THE RE

M000276909

The Science, Politics, and Economics of
Co-Existing with Wolves in Modern Times

Ted B. Lyon & Will N. Graves

With Contributions By
Rob Arnaud, Dr. Arthur Bergerud,
Karen Budd-Falen, Jess Carey,
Dr. Matthew A. Cronin, Dr. Valerius Geist,
Don Peay, Laura Schneberger,
Heather Smith-Thomas, and Cat Urbigkit

Edited by
Linda Grosskopf & Nancy Morrison

Front Cover Photo:
*Patrick Endres with Alaska Photo Graphics, Inc., took this picture
on June 4, 2010, Alaska's Denali National Park.*

Back Cover Photo:
*Chris Ruebusch took this picture in Montana's Gallatin Canyon at Black
Butte during early spring of 2010.*

Written by Ted B. Lyon & Will N. Graves.

With contributions by Rob Arnaud, Dr. Arthur Bergerud,
Karen Budd-Falen, Jess Carey, Dr. Matthew A. Cronin,
Dr. Valerius Geist, Don Peay, Laura Schneberger,
Heather Smith-Thomas, and Cat Urbigkit.

Edited by Linda Grosskopf & Nancy Morrison.
PO Box 85, Billings, MT 59103
www.wordwrightwomen.com

Graphic Design & Pagination by Kara Fairbank.
karaj24@hotmail.com

Softcover version:
ISBN 10: 1-59152-122-X
ISBN 13: 978-1-59152-122-8

Hardcover version:
ISBN 10: 1-59152-127-0
ISBN 13: 978-1-59152-127-3

CIP data is on file at the Library of Congress.

Printed & Published in the United States of America.

For more information, contact Kelley Moore at 1-800-882-5813.
You may order extra copies of this book
by calling Farcountry Press toll free at 1-800-821-3874.

THE REAL WOLF
The Science, Politics, and Economics
of Co-Existing with Wolves in Modern Times

Karl Umbriaco / Shutterstock.com

Authors' Notes

By profession, I'm a trial attorney. I try complex cases that involve death, horrible injuries, toxic torts, and environmental litigation. Over the past 37 years, I've represented clients in more than 150 jury trials as well as in settlements and arbitrations. I've broken nationwide verdict records and received verdicts that ranked in the Top 10 on three different occasions. I win because I stick to the facts.

I came to this approach based on my earlier experiences as a law enforcement officer, as a graduate assistant in college where I taught political science, and as a State Senator and State Representative—all in Texas. While serving as a state legislator, my passions were both law enforcement and wildlife policy based on science. Groups as diverse as the Sierra Club, Texas Farmers Union, Texas Black Bass Unlimited, Greater Dallas Crime Commission, World Wildlife Federation, and Texas Outdoor Writers Association have given me awards, of which I'm very proud.

A second driving force in my life is my love for nature and outdoor sports, especially fishing and hunting. Once upon a time, I was a licensed fishing and hunting guide. Now, while I live in Texas, my wife and I own a home in Montana where we retreat several times a year. I have been on three African safaris and have made a number of trips to Canada and Alaska.

I would like to express my gratitude to the many people who saw me through the process of writing and compiling this book, including those who provided support, allowed me to interview them and quote their remarks, talked things over, read, wrote, offered comments, and assisted in the editing and proofreading:

David Allen, Ray Anderson, Ed Bangs, Benjamin Barmore, Ryan Benson, Toby Bridges, Mark Connell, Harriet M. Hageman, Richard Lyon, Richard Mann, Tracy Stone-Manning, W.R. McAfee, Kelley Moore, Miles Moretti, Justin O'Hair, Governor Bruce Otter, Democratic Majority Leader Senator Harry Reid, Bill Schneider, Governor Brian Schweitzer, Dale Simmons, James Swan Ph.D., Dr. Shannon Taylor, Senator Jon Tester and his staff, Josh Tolin, Skip Tubbs, Stephanie Yarbrough, and the staff of the *Yellow Pine Times*, a valuable resource.

For understanding my countless hours devoted to this book, I'd like to thank my wife, Donna Lyon. I also want to thank Tommy Sellers, who for three years rode around Montana listening to me interview people about wolves and elk, while we were supposed to be hunting birds.

Last and not least, I beg forgiveness from all those who have helped me over the course of the past three years and whose names I have failed to mention.

Ted B. Lyon
September 1, 2013

My first real job started in 1950 when I went to work for the U.S. Department of Agriculture-Bureau of Animal Industry in Mexico. I had the unique experience to work for the Mexican-American Commission for the Eradication of Foot and Mouth Disease, formed to prevent the spread of the foot and mouth disease (FMD) from Mexico into the U.S., as over 15,000,000 head of Mexican cattle had been infected. I became the chief of a livestock inspecting and vaccinating brigade in a horseback-only area. My brigade constantly traveled by horseback inspecting and vaccinating all cloven-footed domestic livestock to prevent them from catching this dreaded disease. Fortunately, there were no active cases of FMD in my sector while I was there. FMD is a highly contagious viral disease. If one animal in a herd catches the disease, within 24 hours, every animal in the herd can be infected. FMD is considered the most costly of all animal diseases, as it is often necessary to conduct a wholesale slaughter of animals whenever there is an outbreak. In 1924, there was an outbreak of FMD in California. The USDA reported that the probable mode of infection of some of the cattle was dogs. An American veterinarian told me that one reason FMD was so difficult to stamp out in Mexico was that dogs and coyotes were spreading the disease. This statement was etched into my mind and had a profound effect on my future interest in livestock and diseases.

In November 1993, I took the opportunity to comment by letter on the Draft Environmental Impact Study about reintroducing wolves into Yellowstone National Park. I wrote that, in my opinion, more research was needed on the potential negative impact wolves would have on bringing and spreading parasites and diseases into the Park. Wide-ranging wolves carry and spread many types of dangerous parasites and diseases. The parasites that wolves carry to wild animals may then be passed on to domestic animals, and then pets may pass them to humans. I believed more research needed to be done about the fact that wolves may cause serious harm by spreading dangerous parasites and diseases over large areas.

I believe that wolves have a legitimate role and place in the ecosystem. I support that their numbers be carefully controlled as the result of scientific research on their impact on given areas. However, after all my years researching Russian wolf behavior, I concluded that, as a general rule, many Western writers and supporters of wolves often over-emphasize the positive role of the wolf in nature and tend to ignore or overlook the negative aspects of wolves in nature.

Will N. Graves
September 1, 2013

Dennis Donahue / Shutterstock.com

Editors' Notes

I came to be part of this project—not because I needed another project, which I most assuredly did not—but because, once I picked the book up, I couldn't lay it down. As the editor of several weekly ag papers for the best part of the last two decades, I have covered and dealt with on a weekly basis the entire spectrum of the "Save the Wolf" campaign—reporting first the blatant falsehood that our native species of wolf was extinct, followed by the nonsensical but catchy banner to "balance the ecosystem" by the reintroduction of an imported—and far larger (read that: far hungrier)—species of Canadian wolf, followed by the inevitable statistical reports that have come in on an annual basis as the ungulate populations have steadily dwindled and even disappeared in direct proportion to the rapidly expanding numbers of wolves.

> *Jake Cummins, Executive Vice President, Montana Farm Bureau Federation, January 1995: "Cattlemen stood by somewhat bewildered that they were cast as villains for trying to protect their cattle from the wolves introduced into Yellowstone Park and central Idaho last week. Soon the killing will start, a little at first, but growing with time. Wildlife will diminish, and hunting with it. Livestock will perish, and ranches will go under. And in 20 or 30 years, the wolf will reign supreme. We'll have a major problem on our hands, and our children will wonder how we could have been so stupid."*

We were *not* stupid. We were steamrolled. And it has *not* been a polite and bloodless takeover of the forests and the plains by any means, but one of blood, guts, and terror experienced by the wild inhabitants of the nation, the domestic livestock of the ag industry, the personal pets of the populace, and yes, in increasingly greater numbers, even the human inhabitants of the wolf's ever-growing territory.

Ted Lyon and Will Graves—together with an impressive array of expert witnesses—have laid the entire saga out for the reader, and they have done so without drama and without any ulterior motive other than to set the record straight. Their goal is a worthy one—to take the forests and the plains back from the wolf and to rebuild our nation's treasured elk and moose herds while giving the human inhabitants of the region who live on and work the land in their chosen avocation to feed the nation and the world the tools to deal with one of Mother Nature's most vicious predators. My goal is to help them.

> *Beware the half truth—*
> *you may have gotten the wrong half.*
> Anonymous

One final note. I long ago came to the realization that this entire issue is NOT about wolves. It's far more basic than that. It's about—plain and simple—private property rights, and as those precious items disappear on a daily basis, it has become a crisis not to be overlooked. It's our responsibility as American citizens to protect the personal property rights that our founders established and thousands of our veterans died to protect.

No four-footed predator shall be held inviolate to the landowners—of both America's treasured private and public property. The responsibility to preserve and safeguard our private property rights is as important to the owner of the small backyard in the suburbs as it is to the owner of the nation's vast ranches and farms.

Linda Grosskopf
September 1, 2013

Whether we like it or not, we now live in an age of global awareness. We are part of the world community, and we do have responsibilities for the greater good, both for humanity and for the world's ecosystems. In an overreaching sense of responsibility, however, it has become part of the environmental dialogue to suppose that legislative control of issues once held at the personal, local, and state levels would provide better governance. In this collective mind-set of trying to see the big picture and trying to take care of everything and everyone, it is easy to be mistaken in what the outcomes will be—especially when those leading the charge for change are possibly uninformed or uneducated in the matter at hand. In the case of wolf introduction, this has proved painfully true.

There are few easy answers to be found without looking to historical evidence for guidance, especially so when speaking of environmental concerns. In the case of Canadian wolf introduction and management, those leading the hugely popular "Save the Wolf" campaign have systematically chosen to ignore historical aspects of the wolf's relationship to its prey and the human race. Some have chosen to ignore the truth because it has been and continues to be financially expedient. Some have not so much ignored the truth as been ignorant of it. One is, of course, more deliberate and one more innocent, but the pair of them combined has been costly and—literally—deadly.

In the business of providing food for the nation, ranchers hold the reasonable position that predators causing significant harm to their personal safety and livestock must be eliminated. This once-common notion of protecting family and personal property from the ravages of predators was bedrock to the development of our country and allowed for the success of settlements throughout the United States. This right to protect personal property from predators, however, has been challenged as the common knowledge base once held by all has been lost, as much of the population has shifted from rural to urban life. When the safety of your children and the success of your livelihood depend on the elimination of dangerous predators, it seems beyond reason to be forced by inappropriate laws to risk both.

In direct contradiction to the bill of goods promised by the wolf introduction proponents, where the introduced wolves have settled, wildlife is not in better balance, and the safety of those living in adjacent areas has been seriously compromised. The promised and promoted "balance" between existing wildlife and the introduced wolves has not happened—rather the opposite. What the Canadian wolf introduction has caused is the exchange of vibrant populations of moose, elk, deer, etc., for that of a predators' paradise—where the main item on the menu more and more becomes domestic livestock. And, in the process, the Canadian wolf introduction has destroyed decades of U.S. Fish and Wildlife Service and wildlife conservationists' efforts to restore and maintain thriving ungulate populations for the benefit of the larger ecosystems as well as the enjoyment of the American taxpaying public who like to vacation in our national parks and public lands to view them.

The ultimate result of the introduction of the Canadian wolf has been to provide in real time both a public menace and the effective, efficient elimination of several revered ungulate populations, so that the wolf then looks to domestic animals as prey. The real irony is that these same groups that promote and romanticize the wolf are blind—either by choice or ignorance—to the ecological tragedies playing out as their pet project tears limb from limb all those species that are smaller and/ or less capable of resisting—not restoring balance in nature but rather destroying it.

The goal of this book is to provide *facts*—neither optimistic nor pessimistic *opinions*—so that you may reassess your understanding of the wolf introduction and management question. In so doing, you may well find yourself on the other side of the fence.

Nancy Morrison
September 1, 2013

Foreword

It has been nearly 30 years since the United States government employees undertook steps to import Canadian wolves into Yellowstone and central Idaho. It has taken nearly 30 years to compile between two book covers all the facts to explain to the American people that they were misled about the wolf. Everything you need to learn about the truth concerning wolves can be found in *The Real Wolf*, which is destined to become the encyclopedia of wolf facts, loaded with resources from some of the most renowned scientists, researchers, investigators, and historians the world has to offer. *The Real Wolf* presents hundreds of pages of documents, facts, and real life stories about gray wolves, including over 460 references, footnotes, and links to sources and facts.

What began in this country at least 100 years ago—a deliberate effort to change the minds of American children—cannot be reversed in one book publication, but as far as correcting the new fairy tales that have been told about gray wolves around the world, *The Real Wolf* is a strong first step. It should be the foundation to understand the wolf and required reading for all wildlife managers and biologists.

For those always wishing they had at their disposal a comprehensive publication to share with others and increase their own knowledge of "real wolves," *The Real Wolf* is certainly something to add to your library.

Some of my favorite researchers and scientists have contributed to this book: Will N. Graves, author of *Wolves in Russia: Anxiety through the Ages*; Dr. Valerius Geist, a leading ethologist and professor emeritus University of Calgary; Dr. Tom Bergerud, the world's leading authority on caribou; and Dr. Matthew Cronin, a research professor of Animal Genetics at the University of Alaska in Fairbanks.

The Real Wolf will teach readers of wolf history across the globe about wolf introduction in the United States, the more than 50 diseases wolves carry, how the U.S. Fish and Wildlife Service alters science to fit agendas, how mongrel mutts are being introduced as pure wolves, the devastation wolves have had on other wild animals and private property depredating livestock, and the unbelievable effect it has had on people, plus a whole lot more.

Tom Remington
Author, Writer, Researcher
tomremington.com
September 1, 2013

Critterbiz / Shutterstock.com

Opening Statement

"May it please the court, ladies and gentlemen of the jury, in this case I will prove the following..." With those time-honored words, I have started hundreds of cases for more than 37 years as a trial lawyer. So...

May it please the court, ladies and gentlemen of the court of public opinion, the case that I am about to make in this book about wolves and how they have been "managed" over the last 30 years is a story about misinformation and myth. It should shock you, like it did me, into understanding that wolves *must* be "managed" differently than they have been. More importantly, they must be managed by the individual states and not the U.S. Fish and Wildlife Service (USFWS). Because of its size, that agency is inefficient, monolithic, and almost impossible to deal with when it comes to wolves. (Besides the Endangered Species Act (ESA), one has to deal with the endless "endangered" lawsuits it triggers—an almost insurmountable task.)

In this book, you will find that a massive disinformation campaign has been perpetrated upon America about wolves. Here are a few examples:

Myth: Wolves do not kill or attack people.
Fact: They do, and regularly. Chapters 3 and 4 will shock you.

Myth: Wolves are the sanitarians of nature and only kill the weak and the sick.
Fact: Wolves kill any and all forms of animals, strong as well as weak.

Myth: Wolves do not destroy game herds.
Fact: Chapters 5 and 8 document how wolves destroy large game herds quickly before moving on when a region can no longer sustain them.

Myth: Wolves are an economic boon to the economies of Idaho, Montana, and Wyoming.
Fact: Chapters 7 and 10 document how and why wolves are a financial disaster for ranchers, farmers, businesses, and states.

Myth: Wolves do not carry harmful diseases.

Fact: Chapter 14 lays out facts, backed by worldwide scientific data, proving that wolves are definite carriers of diseases dangerous to man and animals alike.

This book also shows the reader the enormous amount of taxpayer dollars—more than $200 million and counting—that the federal government and the states have spent to reintroduce wolves in the West, which figure does not include the economic losses that states have had to endure since the wolves' reintroduction.

The preponderance of evidence in this book proves that the Canadian wolf reintroduced into Montana and Idaho is a much larger, different subspecies of wolf than the wolves native to the Western states, *which were not extinct in Yellowstone National Park at the time of the Canadian wolf reintroduction, despite claims to the contrary.*

This book also casts serious doubt over the validity of the Mexican Wolf recovery program that has run up a $26 million tab so far—the concern being whether or not this money was spent trying to reintroduce a wolf-dog cross that does not even qualify as "endangered" under the ESA. A request for a congressional investigation into the validity of this "reintroduction" program has gone forward.

This book also demonstrates the different methodology that surrounds USFWS counts of livestock that are killed by rapidly growing wolf packs each year, versus the number reported by the U.S. Department of Agriculture. A disconnect of several thousand exists.

Further, this book demonstrates the impact that wolves have had on local residents and communities where they have been reintroduced and how these dangerous animals have forced serious lifestyle changes that have cost ranchers, farmers, outfitters, businesses, and states millions of dollars in lost revenue throughout the West.

Finally, this book makes specific recommendations for changes to the Endangered Species Act and to the Equal Access to Justice Act so that states and citizens do not have to endure financial hardships brought on by the USFWS' wolf reintroduction mistake in Montana, Idaho, Wyoming, and Washington. This book documents:

🐾 How millions of dollars have been paid out to government contract "environmental" lawyers under the Equal Access to Justice Act in recent years,

🐾 How the ESA allows any non-governmental organization (NGO) to sue federal agencies for perceived "violations" of the ESA, a practice also extended to any citizen. To date, federal agencies cannot document how much tax money has been paid out under the Equal Access to Justice Act for taxpayer-funded litigation brought on by the EPA, ESA, et al., and

🐾 How the reintroduction of the wolf has been the engine that has driven huge expenditures in the USFWS budget.

In summary, this book strips the mask from the wolf activists' face and calls for changes to the Equal Access to Justice Act and the Endangered Species Act, which changes would move the power to maintain local ecosystems back to the states where it rightfully belongs. Additionally, this books calls for an audit of the millions of dollars spent on "wolf recovery" in the U.S.A. It is also a call for our citizens to educate themselves before accepting on faith the latest environmental activist movement's pet project as necessary or even valid.

Ted B. Lyon
September 1, 2013

Karl Umbriaco / Shutterstock.com

Chapter 1

THE REAL WOLF STORY

By Ted B. Lyon

An attorney specializing in complex litigation with over 37 years of experience, Ted Lyon has represented clients in more than 150 jury trials and was named one of the top 100 lawyers in America by the American Trial Lawyers Association 2007 through 2013. Ted served in both the Texas House of Representatives (1979-1983) and the Texas State Senate (1983-1993). He has also been a police officer, a licensed fishing and hunting guide, and a teacher. Ted has received numerous prestigious and meaningful awards including the 2012 Teddy Roosevelt Conservationist of the Year award.

Editors' Statement of the Issues

The general public's perception of any particular issue is based on access to information. The supporters of Canadian wolf introduction worked hard to put forth their story, but the facts about the wolves' impact on wildlife and affected residents remain largely unheard.

Editors' Statement of the Facts

In order to logically assess the impact to the environment of the reintroduced wolves and to recognize the need to create a workable wolf management plan, the public needs ready access to a concise setting forth of facts on a regular basis. Dissemination of actual wildlife counts as well as documented wolf encounters and depredations will allow the public to become well informed.

> *It became clear to me that the issue*
> *of how the wolves could be controlled was not science at all;*
> *it was pure, unadulterated politics.*
> Ted B. Lyon

Wolves

My first introduction to the wolf issue came in 1999 while my wife and I were staying at a place called Chico Hot Springs just north of Yellowstone National Park. I was soaking in the big hot springs pool when a guy with a beard joined me, and we struck up a conversation. I asked him what he did, and he told me he had been a licensed big game outfitter for more than 15 years and had employed approximately 15 people during hunting season. A native of the area, he said his business had been booming before wolves were reintroduced into Yellowstone in 1995. After the wolves were reintroduced, the elk herd became smaller and smaller and smaller each year until he was forced to shut his business down.

I listened to his story with skepticism. I could not believe a few wolves could cause that much destruction to an incredibly large elk herd—more than 19,000 in 1995.

Montana Real Estate

Time passed, and in 2001, my wife and I bought a piece of property north of Bozeman, Montana. As we drove onto the property that crisp October morning, a whitetail buck ran across the road.

We continued on and saw two ruffed grouse flush off to the side. We stopped, turned off the engine, and got out. An elk bugled to the south. I told my wife the place was speaking to us, and we eventually built a home there.

At the time, the area, which is north of Yellowstone, was full of elk, deer, and moose. To the west of us, between Bozeman and Big Sky, Montana, there was the Gallatin Canyon elk herd with between 1,000 and 1,500 elk.

Horror Stories or Isolated Cases

In the late 1980s and early 1990s, I spent several weeks in the Salmon River Wilderness with local outfitter Brent Hill. We rode horseback throughout the area, camped out, hunted elk and mule deer, and saw moose and bighorn sheep. Before 1995, Brent took approximately 60 hunters a year into this area, which is one of the largest wilderness areas in the lower 48 states. He was there in 1995 when Tom Brokaw, Ted Turner, and Bruce Babbitt, then Secretary of the Interior, watched U.S. Fish and Wildlife Service (USFWS) officers turn Canadian wolves loose onto a nearby airfield.

Before those wolves were released, Brent's hunters and guides took 50-55 elk annually. He told me the new wolves began killing sheep, mule deer, and elk that winter by the river. The next year, only 12 bull elk were taken by hunters, and their success rate went down each year thereafter until the business was forced to close.

I believed at first that Brent's and the other outfitters' stories were isolated cases because the USFWS, Idaho Fish and Game, and Montana Fish, Wildlife and Parks were telling everyone that wolves would *not* affect the wild game populations to any great extent.

I simply could not believe that trained biologists could be so wrong about the wolf and its destructive effect upon wild game. I trusted these agency employees because, as a State Senator and House member in Texas, I spent 14 years on committees that dealt with the Texas Parks and Wildlife Department. That agency continually appeared before my committees and advocated for the preservation of our wild game and fish resources. I used their biologists and staff to support bills pushed through, and I developed a tremendous amount of respect for their scientific knowledge and desire to manage our wild game and fish so that they remained abundant and to properly utilize public resources.

In 2007, I was hunting pheasant with a number of good friends in Choteau, Montana, on a farm owned by my good friend, Skip Tubbs.

3

Skip is an avid sportsman and conservationist, who also owns an art gallery in Bozeman, raises English Setters, and is an avid falconer. All of the people invited to Skip's for opening day had one thing in common: we are, as Southerners say, "dog men." We love to hunt with our dogs and watch them work.

On Saturday night, after bagging our limits, we put some steaks on and broke out some beer and a bottle of wine. It was then that I first became exposed to the strong reaction that Montana hunters as a group have toward wolves as well as toward the USFWS. That evening, I defended the decisions of those who put the 66 wolves into Yellowstone and the Salmon River Wilderness in Idaho in 1995 and 1996—ignorantly, I must say. I also defended the statements made by the Idaho and Montana state wildlife biologists, who parroted USFWS statements.

The men who were crowded around the fire that night were adamant that, under no circumstances, had the introduction of Canadian wolves been a good thing for Montana, Idaho, or anywhere else. Fish and Wildlife statements such as "wolves are not impacting the elk herds" and "hunters just need to work harder to find the elk" were considered "pure B.S."

Again, I just could not bring myself to believe that a USFWS official or an Idaho or Montana state agency wildlife biologist, knowing the economic impact that elk and deer hunting have on Montana and Idaho, would knowingly make misrepresentations about the effects that wolves were having on these states' elk herds or could be that wrong.

That night, as I drove back to where I was staying, I thought that my hunting friends were surely over-reacting. The next thing I expected to hear was that black government helicopters were frequenting the area. But the evening stayed with me, and I recalled my conversation with Brent and the other outfitters and decided to look into what they were saying.

Research and Enlightenment

In 1995 and 1996—at a cost of between $200,000 and $1 million per wolf [1]—the USFWS introduced 32 Canadian wolves from Alberta, Canada, into Yellowstone National Park. An additional 34 Canadian wolves were introduced into the Salmon River Wilderness area in Idaho at the same time. The Salmon River Wilderness area is located some 500 miles to the north of Yellowstone. To get there from Yellowstone, which is located at the southern end of Montana and the northern end of Wyoming, one can follow Interstate Highway 90 up the western side of the Rocky Mountains to Missoula, Montana. From there you can travel the Lolo Pass—made famous by Lewis and Clark in their exploration of the Missouri River—all the way to Idaho. It is a beautiful trip through a place where you expect to see a lot of wildlife.

At the time the Canadian wolves were introduced, Yellowstone was home to some of the healthiest elk, mule deer, and Shiras moose populations in the world. The slopes of the Rockies on the western side of Montana and the Lolo National Forest were also home to thousands of these ungulates (hooved animals)—vibrant populations all and the result of decades of conservation work by sportsmen.

Fast Forward Two Years

When the scenario repeated itself at Skip's annual hunt two years later in 2009, I was better armed, having read a number of articles that said the wolves were *not* impacting the moose or elk herds. There was even an "official scientific study" funded by some groups I had never heard of that said this was true. There was also an economic study that showed that wolves were a positive 35-million-dollar benefit to the Yellowstone area.

That opening night of the 2009 pheasant season, when we all gathered at Skip's house, a new guy was there, Ray Anderson, who had retired to Montana after a successful business career. Ray and Dale Simmons, a website designer, were vocal, articulate, and adamant in their feelings about the wolves. The clincher came when Dr. Shannon Taylor, a professor at Montana State University, and Terry Thomas, a heating contractor, both insisted the moose had all but disappeared from Yellowstone. Each fall, Terry spends the entire elk season camped out next to Yellowstone; he had done so for years. He said the elk herds were severely depleted and the moose were gone.

That night I resolved to thoroughly research the issue. I had to look widely as very little of what these men were saying was in popular print in a comprehensive manner. Frankly, as I got into it, I was shocked. This book is the true story of the greatest destruction of wild game in the United States since the decimation of the bison in the late 1800s and the elimination of the passenger pigeon in 1914 and, more importantly, how people are trying to reverse that unnecessary destruction.

Nothing Harmless

When my wife and I first bought land in Montana, I thought wolves were harmless. The wildlife biologists from the USFWS and the states of Idaho and Montana were telling everyone that wolves:

- were good for the economy,
- brought balance to nature,
- were not now, and never had been, a threat to man,
- could be trained and educated to not attack livestock,
- only ate what they killed,
- were not sport or surplus killers that slaughtered without eating what they killed,
- did not carry deadly diseases, and
- were the sanitarians of nature.

The more I looked into this situation, the more I realized the flood of positive information about wolves was either inadvertently wrong or deliberately deceptive. I had previously believed these statements because they were put out by officials at every level of government. But I found that the people who perpetuated these myths had either failed to research the issue adequately or simply believed the misstatements by the many who either fabricated the scientific data about wolves or deliberately ignored data that had been accumulated since the turn of the century.

Lightbulb Goes On

Then I learned that the environmental and animal rights groups that supported the wildlife officials and the wolf reintroduction program were making *millions* from contributions "to save the wolves," which meant that they were *not* interested in telling any other story, even if it was true.

Irrefutable Data

Finally, in January of 2010, Montana Fish, Wildlife and Parks released a report called "Monitoring and Assessment of Wolf-Ungulate Interactions and Population Trends within the Greater Yellowstone Area, Southwestern Montana, and Montana Statewide—Final Report 2009." Written by Kenneth L. Hamlin, a senior wildlife researcher,

and Julie A. Cunningham, this 83-page report detailed the amazing decline of the Northern Yellowstone herd. It showed that, in 1995 when wolves were first introduced into Yellowstone National Park, the elk herd numbered more than 19,000 animals. The count in 2009 was just a little over 6,000 elk. It also showed there had been a precipitous decline in the moose, which had almost vanished from the same area.

Initially, I simply could *not* understand how a population of 66 wolves could take an elk herd of more than 19,000 down to 6,200 in 14 years and to 4,174 by 2012. Couple that harsh reality with the fact that the Yellowstone moose population dropped from 1,000 to almost zero over the same period, and you will appreciate the seriousness of the situation.

These were *not* isolated cases. Similar devastation has happened to other elk populations throughout the Northern Rocky Mountains since the Canadian wolves arrived. For example, in the Lolo National Forest, where wolves had migrated from Idaho in 1995, the elk herd dropped from 12,000 in 1995 to around 2,000 in 2011. And when my wife and I bought our land near Bozeman in 2001, the Gallatin elk herd numbered from 1,000 to 1,200 elk. By 2011, the Gallatin elk herd numbered just 42 elk. Virtually all of that destruction can be traced to the 66 Canadian wolves and their progeny. Additionally, the Fire Hole elk herd in Yellowstone has been completely wiped out; there are no elk left. I could go on, unfortunately.

More Irrefutable Data

The early "studies" totally "misjudged" the rapid rate by which the wolf population would grow—from 66 in 1996 to, conservatively, 1,700 in 2012—and many scientists believe there are at least five times that many. Unlike other predators like mountain lions or bears, wolves have large litters and can begin breeding by age two.

On top of that, the Canadian wolves are a larger subspecies than the lower 48 states' native wolves and have voracious appetites. Gray wolves can survive on about two and a half pounds of food per wolf per day, but they require about seven pounds per wolf per day to reproduce successfully. A large gray wolf can eat about 22.5 pounds at one time, and the introduced Canadian wolves are much bigger than their native Rocky Mountain brothers.[2] Put larger wolves together with an abundance of prey, and you get a lot of wolves—quickly.

Despite what some people were saying about wolves being nature's sanitarians, wolves—Canadian or native—do *not* care what they eat, healthy or not. Additionally, on occasion, wolves simply go on adrenaline-surging sprees, killing and wounding many times the number of animals that they can eat and leaving the carcasses lying where slaughtered to rot for scavengers.

Side Effects

Predation by wolves is significant, but their impact on wild game herds goes far beyond predation. Research[3] by Professor Scott Creel at Montana State University determined that cow elk in and around Yellowstone were not getting pregnant as a result of the stress caused by wolves. Think about being the fattest animal in the herd and heavy with calf. You run the slowest, and therefore, you become the easiest victim for predators. Creel's later research showed that elk hunted by wolves were actually starving to death in the winter and were either not calving or were having fewer calves.[4]

More Damning Evidence

The damning evidence does not stop there. Before the introduction of Canadian wolves into Montana and Idaho, there was no known incidence of hydatid disease in Montana, Idaho, or Wyoming. Hydatid disease, also known as *hydatidosis* or *echinococcosis*, is a parasitic infection of various animals that can also infect humans. The disease is caused by a small tapeworm living in canids, especially wolves. Tapeworm eggs pass out of the feces of the infected animal. If eaten by a suitable host—ungulates, livestock, and man—these eggs may develop

8

into hydatid cysts in the internal organs of the host, especially the liver, heart, and lungs. This disease did not exist in the moose, elk, mule deer, whitetail deer, mountain goat, or sheep herds before the Canadian wolf introduction. It exists there now, though, and is a serious disease threat to animals and man, as you will see in a later chapter.

Like most people, I was not aware of this disease before I began my research. It was while I was researching wildlife diseases that I found my co-author, former National Security Agency Security Officer Will N. Graves, an incredibly interesting man, who has spent most of his life researching wolves. Will had written a letter in 1993 to Ed Bangs, the USFWS biologist in charge of transporting the wolves to Yellowstone, about his concerns about hydatid disease in the wolves from Canada (see letter in Appendices).

I interviewed Mr. Bangs in Helena, Montana, in June of 2012. A very nice man, Mr. Bangs confirmed that the Canadian wolves were "wormed" twice before they were released. However, the circumstantial evidence is strong, almost overwhelming, that either the wrong type of wormer was used or the parasite existed in Montana and Idaho and was not known. Since the Canadian wolves were introduced, the parasite that carries *hydatidosis* has been transported by wolves throughout the Western states. Over 68% of the wolves in Montana, Idaho, and Wyoming that have been tested have *echinococcosis granulosus* tapeworms; in some areas, the infection rate is 84%. There is also strong evidence that the tapeworm weakens the ungulate, which is an intermediate host, and makes it more susceptible to being preyed upon by predators.[6]

Humans can become infected with this disease simply by petting a dog that has rolled in an area where a wolf has defecated. Contrary to what some spokesmen for U.S. wildlife agencies have reported, *hydatidosis* is a deadly disease to humans with reported deaths all around the world where the tapeworm exists.[5]

Spin Doctors

After realizing that my friends' anecdotal stories about what the wolves had done to the elk were right, I resolved to find out why. One reason that quickly became apparent was that—although the USFWS had declared that wolves were recovered in 2000, easily passing the goal of 10 breeding pairs and 100 wolves in Montana, Wyoming, and Idaho—*neither the states, the hunters, the livestock producers, nor the USFWS were allowed to manage the wolves.*

Endangered Species Act (ESA)

Congress passed the Endangered Species Preservation Act (ESPA) in 1966, providing a means for listing native animal species as "endangered" and giving them limited protection. The Departments of Interior, Agriculture, and Defense were to seek to protect listed species and, insofar as consistent with their primary purposes, preserve the habitats of such species. The ESPA also authorized the U.S. Fish and Wildlife Service to acquire land as habitat for endangered species.

In 1969, Congress amended the ESPA to provide additional protection to species in danger of "worldwide extinction" by prohibiting their importation and subsequent sale in the United States. One amendment to the ESPA changed its title to the Endangered Species Conservation Act (ESCA).

A 1973 conference in Washington, D.C., led 80 nations to sign a treaty called the Convention on International Trade in Endangered Species of Wild Fauna and Flora (CITES), which monitors and, in some cases, restricts international commerce in plant and animal species believed to be harmed by trade.

Later in 1973, Congress passed the Endangered Species Act of 1973 (ESA). It defined "endangered" and "threatened"; made plants and all invertebrates eligible for protection; applied broad "take" prohibitions to all endangered animal species and allowed the prohibitions to apply to threatened animal species by special regulation; required federal agencies to use their authorities to conserve listed species and consult on "may affect" actions; prohibited federal agencies from authorizing, funding, or carrying out any action that would jeopardize a listed species or destroy or modify its "critical habitat"; made matching funds available to states with cooperative agreements; provided funding authority for land acquisition for foreign species; and implemented CITES protection in the United States.

Congress enacted significant amendments in 1978, 1982, and 1988, while keeping the overall framework of the 1973 ESA essentially unchanged. The funding levels in the present ESA were authorized through Fiscal Year 1992. Congress has annually appropriated funds since that time.

I was shocked to find that respected magazines like *Outdoor Life* and *National Geographic* had published stories that did not look into the research that would have pulled open a curtain, detailing how wolves had destroyed elk and caribou herds in Canada. Instead, these respected magazines simply repeated pro-wolf propaganda, quoting anecdotal stories from hunters who said they could not find any elk. In one case, the chief wolf biologist for Montana Fish, Wildlife and Parks, Carolyn Sime, simply said that hunters would have to work harder, that the elk had not disappeared but had simply retreated to the woods, implying that hunters were lazy.

As someone who has been involved in running political campaigns since the age of 21, as well as having served in office and campaigned, I began to realize the wolf issue had been framed by extremely smart, well-funded spin doctors, who were Machiavellian in their approach to the issue of the wolf and masters at manipulating the facts to raise money for their sponsors.

The people on the other side of the issue—hunters, sportsmen, and hunting and fishing groups—were hopelessly outmatched, not from a political power standpoint, but from a politically strategic point of view. The pro-wolf advocates spent their money pumping out propaganda, while the sportsmen and conservation groups conserved habitat and sponsored research (work not well-known to the general public) that was producing monumental successes in restoring big game populations. It was as if the wildlife conservationists were high school baseball players going up against major leaguers; they did not know what to do politically or how to control spin through the media, and consequently, they were, as a group, howling at the moon.

The pro-wolfers also outflanked the wildlife conservation groups and the livestock producers, hiding behind the Endangered Species Act and winning with contract lawyers paid with taxpayer dollars when ranchers, who had to pay their litigation costs out-of-pocket, challenged them in the courts.

One of the issues that I first began researching was how to overcome the court losses that the USFWS kept encountering when they *finally* attempted to delist wolves so they could be managed, not only in those states where Canadian wolves had been introduced in the mid-1990s (Montana, Idaho, and Wyoming), but also in states like Minnesota, Wisconsin, and Michigan, which had long-standing wolf populations.

Lawsuits and Legal Challenges

A lawsuit had been filed in 1993 to keep Canadian wolves from being introduced into the West by one of my collaborators on this book, Cat Urbigkit. She pointed out that there were already wolves in the release area and that the Canadian wolves were a different, larger subspecies that would displace or hybridize the native wolves.

In Minnesota, legal challenges to allow delisting the wolf had been going on since the 1970s. Additional lawsuits by the pro-wolf groups were also filed against the USFWS, which wanted to delist the wolf in the Western states in 2001 when their numbers passed what was required in the initial agreement. In every case, the pro-wolf forces won—postponing, delaying, and stopping state wildlife agencies from managing the wolves.

By the beginning of 2010, wolves in America enjoyed an exalted status over all other species. As a private citizen, you could be walking down a city road in any of the lower 48 states and a pack of wolves could attack your dog, horse, or cow, and you would be committing a felony if you shot the wolves. Wolves were given a free pass to kill sheep, cattle, and horses, and ranchers, states, and citizens were powerless to stop them. Only in defense of human life could a public citizen defend oneself from a wolf attack.

The pro-wolf forces were also manipulating reports that showed how many head of livestock the wolves were killing. Typically, the USFWS or the state wildlife agencies would release at the end of each year the number of "confirmed" kills (and I emphasize the word "confirmed") caused by wolves. For 2010, this number totaled less than 1,000.

Real Names and Faces

In an interview I conducted with Montana rancher Justin O'Hair, he advised me that on one occasion he spotted a young Black Angus calf with its entrails hanging out and a wolf 100 yards away. Initially, the USFWS officer would not confirm that the calf had been attacked by a wolf because, he said, Justin was "not qualified" to confirm the difference between a wolf and a coyote.

Justin and his family own the 80,000-acre O'Hair Ranch located outside of Livingston, Montana, where they run 1,100 head of cattle. Justin has lived on the ranch his entire life, and his family homesteaded the ranch in 1878. They are in the pastures, checking cattle on horseback almost every day.

Justin also related that often ranchers will find just an ear tag lying on the ground at a wolf kill site, but without the corpse, they cannot tell what killed it. Cattle that suffer attacks from wolves almost always die since no amount of antibiotics can overcome the infection that comes with a wolf bite. And wolves often just maim their victims and move on without eating anything.

USDA Report

The U.S. Department of Agriculture (USDA) releases a report every five years as a cooperative effort between the National Agriculture Statistics Service and the Animal and Plant Health Inspection Service-Wildlife Services and Veterinary Services. The report is, and has been, a scientific, validated survey based on producer reports. There is no pro-wolf bias involved in the compilation of these reports. The USDA report published in May of 2011 showed that wolves killed 8,100 head of cattle in 2010 alone—thousands more than the few hundred head that the USFWS reports each year. If wolves are not controlled, this number will climb. As herds of wild ungulates are depleted, wolves turn to livestock and pets for prey and frequent garbage dumps and roads seeking road kills as they spread into other areas.

Handpicked Judges and Other Deck-Stacking Shenanigans

I came to the conclusion that something had to be done about the wolves in North America. They had to be managed by people who truly understood what was going on in the field, and those people needed to be respected and supported.

At first, I approached the problem as a trial attorney. In the past, I have enjoyed tremendous success in using the courtroom as a venue for enforcing justice. I thought that, as a non-governmental lawyer, my fresh perspective and skill as a trial attorney would allow me to succeed where the government's attorneys, going against pro-wolf environmental NGOs (non-governmental organizations), had failed for the past decade.

I wondered about all of the lawsuits filed over the years by groups trying to stop the introduction of Canadian wolves into Montana and Idaho, none of which had been successful. I wondered if they did not have good lawyers, or perhaps, the lawyers representing the USFWS in its attempts to delist the wolf since 2001 were not competent. So I hired two extremely bright, young law students: Ben Barmore, who was at the top of his class at Southern Methodist University, and Richard Mann, a Canadian, who attended the University of Texas

Law School. I asked them to research what could be done from the perspective of a lawsuit to give the states the power to control wolves. Each day we would talk and come up with legal theories to pursue. The next day, they would give me their findings.

What we found is that almost every legal theory that we could come up with to attack the continued protected status of the wolf had already been tried and that the side representing the people who wanted to control the wolves lost at every turn. The government lost, states lost, private citizens lost, and every other outdoor and 501C-3 organization lost every time they tried to allow wolves to be managed by the states. The winners were NGO groups like Defenders of Wildlife, the Center for Biological Diversity, The Nature Conservancy, and others that purported to represent the environmental movement.

Delisting supporters lost in federal courts from Missoula, Montana, to Albuquerque, New Mexico, to Duluth, Minnesota. And in some of the losses, if the pro-wolf people could prove an error by the federal government, even on a technicality, their legal expenses were paid with tax dollars by the U.S. government through the Equal Access to Justice Act.

After reading all the cases, we concluded that the lawyers representing clients that wanted to control the wolves had generally done a very good job. We also found, however, that in some cases the federal judges who heard these cases were handpicked by the pro-wolf groups because their political philosophy was more in line with the groups that wanted no control exerted over wolves. In retrospect, I concluded that it would be almost impossible under the Endangered Species Act to win a legal victory to delist wolves.

Conclusion

In April of 2010, it became clear to me that the issue of how the wolves could be controlled was not science at all. It was, in fact, pure, unadulterated politics. The Canadian wolves had been placed in Montana, Idaho, and Wyoming because of politics. And they could only be controlled and removed by politics. It seemed obvious that the only way to control wolves was by amending the Endangered Species Act, something that had never been done before. I believed in my heart that, if we could just get the truth out to members of the U.S. House and Senate, we could get the Act amended.

🐾 🐾 🐾 🐾 🐾 🐾 🐾 🐾

[1] From the Final Environmental Impact Statement released by the USFWS in May 1994.
[2] http://www.wolf.org/wolves/learn/basic/faqs/faq.asp#19
[3] Funded by the National Institute of Science.
[4] Creel, S., D. Christianson, and J.A. Winnie, "A Survey of the Effects of Wolf Predation Risk on Pregnancy Rates and Calf Recruitment in Elk," *Ecological Applications* 21: 2847–2853, 2010.
[5] www.dpi.nsw.gov.com; "Hydatids—You, Too, Can Be Affected," NSW DPI, February 2007, Australian Government Prime Facts.
[6] http://www.fao.org/docrep/t1300t/t1300T0m.htm

Debbie Steinhausser / Shutterstock.com

S. R. Maglione / Shutterstock.com

Chapter 2

SELLING THE WOLF

By Ted B. Lyon

Stayer / Shutterstock.com

Editors' Statement of the Issues

Advertising campaigns have proved time and again the value of setting forth to the public the benefits received by using their product or service. This same strategy has been used to pave the way for the acceptance of the wolf in modern society, contrary to actual data.

Editors' Statement of the Facts

A clear setting forth of historical data, as well as current information, would provide a counter-balance to this revised telling of the wolf as a beneficial member of the ecosystem. The wolf is a ruthless predator, and no amount of over-selling or soft-soaping can change its nature or the facts. This presentation of the wolf, however, does a major disservice by downplaying its crucial need for management.

> *Environmental battles are not between good guys and bad guys but between beliefs, and the real villain is ignorance.*
> Alston Chase[1]

From Bad Guy to Poster Child

In 1985, Yale sociologist Stephen Kellert conducted a national survey of public opinion about wildlife. He found that wolves were the least liked of all animals in North America; 55% of the people said they were neutral toward wolves or disliked them.[2] Since then, wolves have been reintroduced into the Northern Rockies, the Pacific Northwest, the Southwest, and the Southeast. Wolf populations in the Upper Midwest and New England have grown; wolf populations in Alaska and Canada have increased; and some wolf advocates have set a goal of wild wolves thriving in all 50 states. Similar programs are underway in Europe and Russia. The wolf has gone from a bad guy to a poster child for conservation in less than 30 years.

The unprecedented wolf repopulation program—that brought 66 wolves from Canada to the Northern Rockies in 1995 and 1996 and has since sheltered them, allowing the population to skyrocket to at least 10 times the number called for in the original plan—could only have been accomplished with a massive, multi-faceted promotional sales campaign for, as you will learn, introducing Canadian wolves into a modern social landscape is a serious problem... the intentions may be honorable, but the results can be catastrophic.

The purpose of this book is two-fold: first, to expose the myths about wolves that have been spoon-fed to people in North America and abroad and the falsehoods that have resulted in a war of words and seemingly endless courtroom battles, as well as a war in the woods; and, second, to set the record straight so people on all levels can understand the real issues about living with wolves in modern times and make responsible decisions about the future of our uneasy relationship with *Canis lupus,* the gray wolf.

In Sun Zu's masterful treatise on winning in conflict, *The Art of War*, he insists that, to win, you must understand your enemy. The sad truth is that the "Save the Wolf" campaign is largely based on romantic half-truths, exaggerations, and distortions, which are mixed with negative stereotyping and the stigmatizing of anyone who questions the wolf restoration program. Nonetheless, it has been extremely successful. So, let's see how and why this is so.

A Brief History

The ancestors of the modern gray wolf, the largest living member of the wild dog family *Canidae*, trace back to the Pleistocene era, perhaps as far back as 4.75 million years ago. The gray wolf was once the most widely distributed large mammal on earth. Everywhere that wolves and people are found together, there is a history of respect, distrust, and mutual predation. This is the primary reason why wolves are not as common today as they once were.

When European settlers arrived in the U.S., they found wolves, as their ancestors had known in far-distant lands for thousands of years. American Indians lived with wolves, which were integrated into their spirituality, mythology, and rituals, but the Indians also trapped and killed wolves, using their skins for clothing. Eating them was a delicacy. While there were no newspapers or written records in those days, there are many tales of people being attacked, killed, and eaten by wolves. In the 1800s, as the buffalo were nearly exterminated by market hunters and the Indians were driven onto reservations through a planned military strategy, elk and deer were killed in large numbers, and the natural habitat declined dramatically due to logging and farming. In response to the lack of prey, wolves switched their predation to livestock. This triggered a war on wolves—bounties, trapping, hunting, and poisons—that was supported by the U.S. government. This was the first educational wolf campaign—Get Rid of Wolves—and Congress supported it.

The second wolf educational campaign—still "Get Rid of Wolves"—was again backed by the U.S. Congress when, in 1914, it passed legislation calling for the elimination of predators from all public lands, including national parks, as wolves and other predators kept down the numbers of elk, deer, moose, and antelope. Aided by modern weapons, traps, and poisons, by 1930, wolves were all but gone from the lower 48 states, except for small numbers in the Northern Rockies and northern Minnesota and a handful of Mexican wolves in Arizona, New Mexico, and Mexico. Remaining wolves became very wary of man and were seldom seen, except in the far north.

Poison baits, which were heavily used on coyotes after the wolves were nearly eliminated, were not banned until 1972, and then in large part due to Earth Day 1970, when banning the poison 1080 was a hot issue at teach-ins across the U.S.

With an absence of wolves and diminished numbers of bears and mountain lions, as well as habitat conservation programs supported by many groups, by the 1960s, elk, deer, moose, and antelope numbers in Yellowstone National Park and elsewhere across the U.S. mushroomed to record high numbers. The wild game restoration campaign was spearheaded by conservation and sportsmen's organizations with support from state and federal resource agencies.

The concept of restoring wolves to the lower 48 states as a way to control big game herds was first introduced to Congress in 1966 by biologists.[3] Support for this strategy came from years of studying wolves on Mt. McKinley in Alaska by Adolph Murie and studies of wolves and moose on Isle Royale in Lake Superior by Purdue University wildlife biologist Durward Allen and his students, including David Mech and Rolf Peterson.[4] This research concluded that wolves are shy creatures of the wilderness that do not attack people or seriously reduce large ungulate populations and that wolves are nature's sanitarians, attacking only the old, the lame, and the diseased. That perspective became the gospel in wildlife management for decades. This isolated and specific research kicked off another wave of wolf education, but for the first time, it was in favor of wolves. The problem is, as you will soon learn, that wolves are very adaptable, and in different situations they behave very differently.

The "harmless wolf" research was woven into the 1963 "The Leopold Report," otherwise known as "Wildlife Management in the National Parks," written by Aldo Leopold's son, Starker, a renowned wildlife biologist.[5] The Leopold Report called for active management of wildlife to insure that "a reasonable illusion of primitive America (what things looked like when white men first arrived there) … should be the objective of every national park and monument."[6]

Following Earth Day 1970, support for restoring wolves began rising. "Wolfism" joined racism, sexism, ageism, and pollution as another form of oppression. Riding on the wave of the first Earth Day, the Endangered Species Act was passed in 1973. One year later, the gray wolf was added to the list of endangered species in the lower 48 states. Saving the wolf became a growing rallying cause for environmentalists, who were joined by animal rights groups, resulting in a "Save the Wolf" movement. But, as the Kellert study found, even by 1985, the general public was still not too keen on wolves. To bring back the wolf, an unprecedented massive public education program was needed to change the prevailing negative opinion of wolves.

With the only wolves found in zoos or remote areas, media became the new sense organ of urban Americans, as a swarm of books, articles, lecture tours, exhibits, public meetings, films, toys, and TV shows in support of wolf restoration exploded. "Wolf experts" were suddenly everywhere. The wolf became a symbol of green ecological action, along with stopping pollution, recycling, promoting sustainability, and fighting global warming.

The new wolf emerged as a romantic, mythic image of wilderness that urbanized Americans—clustered in concrete, steel, plastic, and wood canyons—longed for in their soul. Reviewing 38 quantitative surveys conducted between 1972 and 2000, Williams, Ericsson, and Heberlein found that attitudes toward wolves consistently showed that *the farther one lives from wolves, the more likely public opinion is in favor of wolf restoration.*[7] In contrast, Williams, Ericsson, and Heberlein found, as did Kellert, that *people who have the most first-hand contact with wild wolves—ranchers, farmers, outfitters, and hunters—held the most negative views of wolves,* and despite the pro-wolf campaign, positive attitudes about wolf restoration have not continued to increase over time.

In the United States, Williams et al. found that 55.3% overall are favorable to wolf restoration. In Europe, where wolves have a history of contact with people, attitudes about wolves are less favorable—37% are favorable to wolves in Western Europe, and 43% are favorable in Scandinavia. While one result of the "Save the Wolf" movement has been wolf restoration programs, a second consequence is growing antagonism between pro- and anti-wolf groups.

Owning the Truth

Unfortunately, in the flood of wolf media, there has been very little accurate information about the problems associated with wolf restoration. Setting the record straight is a major goal of this book.

There are at least four major problems with the "Save the Wolf" movement's educational campaign. The first is that wolf behavior around people is heavily influenced by human behavior. Wolves are intelligent and adaptable, as well as unpredictable. In localities in Europe and Asia where people are commonly armed, as they are in North America, wolves are shy and reclusive. Where the populace is not heavily armed, wolves adapt, become habituated, and act much more boldly, preying on livestock and venturing into towns where they feed on garbage and attack pets and people. Dr. Valerius Geist's chapter shows a predictable behavior pattern of habituation that happens when wolves contact people and meet little or no opposition.

The "Save the Wolf" campaign has largely avoided reporting the fact that, in addition to rabies, wolves may carry over 50 diseases, some of which can be fatal to humans and livestock, such as *hydatidosis*. That we have little record of these diseases in the U.S. is simply due to the previous 50-year absence of wolves and, in some cases, a lack of reporting of wolf-borne diseases. Warnings about such diseases are, at best, a footnote in the many "Save the Wolf" messages. Educating the public about wolf diseases is bad, after all, for the "Save the Wolf" business.

A third major problem is that the so-called economic benefits of a wolf restoration on a large scale are far outweighed by its costs, but again, the costs are not given anywhere near full coverage.

A fourth major problem is that the pro-wolf media has not only sold us a harmless wolf, but for the first time ever, it has sought to discredit— as pure superstition, wrong, and worthless—the rich legacy of myths, fables, folklore, and fairy tales about wolves that originates from Europe and Asia. This attack fails to account for how and why these tales came about, for they represent the earliest wolf educational campaign.

Fairy Tales, Mythology, and Folklore

Fables, folklore, and mythology of Europe and Asia were the first wolf educational campaigns; most all teach that the wolf is dangerous, a lesson that is far from wrong. In Europe and Asia for thousands of years, wolves have attacked and killed big game, livestock, pets, and people. From centuries of study in Europe and Asia, it is known that wolves are adaptable, intelligent, mysterious, and unpredictable predators. Unlike most other predators, occasionally wolves run amok and engage in mass spree killings for sheer joy, such as the pack of wolves that killed 120 sheep on one August 2009 night in Dillon, Montana. Put those qualities together with the possibility of a rabid wolf, and you have an animal that people should fear with good reason.

The Moral of the Story

After reviewing the history of man-wolf relations, Barry Holston Lopez, in his award-winning book, *Of Wolves and Men*, arrives at this conclusion: "No one—not biologists, not Eskimos, not backwoods hunters, not naturalist writers—knows why wolves do what they do."[8] That is a very good reason why folklore about wolves carries warnings. If an animal is unpredictable and carnivorous, you have a dangerous situation. This is why a terrorist acting alone is often called "a lone wolf."

It is understandable then—in cultures where a significant number of people do not own firearms and where children may venture into areas where wolves are present—that fairy tales and folklore (such as Aesop's fables, "Little Red Riding Hood," "The Three Little Pigs," Shakespeare, Grimm's fairy tales, as well as holy books including the *Bible*, the *Rig Veda*, etc.) cast *Canis lupus* in a negative light. These stories are warnings, especially to children and shepherds, designed to keep people alive.

We know the wolf on an unconscious level, too. Animals in our dreams are symbols of instincts in the roots of the psyche—"archetypes." The wolf is an archetypal symbol of pure wildness, both in nature and human nature, and a reminder of one's own inner wolf-like qualities—positive in terms of being a skillful hunter and a family protector, and negative in terms of the wolf's dark side of lust, violence, greed, killing, unpredictability, etc., that can make a wolf seem like a sociopath. The call of the rogue male out on the prowl is the "wolf whistle," and calling a person a "wolf" means that he/she is not trustworthy and can, in fact, be dangerous. Among the Navajo, the word *mai-coh* means both wolf and witch, which the Navajo see as a werewolf, a person who is most likely to perform evil acts during twilight or at night while wearing a wolf skin.

The belief in half-human and half-wolf creatures (the werewolf) and the rare mental disease of *lycanthropy* (where a person goes berserk with almost superhuman strength, howling, making wolf-like sounds, and attacking people as if they are prey) all speak of our fear of raw human instinctual emotions that make people behave like wolves.

Referring to the Wolf

The meanings of "wolf" are many. In medieval times, famine was called "a wolf," and still during hard times, people refer to "keeping the wolf from the door." Werewolves were a principal target of the Inquisition. In Dante's *Inferno*, the wolf presides over the eighth circle of hell where punishment is meted out to those who have committed the "sins of the wolf" in their lives—religious hypocrites, magicians, thieves, and seducers. In the fairy tale "Little Red Riding Hood," the wolf uses trickery to try to lure a young girl into his clutches.

In myth, magic, and medicine in earlier times, wolves have a strong association with the supernatural. Wolves prowl at night and at twilight, a time when the imagination grows larger. Latin for "dawn" is *interlupum et canum*, which translates as "the time between the wolf and the dog."

There are 13 references to wolves in the *Bible*, almost all as metaphors for destructiveness and greed. It should not be surprising then that *The Book of Beasts*, a medieval bestiary derived from a chain of Christian monks adapting earlier natural histories that date to Pliny and Aristotle, states: "The devil bears the similitude of a wolf: he who is always looking over the human race with his evil eye, and darkly prowling round the sheepfolds of the faithful so that he may afflict and ruin their souls... Because a wolf is never able to turn its neck backward, except with movement of the whole body, it means that the Devil never turns back to lay hold on repentance."[9]

In his early days, Adolph Hitler referred to himself as "Herr Wolf" and changed his sister's name to "Frau Wolf." His name "Adolf" is a derivative of Athalwolf, meaning "Nobel Wolf." He called his retreat in Prussia "The Wolf's Lair," and he named three of his military headquarters *Wolfsschanze*, *Wolfsschlucht*, and *Werwolf*. His favorite dogs were wolfshunde, and he referred to his S.S. troops as "my pack of wolves." Little wonder then that journalists spoke of groups of German submarines patrolling the North Atlantic as "wolf packs."

There are exceptions to the negative wolf mythology, as in Roman mythology with the she-wolf raising Romulus and Remus. In Rudyard Kipling's *The Jungle Book*, the boy Mowgli is adopted by wolves. Japanese farmers once left offerings to the wolf *kami* (spirit) to ask his help in protecting their fields from deer and wild pigs. However, wolves were eradicated in Japan in the early 1900s when a rabies epidemic broke out.

The Bottom Line

The bottom line is that, in Asia, wolves attack and kill children far more often than they adopt them, and in European history, there are many cases of fatal wolf attacks. *This is why myths, folklore, and fairy tales almost always portray wolves in a negative light abroad.*

Many American Indian tribes respect the wolf's prowess as a mighty hunter, and they too have many legends, rituals, and myths about wolves, but Indians—and Eskimos, too—also kill wolves for their fur, for food, and in self-defense. Barry Lopez writes: "It is popularly believed that there is no written record of a healthy wolf ever having killed a person in North America. Those making the claim ignore Eskimos and Indians who have been killed..."[10]

Psychiatrist James Hillman found that, in the dreams of most modern people, either animals are pursuing us or we are trying to kill them.[11] Hillman interpreted this as the result of suppression of our own primal instincts, which Hillman and many others believe is a primary cause of the epidemic of anxiety that afflicts our age. In the same vein, psychologist Aniela Jaffe observes: "Primitive man must tame the animal in himself and make it his helpful companion; civilized man must heal the animal in himself and make it his friend."[12]

Clarissa Pinkola Estes' best-selling book about the wild woman archetype, *Women Who Run with the Wolves*, is an example of the power of a symbolic association with wolves. People who live far from wild wolves have an unconscious desire to reconnect with nature, more than to conserve the actual wild wolf, which few have even seen. Estes' book is really not about wolves at all, but rather her biased symbolism about wolves.

To discredit the rich legacy of wolf folklore and mythology was to invite disaster. The modern myth of the "harmless wolf" is not only inaccurate but may also have contributed to attacks and deaths by wolves in North America in recent years by downplaying wolves' inherent viciousness.

As this book is being written, hungry wolves—hunting for garbage, killing pets, and testing humans—have been showing up in broad daylight in the city limits of Sun Valley, Idaho; Jackson Hole, Wyoming; Anchorage, Alaska; Reserve, New Mexico; and Kalispell, Montana. There are documented deaths of three unarmed people killed by wolves in the last decade, and many others have been attacked; some attacks were reported, and others not.

People need to distinguish fact from fiction and to appreciate wolves for what they really are.

The Wolf as a Cash Cow

> **Every great cause begins as a movement,**
> **becomes a business, and ends up as a racket.**
> Eric Hoffer, *The True Believer*

The leader of the pack of environmental and animal rights groups promoting saving the wolf is Defenders of Wildlife, an organization that idolizes wolves so much that the wolf is its logo. Founded in 1947 as Defenders of Furbearers, the group's initial target was banning steel-jaw leg-hold traps and poisons. Defenders began with one staff person and 1,500 members. Today, Defenders' mission statement contains language about promoting "science-based, results-oriented wildlife conservation," saving "imperiled wildlife," and championing the Endangered Species Act. It does these things with a staff of 150 and, it says, over one million members.[13]

According to Charity Navigator, in fiscal year 2010, Defenders of Wildlife had an annual budget of $32,595,000 and its president received an annual salary of $295,641.[14] Much of this money is the result of its "Save the Wolf" campaign. The American Institute of Philanthropy gives Defenders of Wildlife a "D" for the percentage of its budget spent on charitable purposes (43%) and notes that the organization sends out 10-12 million pieces of direct mail each year to draw in about $29 million.[15] This U.S.P.S. tidal wave hardly seems "green." Appeal letters are written by special direct mail and telemarketing firms that crank out the same types of letters for all kinds of causes and use focus groups to determine the most emotionally-engaging pitch.

Sometimes the science behind such appeals is questionable or wrong, but what you read is crafted to have the greatest potential for drawing in donations. For example, a common emotional hook is a crisis—fear that, if you don't give, something terrible will surely happen. The opening line for the 2012 Defenders' "Campaign to Save America's Wolves" on its website was this: "America's Wolves Need Our Help!" which was followed by this: "America's wolves were nearly eradicated in the 20th century. Now, after a remarkable recovery in parts of the country, our wolves are once again in serious danger."[16] Of course, if something bad does occur, then they can make another appeal based on guilt—if you had donated more, this would not have happened.

Another popular appeal is sentimentality, such as this Defenders' ad: "Won't you please adopt a furry little pup like Hope? Hope is a

cuddly brown wolf... Hope was triumphantly born in Yellowstone."
For the record, the U.S. Fish and Wildlife Service does *not* name
wolves; it gives them numbers, and it does *not* put them up for adoption.
Nonetheless, the World Wildlife Fund also offers donors the chance to
"Adopt a Wolf."[17]

Another popular appeal is to identify a dastardly, cruel enemy,
who, if not stopped, will surely cause great damage or extinction of a
species or already is doing so. The American Farm Bureau has been a
favorite target for its stance on wolves. In short, from a psychological
standpoint, the wolf-loving organization must operate as a crisis addict
to keep itself in business, for the new wolf is its cash cow.

To Defenders' credit, it initially had a Wolf Compensation Fund
to pay ranchers for livestock lost to wolves. However, on August 20,
2010, Defenders announced cancellation of the wolf compensation
fund and declared that states and tribes could take over the cost while it
(Defenders) would work with farmers and ranchers on non-lethal means
of wolf control.[18] This decision has placed a heavy burden on states,
diverting funds that could have served more critical wildlife needs.

Ranchers and farmers, additionally, report that compensation,
when paid, is paid only for confirmed wolf kills, not for kills where
what killed the cow or sheep in the first place cannot be determined.
(See Chapter 12: Collateral Damage Identification.)

Pro-Wolf Organizations

A Google Search for "Save Wolves" today comes up with 35.9
million results; many organizations and petitions joined the pack
when they saw that wolves were cash cows. In addition to Defenders
of Wildlife, some of the best-known "Save the Wolf" groups, which
use both "educational" campaigns and litigation, include the Center for
Biological Diversity, Earthjustice, Friends of Animals, Humane Society
of the U.S., World Wildlife Fund, and the Sierra Club.

A major theme running through "Save the Wolves" appeals is that
wolves are in danger of extinction. This, of course, is false. There may
be as many as 100,000 wolves in North America, and despite USDA
Wildlife Services, USFWS, U.S. Park Service, state natural resource
departments, and agricultural agencies removing problem wolves,
road kills, natural mortality, and legal and illegal hunting, the North
American wolf population is growing appreciably and spreading.

Defenders of Wildlife also uses public opinion polls to support its
advocacy, but not the same polls that unbiased researchers conduct. For
example, on the Defenders' website, it reports that, in response to an

NBC Dateline segment on wolves, more than 1,500 viewers responded, with less than 11% saying they were opposed to wolf reintroduction.[19] Defenders selectively draws on a few surveys to show that people "everywhere" favor wolves, although they acknowledge that people in rural areas are more likely to feel negatively about wolves.

The Natural Resources Defense Council (NRDC), which funds "wolf advocates" in the field, states that "persistent intolerance among humans..." is one of the two greatest threats to wolves, the other being loss of habitat.[20] Intolerance is ironically characteristic of some of the NRDC advocates, who engage in vitriolic Internet media campaigns against people with different points of view. Wolf advocates often discredit and attack those people who want wolves managed, portraying them as intolerant, fearful, uninformed, naive, somehow inferior, mentally unsound, unethical, and even a threat to society. And if someone targeted does lose his temper, it only helps the pro-wolf advocate organizations raise money as they can say, "These people are so unreasonable."

The Humane Society of the U.S. (HSUS), the nation's largest animal rights group, said on its website in November 2011: "Social, family-oriented, and highly adaptable, wolves have a lot in common with humans. And while there's no record of a healthy, wild wolf ever attacking a person in the United States, old myths and fears plus competition for land and prey threaten the survival of this wild canine."[21] The April-May 2012 issue of Charity Watch's Charity Rating Guide and Watchdog Report, gave HSUS a "D" (unsatisfactory) rating for the second year in a row, based on how much money it spends to raise money. In contrast, PETA gets a "C+", and the American Red Cross, and the Wildlife Conservation Society get an "A".[22]

The Earthjustice website's wolf page, entitled "Wolves in Danger," proclaims: "For the past decade, Earthjustice was instrumental in protecting the gray wolves in court. Our work is now shifting to Congress where there have been legislative attempts to derail wolf recovery and push these animals to the brink of extinction." This statement is hardly accurate, but that does not prevent them from using it.

None of the major pro-wolf groups say much about wolves representing a danger to people and livestock due to about 50 diseases they may carry. Instead, the Sierra Club uses the slogan "Those faithful shepherds" on their wolf campaign page. The Sierra Club states that they "Educate the public about wolves and their biology to dispel negative stereotypes."[23] The Sierra Club calls the wolf a "Species at

Risk" when, in reality, wolves are plentiful in many parts of North America as well as abroad.

"Save the Wolf!" messages ultimately are picked up by the general media, including PBS, which further inflames polarization.[25] Often the messages of the wolf advocates are misleading or simply wrong. For example, a common campaign message of many pro-wolf groups is that browsing elk are destroying aspens in Yellowstone National Park. Introducing wolves, they say, is the best way to restore the aspens by establishing a "landscape of fear" that keeps elk away from aspens, which results in habitat improvement that benefits many other species. In the September 2010 issue of *Science Daily*, 15 years after wolves were released into Yellowstone, USGS scientist Matthew Kauffman reported that elk were continuing to browse on aspens, regardless of wolves. Kauffman stated: "This study not only confirms that elk are responsible for the decline of aspen in Yellowstone beginning in the 1890s, but also that none of the aspen groves studied after wolf restoration appear to be regenerating, even in areas risky to elk."[26]

In response to the 2012 arrival in California of one wolf from Oregon, the Center for Biological Diversity sent out an e-mail message that began: "Wolves are smart, fast, curious, and strong. It was inevitable that they'd find their way to California. It is not inevitable, though, that they'll survive. The livestock industry has already vowed to kill any wolf it sees and is gearing up its lobbying machine to keep them out of the state." The message ended with this plea: "Make a generous gift to support our California Wolf Fund."[27] That email message is a perfect example of negative stereotyping and polarization.

USFWS Efforts

There could be no successful campaign to bring back wolves without the support of the U.S. Fish and Wildlife Service (USFWS), which has jurisdiction as wolves are an endangered species. Since the 1980s, USFWS has promoted wolf recovery programs all around the U.S. through news media, public hearings, interviews, websites, exhibits, and personal appearances. One of the most visible parts of this program was the widespread public review of the Environmental Impact Statement (EIS) that led up to the 1995-96 relocation of Canadian wolves into the Northern Rockies. Ten years after the relocation took place, the Wyoming Game and Fish Department (WG&F) did a review of the predictions made by the USFWS in that EIS. This is what they found: "Despite research findings in Idaho and the Greater Yellowstone Area and monitoring evidence in Wyoming that indicate wolf predation is having an impact on ungulate populations that will reduce hunter opportunity if the current impact levels persist, the Service continues to rigidly deny wolf predation is a problem."

The 1994 EIS predicted that the presence of wolves would result in a 5-10% increase in annual visitation to Yellowstone National Park. On this basis, the EIS forecast wolves in the region would generate $20 million in revenue to the states of Idaho, Montana, and Wyoming. However, WG&F reports that annual park visitation has remained essentially unchanged since wolf introduction. (A later chapter will examine in detail the real economics of the wolf reintroduction.)

WG&F stated: "Wolf presence can be ecologically compatible in the Greater Yellowstone Area only to the extent that the distribution and numbers of wolves are controlled and maintained at approximately the levels originally predicted by the 1994 EIS—100 wolves and 10 breeding pairs. USFWS … has a permanent, legal obligation to manage wolves at the levels on which the wolf recovery program was originally predicated, the levels described by the impact analysis in the 1994 EIS."[28]

In a September 2010 interview with the *Bozeman Daily Chronicle*, two leading federal wolf biologists, Ed Bangs of the USFWS and Doug Smith of the U.S. National Park Service, state that, from the beginning, their long-term goal has been to delist wolves so they can be managed, which would mean controlled hunting.[29] A number of wildlife biologists, including Dr. L. David Mech, Chair of the World Conservation Union Wolf Specialist Group, support that goal.[30] Nonetheless, a lot of the general public believe that any hunting of wolves is wrong, largely because the general public has been sold a bill of goods. For example, a 1999 poll in Minnesota showed that, while people favored wolf management, they preferred non-lethal methods.[31] A 2004 poll in Ontario, which does have a native wolf population but not near any population centers, found that 70% of the public opposed hunting wolves, 88% opposed sport hunting of wolves, and 82% did not support killing wolves for their pelts.[32] Such opposition also holds today for Scandinavia, where a 2003 study found that, while a majority supported hunting wolves if livestock were being harmed or wolves were entering cities, they did not favor a general wolf hunt.[33] The problem is that nonlethal wolf controls seldom work, and if they do, it is a short-term, costly fix. Issuing wolf sport hunting licenses is one way for state agencies to try to pay for the economic burden of wolf management, which is considerable; however, when this happens, it raises the hackles of anti-hunting groups.

Ohio State University researcher Jeremy Bruskotter reported in *Bioscience* in December of 2010 that the USFWS is essentially supporting the wolf advocates by suppressing research on the real and potential negative consequences of wolf populations and that this is contributing to the polarization of public opinion about wolves as well as misleading people about the negative consequences of expanding wolf populations.[34]

Selling Wolves to the General Public

Exhibits. In 1986, writer Rene Askins launched a traveling exhibit, the Wolf Fund, whose primary purpose was promoting the release of wolves into Yellowstone National Park. To her credit, Askins closed the Wolf Fund when the first wolf was released into Yellowstone.[35]

Defenders of Wildlife at one time had a traveling wolf exhibit that was the largest such wildlife exhibit in the U.S. Many parks in the U.S. and Canada have educational exhibits about wolves, especially Yellowstone National Park and Algonquin Provincial Park in Ontario, where they also feature "howl-ins." The same is true for USFWS refuges where wolves are found. It's never pointed out, however, that visiting a park and howling with wolves is a much different experience than living with them day to day and suffering loss of property from their killing.

Wolf Education Centers. In both the U.S. and Canada, there are a number of "Wolf Education Centers" where people may see wolves in captivity and be exposed to exhibits and educational programs about wolves. Some of the most popular centers include the following: Wolf Park in Battle Ground, Indiana[36]; Wolf Education Research Center in Winchester, Idaho[37]; Colorado Wolf and Wildlife Center, Divide, Colorado[38]; Wolf Song of Alaska Education Center, Eagle River, Alaska[39]; Wolf Conservation Center, South Salem, New York[40]; and Northern Lights Wolf Center, Golden, British Columbia.[41]

One of the most grounded wolf education centers is the International Wolf Center in Ely, Minnesota. Located in the center of the Minnesota wolf population, it was launched in 1989 by Dr. L. David Mech. Mech was once a firm believer that wolves in North America did not attack people, but he realized the limits of his early statements on human attacks. He has reversed his position and has made Mark McNay's landmark study on attacks available on his website, has sponsored a conference on attacks in Europe and Asia, and supports managing wolf populations with hunting.[42]

The International Wolf Center serves about 50,000 visitors a year and conducts classes, workshops, and lectures in the U.S. and Canada. Such a facility near Yellowstone National Park might enable visitors to see wolves and learn about them without the need to try to keep large packs of wolves running free in the Park, especially near the highway where they can become habituated or spill over into nearby areas, reducing elk and moose populations, not to mention off-park domestic animals.

Books. There have been hundreds of books written about wolves, especially since Earth Day 1970.[43] Here we will spotlight a few that have gained recognition and contain misleading information.

White Fang by Jack London (1906). Following London's very successful novel, *Call of the Wild*, about a domestic dog that returns to a wild state, his 1906 novel *White Fang* is about a wild, three-quarter wolf wolf-dog that becomes domesticated. Following its publication, Theodore Roosevelt declared that London was a "nature faker" and that some of the scenes in *White Fang* were "the sublimity of absurdity."[44] Nonetheless, *White Fang* has been made into several films, including a 1991 adaptation starring Ethan Hawke.

A Sand County Almanac by Aldo Leopold (1949). When he was just out of college and working in the Southwest for the U.S. Forest Service, Leopold believed that predators—bobcats, wolves, cougars, and black and grizzly bears—should be removed from special areas so that big game populations like deer and elk could build up reserves and spill over into adjacent lands, increasing recreational opportunities for hunters and wildlife watchers. Teddy Roosevelt supported Leopold, as did ranchers, farmers, and hunters. The Kaibab Plateau, adjacent to the Grand Canyon, was chosen as a place to implement Leopold's plan. When predators were eliminated at the Kaibab Plateau, the resident deer population exploded from less than 10,000 to nearly 100,000. But the deer stayed on the reserve and ultimately ate all the food, leading to massive starvation, a horrific crash in deer population, and destruction of habitat that lasted for years after. That incident changed Leopold's attitude toward predators.

One of the most commonly quoted pro-wolf passages appears in Leopold's masterful treatise on man and nature, *A Sand County Almanac*, where he recounts an incident in 1909 when he shot a female wolf, but didn't immediately kill her. Approaching the wounded animal to administer the final shot, he looked at the old she-wolf and watched "a green fire dying in her eyes. I realized then, and have known ever

These photos are included in an effort to give the reader an accurate visual representation of the significant size difference between the imported Canadian wolf and the native Mexican wolf.

When this wolf was legally shot on February 15, 2010, he weighed 127 pounds and was approximately eight years old. The wolf had been collared in 2006 by Idaho Fish and Game, at which time he weighed 127 pounds, and they estimated he was four. The largest wolf

Starker Leopold and his wolf trophy. 1948.

weighed by Douglas W. Smith, wolf biologist for Yellowstone National Park, weighed 148 pounds. For reference, the woman in the photo is 5'4" tall. [47]

The photo on the left needed no photoshopping to show the major difference in size between the two species of wolf. The untouched photo tells the story. However, a signed affidavit from the woman in the the photo is on file.

since, that there was something new to me in those eyes, something known to her and the mountain."[45]

This story fits perfectly into the mindset of "Save the Wolf" people, but it would be a mistake to think that Leopold believed that predatory animals like bears, cougars, wolves, bobcats, and coyotes should not be managed. In 1933, Leopold, the first professor of wildlife management, wrote in his textbook *Game Management* (the first college wildlife management text, it remains in use by many colleges today) in the chapter on Predator Control:

Predatory animals directly affect four kinds of people: (1) agriculturists, (2) game managers and sportsmen, (3)

*students of natural history, and (4) the fur industry. There is
a certain degree of natural and inevitable conflict of interest
among these groups. Each tends to assume that its interest is
paramount. Some students of natural history want no predator
control at all, while many hunters and farmers want as much
as they can get up to complete eradication. Both extremes are
biologically unsound and economically impossible. The real
question is one of determining and practicing such kind and
degree of control as comes nearest to the interests of all four
groups in the long run.[46]*

Aldo Leopold and his entire family were life-long avid hunters.
Estella, his wife, was the Wisconsin state women's archery champion.
In the Aldo Leopold Archives at the University of Wisconsin-Madison,
there is a photo of Aldo's son, Starker (who became a very prominent
wildlife biologist and university professor, and who edited *A Sand
County Almanac* after his father's death) proudly standing beside a
Mexican wolf that he shot in New Mexico in 1948.[47]

Wild Animals I Have Known by Ernest Thompson Seton (1898).
Naturalist, artist, writer, and predator bounty hunter, Seton was an early
pioneer of the modern school of animal fiction writing. His most popular
work, *Wild Animals I Have Known*, contains the story of his killing of
a renegade Mexican wolf named Lobo, "The King of the Currumpaw."
Seton later became involved in a literary controversy about writers
who fictionalized natural history and distorted wildlife biology and
behavior—"nature fakers" as Teddy Roosevelt called them. In a 1903
article in the *Atlantic Monthly*, John Burroughs charged Seton with
purposefully deceiving people with his writing. The controversy lasted
four years and involved many important American environmental and
political figures of the day, including Teddy Roosevelt, who negotiated
a deal with Seton to clean up his act.[48] There will be more about this in
the section on the documentary film made about Seton's wolfish tale,
"The Wolf That Changed America."[49]

Never Cry Wolf by Farley Mowat (1963). Mowat's humorous
account features a young government biologist, who in 1958 is flown to
the tundra plains of Northern Canada to study the area's wolf population
and gather proof of the ongoing destruction of caribou herds by wolves.
After locating on the remote tundra, the biologist contacts wolves as he
discovers a den with pups and three devoted protectors of the young. In
the absence of caribou, the wolves happily feed on mice and lemmings.
The biologist meets two Inuit, who tell him their own stories about the

wolves. As he learns more and more about the wolf, he comes to fear the onslaught of hunters out to kill the wolves for their pelts. Ultimately, he runs naked with the wolves as they chase a herd of caribou.

The book is supposedly based on a real-life experience. However, Eskimos refer to Farley Mowat as "Hardly Know It." Scientists agree. Writing a review in *Canadian Field-Naturalist* in 1964, Canadian Wildlife Federation officer Alexander William Francis Banfield, who supervised Mowat's field work, accused Mowat of blatantly lying; for starters, Mowat was part of a team of three biologists and never alone. Banfield also pointed out that a lot of what was written in *Never Cry Wolf* was not derived from Mowat's first-hand observations, but was lifted from Banfield's own works, as well as those of Adolph Murie's studies of wolves on Mount McKinley. Ultimately, Banfield compared *Never Cry Wolf* to "Little Red Riding Hood," stating that "both stories have about the same factual content."[50]

Wolf biologist L. David Mech writes about the book: "Whereas the other books and articles were based strictly on facts and the experiences of the author, Mowat's seems to be basically fiction founded somewhat on facts."[51] And animal behaviorist Dr. Valerius Geist calls *Never Cry Wolf* "a brilliant literary prank."

Of Wolves and Men by Barry Holston Lopez (1978). Written before Mark McNay's documentation on wolf attacks and the deadly wolf attacks of 2005 and 2010, Lopez explores many aspects of the relationship between people and wolves through history and clearly states that wolves can and do kill people, but he does not acknowledge any attacks on white people in North America in the 20th century. After surveying literature, mythology, and folklore from around the world, and trying to raise two red wolves, in the *Of Wolves and Men* epilogue, Lopez concludes: "Wolves don't belong living with people. It's as simple as that."[52]

Vicious: Men and Wolves in America by Jon T. Coleman (2004). An Associate Professor of History at Notre Dame, Coleman admits at the outset that he is an "animal person." *Vicious* begins with a description of John James Audubon watching an Ohio farmer catch three wolves in a pit trap and then slowly kill them. He explains the farmer's behavior as an example of "theriophobia," an excessive fear of wild animals. "Oblivious to the actual behavior of wolves, anti-wolf people based their hatred on 'myths, tales, and legends."[53] To generalize all people who do not like wolves as being the same is engaging in cheap stereotyping, which is hardly scholarly or accurate. Coleman fails to say whether the farmer had lost livestock, pets, or even friends to wolves prior to this time, which would mean the farmer was simply very angry and seeking revenge.

On the same page, Coleman states, "There is no record of a non-rabid wolf killing a human in North America since the arrival of the Europeans." The irony is that Audubon personally investigated and confirmed the death of a man by three wolves in 1830.[54]

Two problems with finding information about wolf attacks is that, until very recently, there was no reliable recording system; you had to look for newspaper reports. And when people are lost in the woods and never found or their remains are found later, confirming if they were killed by wolves or not is very difficult if not impossible. However, when Young and Goldman looked for records of wolf attacks on humans before 1900 in North America, they found 30 accounts of attacks and six possible human kills.[55]

Coleman's book came out in 2004, two years after Mark McNay's report on wolf attacks, yet he gives no mention of this research. Could the wolf attack deaths of Kenton Carnegie in 2005, Taylor Mitchell in 2009, and Candice Berger in 2010 been averted if people like Coleman were not spreading misinformation about potential dangers of wolves? None of these three young people were carrying firearms or even pepper spray when they were attacked.

On page 14 of his book, Coleman states, "In 1995, the Fish and Wildlife Service set 14 wolves from Alberta, Canada, loose in Yellowstone. Nine years later, the population had grown to 148 predators, and packs of 'non-essential' gray and Mexican wolves loped in Idaho, Arizona, and New Mexico." Actually a total of 66 Canadian wolves were released into Yellowstone and Idaho by the USFWS, establishing "experimental, non-essential" populations, according to Article 10(j) of the Endangered Species Act. In 2005, the wolf population of the Greater Yellowstone Area alone was estimated at a minimum of 325 wolves, not 148. Adding the wolves in Idaho with those in the Yellowstone area and those in Montana, some of which were already living in the state before the 1995-96 releases, results in at least 1,500 wolves in the Northern Rockies, and some say the number is at least 3,000.

In March 1998, the USFWS released three packs of Mexican wolves, bred in captivity, into the Apache-Sitgreaves National Forest in Arizona, and 11 wolves into the Blue Range Wilderness Area of western New Mexico. There are currently approximately 75 Mexican wolves in the wild in both Arizona and New Mexico, three-fourths of the original goal.

Coleman goes to great lengths to deride the negative mythology, folklore, and fairy tales that originated from Europe, but fails to report the numerous well-documented attacks and killings of people by wolves throughout Eurasia and the importance of folk tales, fables, and fairy tales in educating children to avoid becoming victims.[56]

Despite the bias of the author, the many inaccuracies of the work, and his lack of understanding of psychology, folklore, and mythology, *Vicious* won awards from the American Historical Association and the Western Historical Association. A falsehood often repeated does not, however, make it true.

Magazine Articles. There have been thousands of articles about wolves in major national magazines and newspapers including *National Geographic*, *New York Times*, *Audubon*, *Sierra*, *Newsweek*, and *Wall Street Journal*, as well as numerous local newspapers and magazines. The majority of these favor wolf restoration, and most are written by journalists parroting what they are told by wolf advocates from environmental groups and federal and state agencies charged with wolf management. An article worth noting illustrates one way the "Save the Wolf" campaign has often used self-righteousness to create enemies to hate. "Cry, Wolf: How A Campaign of Fear and Intimidation Led to the Gray Wolf's Removal from the Endangered Species List" was written by James William Gibson and printed in the Summer 2011 issue of *Earth Island Journal*.[57]

Gibson, a professor of sociology at California State University at Long Beach, claims that "an extreme right-wing culture that celebrates the image of man as 'warrior' recognizes only local and state governance as legitimate and advocates resistance, even armed resistance, against the federal government. ... Montana's anti-wolf alliance of ranchers, hunters, and militia sympathizers is built around some shared myths that focus on the evils of wolves in general and the Rockies' wolves specifically" and is engaging in "fear-driven demagoguery" targeting pro-wolf groups and individuals. According to Gibson, this is the real reason why wolves were delisted—and not biology. In a December 8, 2011, op-ed article in the *Los Angeles Times*, Gibson describes proponents of hunting wolves to manage their population as "paramilitary militia advocates." He goes on to say: "Afraid for their lives and their families, regional wolf advocates stopped participating in public hearings held by fish and game agencies and legislative committees and retreated to the relative safety of the Internet to spread their message." He charges that "a dysfunctional political system in which fear—both irrational fear and fear harnessed for political gain—determines policy."

Gibson claims that, "In the entire 20th century, wolves attacked about 15 people in North America, killing none." He does admit that, "In 2010, wolves did kill a woman jogging on the outskirts of her Alaskan town," but he makes no mention of other attacks and fatalities. Little or no mention is made of the numerous professional wildlife biologists and the many present and former state and USFWS biologists that support wolf management and controlled hunts—none of which belongs to right-wing paramilitary organizations (nor do any of the contributors to this book).

This is not to say that some ranchers, hunters, and others have not written or said negative things about wolf supporters. Gibson, however, makes absolutely no mention of the swarm of threats in electronic media and the direct personal attacks that have been made by wolf advocates to anyone who differs with their view, including many contributors to this book. He completely fails to acknowledge how the whole situation is made worse by "wolf advocates," who are paid to carry on campaigns against anyone who questions their agenda, to keep controversies going, and to fuel crises that can be used by their employers to raise money. (For an interesting case study, Google the five-part series "Environment Inc." by Pulitzer Prize-winning *Sacramento Bee* investigative journalist Thomas Knudsen, documenting how large environmental groups use inflammatory articles, exaggeration, and even purposeful lying to keep themselves visible and money flowing in.[58])

A most important book about the politics of the wolf wars of the Northern Rockies is *Yellowstone Wolves: A Chronicle of the Animal, the People and the Politics*, by Cat Urbigkit.[59] Urbigkit, a Wyoming journalist and sheepherder, traces the history of wolves in the Yellowstone area, clearly showing that—while hunting, trapping, and poisoning wolves knocked the population down in the mid-1900s—wolves never were eradicated from that area. She traces the history of groups, especially Defenders of Wildlife, which began sending out Action Alerts as early as 1992, and the federal government, which declared at public meetings about wolf reintroduction in the early 1990s: "As the wolf population grows, support can grow along with it..." Urbigkit's story of how she and her husband filed suit to block the USFWS wolf relocation program as wolves were already present in the Northern Rockies is a later chapter in this book.

In response to Gibson's inflammatory and negative stereotyping, Urbigkit quotes a January 1999 editorial in the national newsletter of the American Farm Bureau by AFB President Dean Kleckner that describes what happened when the AFB filed a lawsuit to have the introduced Canadian wolves removed from the Northern Rockies and returned to Canada:

> *Defenders of Wildlife launched a nationwide campaign against Farm Bureau in the press, television, radio, and Internet, falsely describing our organization and our lawsuit. (Note: AFB did not call for killing off wolves. It simply wanted them removed from the area.)*
>
> *Wolf stocking advocates, incited by Defenders of Wildlife, organized a campaign of harassment and intimidation against the Bureau with the aim of forcing us to abandon our own farmer-written policies and drop the lawsuit. We have received several bomb threats and threats against the lives of Farm Bureau officers and their families. Even a federal judge's life was threatened. (Note: The judge found that wolves could be captured humanely and returned to Canada.)*
>
> *Working Assets (in support of Defenders of Wildlife) brags about sending us 34,000 letters and calls. ... Defenders of Wildlife and Working Assets crossed a line when they intended to shut down our phone system and encouraged callers to harass us into dropping our case against the Department of Interior's illegal program.*

The American Farm Bureau filed a complaint with the Federal Communications Commission and the FBI about what they had been subjected to, which is how such threats should be handled.

Films. There are many wolf films—documentaries and feature films.[60] Nearly all support the "Save the Wolf" movement, and many have inaccuracies, such as "Wild Wolves" on PBS NOVA, which includes an interview with USFWS wolf biologist Ed Bangs, where Bangs stated:

> *The studies that we've done and that other people have done indicate that wolves normally kill less than one-tenth of one percent of the livestock available to them. To date, in the past 15 years in the Northern Rocky Mountains, we've lost an average of about five cattle and five sheep per year to wolf depredations.*

Those statistics do *not* represent what is actually happening. For example in 2010, wolves killed 65 livestock (cattle and sheep), 3 horses, and 1 dog in Wyoming alone.[61] Bangs then made three incorrect statements:

> *You know when you teach your dog to not go out of the yard or not go in the flower bed—and your dog learns that for the rest of its life? It's just something it won't do? That's the same reason that wolves never attack people. Behaviorally, they just don't recognize people as anything they want to screw with. And they live their entire lives without ever trying it.*[62]

Not all films support the "Save the Wolf" movement, however. Well researched and based on irrefutable facts is J.D. King's most excellent 2011 documentary, *Crying Wolf.* King described his hour-long film as follows: "My three loves—cameras, wildlife, and truth—met in *Crying Wolf.* The goal of this film is to shed light on a topic encompassed by lies, myths, and hidden agendas; assert the rights of the people and the states over the central government; and largely change the way our broader culture looks at wolves, wildlife management, and ecology." King told this book's editor that, after interviewing numerous people for the film, the most important thing he learned was that the wolf reintroduction was NOT about wolves at all. "It's about an agenda of the environmentalists and the U.S. Fish and Wildlife Service to, basically, eliminate hunting, trapping, logging, etc. They are moving towards corridors and connectivity, and they want to create more wilderness areas and then remove man from them. One of my goals is to educate the general public about the wolf program, which is surrounded by misconceptions."

In feature films, we sometimes find a more balanced coverage of wolves, although that was not the case with *Never Cry Wolf* based on Farley Mowat's 1963 best-selling book, which is an excellent example of misrepresenting the facts with emotion. The book was adapted for the big screen in 1983 as a featured film by Carroll Ballard. However, in the May 1996 issue of the popular Canadian magazine *Saturday Night*, John Goddard wrote a heavily researched article entitled "A Real Whopper," in which he poked many holes in Mowat's claim that the book was non-fictional. Goddard also reported that Mowat told him, "I never let the facts get in the way of the truth." While Mowat called Goddard's article, "bullshit, pure and simple," he refused to refute Goddard's main claims.[63]

Films often, like books, distort the behavior and tendencies of wolves in order to romanticize them, which does injustice to the perception of wolves and, consequently, does injustice to the people who have to live with wolves. An excellent example of that was the 1990 movie *Dances With Wolves* featuring Kevin Costner and based on the 1988 book of the same name by Michael Blake. Costner, as a First Lieutenant, gets posted to an isolated fort on the Western frontier where he not only becomes friends with the Lakota Sioux but also with a wolf. Memorable for its poignant ending, *Dances With Wolves* has Costner ultimately losing not only his Indian friends, but also his horse and, even worse, his wolf friend when they are coldheartedly shot by the troops.

Conclusion

As the federal government seeks to turn wolf management over to states and habitat conservation organizations seek to have the wolf delisted and managed by hunting, pro-wolf groups counter with a seemingly endless and escalating barrage of legal challenges accompanied by mass mailings and other expensive propaganda. Already in 2013 as I finish this book, three different lawsuits have been filed, challenging Wyoming's plan to manage wolves, one in federal court in Colorado and two in federal courts in Washington, D.C.

However, people *do* want to hear the truth about wolves. In the following pages, you will learn more about what that is.

🐾 🐾 🐾 🐾 🐾 🐾 🐾 🐾

[1] Chase, Alston, *In a Dark Wood: The Fight Over Forests and the Rising Tyranny of Ecology*, New York, NY: Houghton-Mifflin, 1995, xiii.
[2] Kellert, R.S., "Public Perceptions of Predators, Particularly the Wolf and Coyote," *Biological Conservation* 31, 1985, 167-189.
[3] Urbigkit, Cat, *Yellowstone Wolves: A Chronicle of the Animal, the People, and the Politics*, Ohio: McDonald and Woodward Publishing Company, 2008, 28.
[4] Murie, Adolph, *The Wolves of Mount McKinley*, University of Washington Press, 1985.
[5] Allen, Durwood and L. David Mech, "Wolves Versus Moose on Isle Royale," *National Geographic*, February 1963, 200-219.
[6] Leopold, A. Starker, et al., *Wildlife Management in the National Parks,* National Park Service, 1963.
[7] Williams, C.W., Gordan Ericsson, and Thomas Herbelein, "A Quantitative Summary of Attitudes Toward Wolves and Their Reintroduction, 1972-2000," *Wildlife Society Bulletin* 30, 2002, 1-10.
[8] Lopez, Barry Holston, *Of Wolves and Men*, Scribners, 1978, 4.

9   White, T.H., *The Book of Beasts*, New York, NY: Putnam, 1954, 59.
10 Lopez, Barry, *Ibid.*
11 White, Jonathan, *Talking on the Water: Conversations about Nature and Creativity*, Sierra Club Books, 1994, 121-136.
12 Jaffe, Aniela, "Symbolism in the Visual Arts," *Man and His Symbols,* edited by Carl Jung, New York, NY: Dell, l968, 266.
13 http://www.defenders.org/about_us/history/index.php
14 http://www.charitynavigator.org/index.cfm?bay=search. summary&orgid=3605
15 http://www.charitywatch.org/articles/defendersofwildlife.html
16 http://action.defenders.org/site/PageServer?pagename=savewolves_ homepage
17 http://www.worldwildlife.org/gift-center/gifts/Species-Adoptions/Gray-Wolf.aspx?gid=13&sc=AWY1000 WCGP1&searchen=google&gclid=CLz rv6rKw6ACFR6kiQodFmSybA
18 http://www.defenders.org/programs_and_policy/wildlife_conservation/ solutions/wolf_compensation_trust/
19 http://www.defenders.org/programs_and_policy/wildlife_conservation/ imperiled_species/wolves/america_votes_yes!_for_wolves.php
20 http://www.nrdc.org/wildlife/habitat/esa/rockies02.asp
21 http://www.humanesociety.org/animals/wolves/
22 http://humanewatch.org/index.php/site/post/hsus_earns_some_detention/
23 http://www.sierraclub.org/lewisandclark/species/wolf.asp
24 http://www.biologicaldiversity.org/publications/earthonline/endangered-earth-onlineno605.html
25 http://www.pbs.org/now/shows/609/index.html
26 http://www.sciencedaily.com/releases/2010/09/100901111636.htm
27 http://www.pinedaleonline.com/wolf/wolfimpacts.htm
28 http://www.bozemandailychronicle.com/news/article_32c72a40-3c5b-11df-91e5-001cc4c002e0.html; http://digitalcommons.unl.edu/cgi/viewcontent.cgi?article=1040&context= vpc16
29 http://www.wolf.org/wolves/learn/scientific/challenge_mech.asp
30 http://www.wolf.org/wolves/learn/intermed/inter_human/survey_shows.asp
31 http://www.wolvesontario.org/wolves/Full%20Wolf%20Poll%20March%2 011%202004%20revised.pdf
32 http://www.wildlifebiology.com/Downloads/Article/477/En/10_4_ericsson. pdf
33 http://www.sciencedaily.com/releases/2010/12/101206093703.htm; http://bruskotter.wordpress.com/
34 http://www.people.com/people/archive/article/0,,20108666,00.html
35 http://www.wolfpark.org/
36 http://www.wolfcenter.org/default.aspx
37 http://www.wolfeducation.org/

[38] http://www.wolfsongalaska.org/education_center_tour.html

[39] http://www.nywolf.org/

[40] http://www.northernlightswildlife.com/

[41] http://www.wolf.org/wolves/index.asp

[42] http://www.inetdesign.com/wolfdunn/wolfbooks/

[43] Carson, Gerald, "T.R. and The Nature Fakers," *American Heritage*, February 1971, 22.

[44] Leopold, Aldo, "Thinking like a Mountain," *A Sand County Almanac*, Oxford University Press, 1949, 130.

[45] Leopold, Aldo, *Game Management*, University of Wisconsin Press, 1933, 230.

[46] Holding the wolf is Heidi Leavitt of North Fork, Idaho, who sent Ted Lyon an email dated April 21, 2012, that read: "I received your letter requesting more info. That wolf weighed 127 pounds, approximately 9 years old. The wolf had been collared in 2006 by fish and game. They figured he was 4 at the time and weighed 127 then. I am 5 ft. 4 in. and weigh 140. Heidi Leavitt."

[47] http://digicoll.library.wisc.edu/WebZ/FETCH?sessionid= 01-43349-194506 3810&recno=2&resultset=3&format=F&next=html/nffull.html&bad= error/ badfetch.html&&entitytoprecno=2&entitycurrecno=2&entityreturnTo=brief

[48] Carson, Gerald, *Op. Cit.*

[49] http://factoidz.com/lobo-king-of-the-currumpaw/

[50] Banfield, A.W.F., "Never Cry Wolf," *Canadian Field Naturalist* 78, January-March 1964, 52-54.

[51] Mech, L. David, *The Wolf: The Ecology and Behavior of an Endangered Species*, Doubleday Publishing Co., N.Y, 1978, 389.

[52] Lopez, Barry Holston, *Of Wolves and Men*, Charles Scribner's Sons, 1978, 280.

[53] Coleman, Jon T., *Vicious: Men and Wolves in America*, Yale University Press, 2004, 3.

[54] Audubon, J.J., and J. Bachman, *The Quadrupeds of North America*, New York, NY: Wellfleet Press, 1851 – 1854.

[55] Young, S. and E. Goldman, *Wolves of North America,* Vol. 1 and 2, Dover, 1944.

[56] Von Franz, Marie-Louise, *Shadow and Evil in Fairy Tales*, Shambala, Boston: 1995.

[57] http://www.earthisland.org/journal/index.php/eij/article/cry_wolf

[58] *icecap.Us/images/uploads/ENVIRONMENTINC.doc*

[59] Urbigkit, Cat, *Yellowstone Wolves: A Chronicle of the Animal, the People, and the Politics,* McDonald and Wordwood, 2008.

[60] http://www.bullfrogfilms.com/catalog/wolf.html; http://wolves.biginterest2u.com/docs.html

[61] http://www.pinedaleonline.com/news/2012/02/Wyomingwolfpopulation.htm

[62] http://www.pbs.org/wgbh/nova/wolves/

[63] http://www.salon.com/1999/05/11/mowat/singleton/

Chapter 3

THE MYTH OF THE HARMLESS WOLF

By Ted B. Lyon

Holly Kuchera / Shutterstock.com

Editors' Statement of the Issues

"Save the Wolf" dogma indicates that wolves are not dangerous to humans, despite centuries of historical evidence to the contrary. This willingness to ignore historical records accompanies a propensity to rewrite the wolf's history and habits, leaving people unprepared.

Editors' Statement of the Facts

Centuries of historical evidence clearly and specifically recounts fatal wolf attacks. Wolves hunt in packs, thereby making them very dangerous to the relatively defenseless human. Unless wolves are routinely hunted, trapped, or shot at, they do not recognize or fear humans as the dominant species and instead view them as potential prey. Wolves are dangerous predators and should be viewed as such.

> ### There are no known gray wolf attacks on humans in modern times in North America.
> U.S. Fish and Wildlife Service website[1]

Predators

There are five species of animals in the United States that engage in the act of predation (one animal killing another for food). Let's break these American predators down by species.

Bears. There are an estimated 750,000 black bears in the U.S. According to wildlife biologist and bear expert Dr. Gary Alt, bears operate as lone individuals and kill an average of one to two people a year, primarily with the intent of predation.

There are less than 1,000 grizzly bears in the lower 48 states. Grizzlies primarily attack in defense of food and cubs. Attacks on humans average three to five per year, and occasionally they are deadly, as in the summer of 2011 when two people were killed in the Northern Rockies.[2]

Mountain Lions. There are an estimated 50,000 mountain lions—also called cougars, pumas, and panthers—in the lower 48 states; they are increasing in numbers and spreading eastward to feed on the mushrooming whitetail deer population. In August of 2011, a mountain lion was killed by an SUV in Connecticut, the first such sighting in a century.[3] Mountain lions are solo hunters that shy away from people, unless they run out of food or have never been hunted. Attacks per year for North America run from two to nine, with up to two deaths resulting. These attacks are almost always predatory, with cougars targeting runners or hikers, who must look like deer to mountain lions.

Dogs. There are 75 million dogs in the U.S. The Center for Disease Control's "Dog Bite: Fact Sheet" says that, each year, 4.7 million Americans are bitten by domestic dogs and that as many as 50 Americans die every year from dog attacks. Not surprisingly, 75% of the attacks and over 50% of the fatal attacks are by three breeds—Pit Bull, Presa Canario, and Rottweiler, all bred for aggressiveness. Most importantly, dog attacks are almost always defensive, unless the dogs have been bred to attack.[4]

Coyotes. Once a Western species, coyotes are now found in all states but Hawaii. No one is quite sure how many coyotes there are in North America; estimates widely range from 10 million to 100 million. The number of coyote attacks per year in the U.S. averages around 14, and it is rising as more and more coyotes move into urban areas including Chicago, San Francisco, and Los Angeles.[5] Human deaths from coyote attacks are rare, but on October 28, 2009, Canadian folk singer Taylor Mitchell was hiking in a provincial park in Nova Scotia when she was attacked and killed by two wolf-coyote hybrids.[6] Coywolves, or Eastern coyotes, are larger than Western coyotes, and most scientists believe they are hybrids between Western coyotes and gray wolves or red wolves.[7]

Wolves. Wolves are ancestors of dogs, but the "pure" wolf remains a very different species of wild canid. Estimates indicate that there are at least 6,000 gray wolves in the Northern Rockies and Northern Great Lakes states, 40,000 to 50,000 wolves in Canada, and 7,700 to 11,200 in Alaska—65,000-75,000 wolves for all of North America, possibly as many as 100,000.

There also are about 75 Mexican wolves in New Mexico and Arizona and less than 100 red wolves in the southeastern section of the United States.[9] Both Mexican and red wolves were raised in captivity and released into the wild. There is considerable debate if these animals are truly wild or not.

As the wolf genetics chapter in this book, "Canis Stew," explains, crossbreeding among wolves, coyotes, and domestic dogs will inevitably increase. In hybridizing with other canid species, especially domestic dogs, the wolf loses some of its instinctual learning and wildness, seemingly born semi-habituated, which is the case with the Mexican wolves of the Southwest. There are between 300,000 and 500,000 wolf-dog hybrids in the U.S., and their numbers are rising, even though such animals are illegal in 40 states. Typically, such animals are unpredictable, and people find that what started out as a cute puppy becomes a dangerous adult. All too often, wolf-dogs are responsible for attacks on livestock and people.[8] One might predict then that an increasing number of attacks on people, their livestock, and their pets is inevitable, not only due to habituation, but also hybridization.

Do North American Wolves Attack People?

As the quote from the U.S. Fish and Wildlife Service (USFWS) website that opens this chapter shows, for decades we have been told—in colleges across North America and in countless popular books, articles, and films—that wolves do not attack and kill people in North America and that wolves have not attacked people in North America for almost a century.[10] The reality is that this is *not* correct.

Wolves are voracious predators—they kill other animals to eat and for pleasure. They prefer large ungulates (hoofed mammals)—deer, elk, moose, bison, and caribou. A pack of three to five wolves will kill an average of two reindeer or caribou every three days. They can eat six to seven pounds of meat per day and over 10 pounds if they have not eaten for a while. That translates into between 2,200 and 3,600 pounds of meat per wolf per year. When wolves run out of wild meat, domestic livestock come next.

Always, wolves test conditions. If hungry and especially if humans do not appear to be a threat, wolves move closer and closer—killing pets and livestock, eating garbage, and ultimately stalking and attacking people. This is proven by the historical records of Europe and Asia, as well as North America.

Let us be clear at the outset that the record of wolf attacks on people does not reflect the total number of attacks for several reasons. The first is that there was no written record of wolf attacks in North America until the 1700s. In the oral tradition of American Indians, there are many stories of attacks, but there is no way to authenticate them. The second is that there has never been a central system for wolf attack reports. In general, newspaper reports are all that researchers have to go on. And third, if people are lost in the wilderness and not ever found or at least not for some time, they may well be victims of wolf attacks, but we will never know for sure. Thus, anyone who claims that there has never been a fatal wolf attack in North America is basing this statement on wishful thinking and opinion, not realism.

Based on what we can learn from scattered reporting, it is true that there have not been many wolf attacks in North America in the 20th century. A primary reason why is that wolves in the U.S. and Canada were relentlessly hunted, trapped, and poisoned beginning in the 1800s and continuing through the 1930s. Those few wolves that survived along the Canadian border in the Northern Rockies and Minnesota learned that they needed to retreat to wilderness areas and stay away from people to survive. Wolves are intelligent, and North American wolves became seldom seen out of the necessity to survive. What the numbers will look like in the 21st century if the wolf populations are allowed to grow unchecked is daunting, given current growth.

Another reason that more attacks have not occurred on people in North America is that the settlers arriving from Europe were heavily armed. As we shall see, in countries where people are not heavily armed, wolves have periodically killed many people. And, as this chapter will show, wolves do attack people in North America, and the number of attacks is increasing as the wolf population grows and these largest of the wild dogs increasingly come into contact with people, ranches, and towns, as well as coyotes and dogs with which to breed.

Wolf Attacks around the World

Originally, gray wolves were found throughout the northern hemisphere in every habitat where large ungulates were found. Saturating most of the region between 20°N latitude (mid-Mexico and India) and the North Pole, in temperatures from -40° to +40° C, the wolf inhabited areas as diverse as Israel, North Africa, China, Ireland, and Greenland. In North America, they were found from Mexico City north to the Bering Sea.[11] Wolves were once the most widely distributed mammal living north of the 15°N latitude in North America and 12°N in Eurasia, other than man.[12]

There may have been as many as two million wolves worldwide in earlier times, but today the global population is approximately one-eighth of that, because man has trapped, hunted, and poisoned wolves—not just in the United States, but worldwide—for centuries. A primary reason for the campaign against wolves has been the fear of attacks on livestock and people. Let us look at what history tells us about the propensity for wolves to attack prey other than wildlife.

Wolf Attacks on People in Europe and Asia

There is a long and well-documented history of wolf attacks in Europe and Asia. This is where most of our Western fables and fairy tales originated and, as you shall see, for good reason. Let us briefly look at some statistics about wolf attacks on people abroad.

France. Between 1580 and 1830, 3,069 people were killed by wolves in France; of this number, 1,857 of these victims were killed by wolves that were non-rabid.[13] In the winter of 1455, 40 people were killed by wolves in Paris. These wolves came to be known as the

"Wolves of Paris."[14] Between 1763 and 1767 in Gévaudan, Auvergne, and Languedoc, France, 99 people were killed by a wolf known as "The Beast of Gévaudan" and its offspring.[15]

Sweden. From December 30, 1820, to March 27, 1821, around Gysinge near the border of central Sweden, a wolf attacked 31 people, killing 11 children between 3½ and 15 years of age and one 19-year-old woman; most of the victims were partially consumed by the wolf. The wolf had been captured as a pup and raised in captivity for three to four years before it was released prior to the attacks.[16]

Italy. Between the 15th and 19th centuries, 440 people were killed by wolves in the Po Valley of Italy. For example, in 1704, 16 people were killed by wolves in what is today the popular tourist region of Varesotto in northern Italy.[17] Today there are between 500 and 1,000 wolves in Italy. According to Luigi Boitani, who heads the La Sapienza University Animal and Human Biology Department, "The problems start when wolves return to an area after decades, and the ability to coexist has been forgotten."[18]

Russia. There are more wolves in Russia than any other country in the world today, and those wolves have evolved in a culture that, traditionally, had very few firearms owned by the general populace and scant few hunters. That we have not heard more about wolves in Russia since the fall of Czarist Russia, author and scientist Will Graves says is a function of the Communist government suppressing such information because, if people knew how really dangerous wolves were, the people would demand to be well-armed, and an armed populace could lead to revolution.[19]

During the last century, wolf numbers in Russia have risen and fallen through five cycles. When wolf populations became large, special hunting efforts were organized to reduce them. When the population was reduced, the hunting efforts were disbanded, and the wolf populations rose again. The greatest number of wolves occurred during World War I and World War II, when men were called off to war. The general populace of Russia has few firearms, and so in the absence of men and without firearms, the wolf population grew quickly, as much as 30% per year.

In contrast to the United States' currently popular standard line that wolves do not attack people, Russian wildlife biologist Mikhail Pavlov states in his 1982 book, *The Wolf in Game Management*:

> *Cases of severe, unprovoked aggression by wolves toward humans are numerous. In 1988, the editorial board of the magazine* Hunting and Hunting Economics *sent me*

50

information on aggressive wolves in Kaluzhskaya Oblast. Ex-chairman of the hunter's society, S. Semiletkin, informed the magazine that, in 1943-1947, there were 60 victims of wolf aggression, including 46 children. ... In Viatskaya in 1896-1897, 205 people became victims of predation, while there were only 10 in Vologodskaya, 18 in Kostromskaya, 1 in Arkhangelskaya, and 9 in Yaroslavskaya.[20]

Will Graves' exhaustive study, *Wolves in Russia: Anxiety through the Ages*, builds upon Pavlov's work. Graves reports that, from 1804 to 1853, non-rabid wolves killed 111 people in Estonia, of which 108 were children. Another 266 adults and 110 children were killed by wolves in the period from 1849 to 1851 in the European sector of the Russian Empire. Another 1,445 people were killed by wolves in the period from 1870 to 1887 in the European sector of the Russian Empire.[21]

Rabid wolves have long been a problem in Russia, with wolves contracting rabies from foxes, raccoons, dogs, and other mammals. According to Pavlov, in 1975-1976, attacks of rabid wolves on humans were recorded in Ulianovskaya Oblast (15 cases), Kaluzhskaya Oblast (seven cases), Orenburgskaya Oblast (six cases), and Orlovskaya Oblast (four cases). In Gorkovskaya Oblast, during the 10 years from 1929 to 1939, 40 people bitten by a rabid wolf were treated in hospitals. In the same area, after a rise in numbers of wolves in 1978, some 24 attacks on humans were recorded. In the 1980s, there also were numerous stories in newspapers on battles between humans and rabid wolves. In July 1976, in three days, a rabid wolf bit 16 people in the Lubomil district of Volinskaya Oblast.[22]

India and Pakistan. India typically has 2,000 to 3,000 wolves in rural areas. Pakistan and neighboring Kazakhstan traditionally have had many more wolves. Although an imaginary friendly wolf pack raised Mowgli in Rudyard Kipling's classic children's tale, *The Jungle Book*, attacks on people in this area are all too common, especially where there are few weapons available for self-defense and little or no police or military presence. A sample of what has taken place there includes 721 people killed by wolves in 1875 in the Northwestern Province and Bihar State, British India.[23] One of the worst cases ever recorded of wolves attacking people occurred in 1878 in British India when 624 people were killed by man-eating wolves.[24]

These tragic events might seem like a thing of the past, but wolves killing people in India has continued over the years. In 1926, 95 people were killed by wolves in the Districts of Bareilly and Pilibhit, United Provinces, India.[25] Fast forward to 1993-1995 in the Bihar State of

India when wolves killed 60 children, and another 20 children were rescued from what would surely have been death. All the children were taken from settlements primarily during March to August between 5 p.m. and 9 p.m. There were more female victims (58%) than males, and 89% were three to 11 years old. Of the 80 child casualties, only 20 were rescued.[26]

Yadvendradev Jhala, a scientist at the Wildlife Institute of India, reports that, in Uttar Pradesh during a two-year period (1996-1997), a wolf or wolves killed or seriously injured 74 humans, mostly children under the age of 10 years.[27]

An August 29, 2013, newspaper from Kashmir, India, reported: "Wild animals, particularly wolves, have unleashed a reign of terror in a number of villages in the Kreeri area of northern Kashmir's Baramulla district. The locals said that three minors have been mauled to death by wolves in the past seven months in the area. They said that the presence of wolves in Kreeri villages has created a sense of fear among the people. Locals said that the killing of their children by wolves has triggered panic in the villages and that the Wildlife Department is doing very little to tackle the problem. Villagers said that, on February 2, 2013, an 11-year-old boy—Tariq Ahmad Shiekh, son of Abdul Rahim Shiekh of Kharchek Watergam—was mauled to death by wolves while playing in a field. On July 11, an 8-year-old boy—Yasir Khan, son of Manzoor Ahmed Khan of Harushnar Kreeri—was mauled to death by wolves. It was reported that two wolves caught hold of the boy and dragged him into the fields where they tore him apart. And on July 13, 9-year-old Aadil Hameed Sheikh, son of Abdul Hamid Sheikh of Hail Kreeri, was killed by wolves while walking home.[28]

Iran. Two recent fatal attacks by wolves in Iran suggest why wolves continue to be feared in the Middle East, even when the land is so heavily developed. On January 4, 2005, an Iranian homeless man was eaten alive by wild wolves in the village of Vali-Asr, near the town of Torbat Heydariya in northeast Iran.[29] In November 2008, a wolf attacked an 87-year-old woman in the village cemetery of the central Iran town of Kashan. The wolf bit one of her fingers, but she fought back and suffocated the wolf to death.[30]

Historical Accounts of North American Wolf Attacks

Wolves were largely exterminated from heavily-armed North America 100 years ago when cattle and sheep replaced buffalo and other wild game as the wolves' primary sources of meat and ranchers responded with traps, guns, and poison. In direct contradiction to historical evidence, a 1997 CNN article states, "While the wolves pose some threat to humans' domesticated animals, there is little risk to people themselves. And while humans have killed an estimated two million wolves in this century, there is not a single documented case of a human being killed by a healthy wild wolf."[31] That statement is blatantly false.

Let us begin a discussion of wolf attacks in North America by clearing up a common misconception. Often we hear from pro-wolf people that American Indians and other traditional cultures considered the wolf to be a spiritual being, implying that Indians revere wolves and want them protected. Among traditional cultures all around the world, all parts of nature—animals, stones, clouds, bodies of water, plants, places, and heavenly bodies—have a spiritual dimension, and each part of nature has a unique spirit and power that man may be able to access. That an animal like a wolf is believed to have a spiritual aspect or power does not mean, however, that a native person will not kill it, wear its pelt, or even eat it. In fact, in many cultures, killing a wolf and wearing its skin are considered the means to gaining the wolf's powers as a brave hunter. This is why the Mongolian Winter Olympic Team wears parkas with wolf fur.

Additionally, one should never assume that all Native American tribes hold the same beliefs about wolves. There are 562 federally recognized Indian tribes in the U.S., as well as Aleuts and Eskimos. While some tribes regard wolves positively, others, such as the Navajo (like some European people), see wolves as witches—people taking the shape of wolves—and people who wear wolf skins become like werewolves.[32] The Chilcotin tribe of British Columbia believes that contact with wolves can lead to mental illness and death. The Blackfeet, while respecting wolves as great hunters, believe that wolves can never be trusted.[33] Among the Koyukon tribe of Alaska, the wolf's spirit is seen as third in ranking of power after the bear and the wolverine, and all three mammals may be trapped or shot, providing that proper ceremonies are conducted to honor the animal's spirit.[34]

In present times, much has been made of the Nez Perce tribe in Idaho supporting wolf reintroductions, and the tribe has made a considerable contribution to wolf conservation along these lines. However, in 2009, when hunting licenses for wolves were issued in Idaho, 35 licenses were sold specifically for killing wolves on Nez Perce lands. In 2011, the Blackfoot tribe in Montana issued 10 wolf tags for hunters on the reservation. Farther south, while the White Mountain Apaches support Mexican wolf restoration, the San Carlos Apaches oppose the Mexican wolf restoration program in New Mexico and Arizona.[35]

American Indians and Eskimos have co-existed with wolves for thousands of years. They respect wolves as great hunters, but they also fear attacks by both rabid and healthy wolves. According to Barry Lopez, "It is popularly believed that there is no written record of a healthy wolf ever having killed a person in North America. Those making the claim ignore Eskimos and Indians who have been killed and are careful to rule out rabid wolves."[36]

Early Europeans encountered wolves upon landing in the New World. Recalling what they had known about wolves in Europe, the settlers responded with firearms, which drove wolves back into the wilderness where there was abundant wildlife, especially buffalo, which once numbered 60 million in the U.S. Accounts from the Lewis and Clark Expedition reported large packs of wolves following herds of buffalo. Even though wild game was abundant, wolves were considered fearless, and it was thought they might attack people, including the Expedition.[37] In 1807, Clark's squad was camped along the Yellowstone River near where Billings, Montana, was later established, when a wolf stole into camp and bit a sleeping Sergeant John Pryor "through the hand." The wolf then turned on Private Richard Windsor before being dispatched by another squad member.[38]

The noted naturalist and artist John James Audubon, who was traveling through Kentucky near the Ohio border in 1803, reported that wolves attacked two men, who defended themselves with axes. Three wolves were killed, and the wolves killed one man.[39]

In his book *The Great American Wolf*, Bruce Hampton reports that, in 1833 at a fur trapping rendezvous for trappers in Wyoming, a rabid wolf wandered into camp on two evenings and bit 13 people.[40]

The noted naturalist George Bird Grinnell, who was skeptical about wolf attacks on humans, confirmed that, in the summer of 1881, an 18-year-old girl in Colorado was attacked and bitten on the arms and legs by a wolf.[41]

Rabies is possible in any warm-blooded animal, but in Asia, Africa, and parts of North America, dogs are the primary host, especially feral dogs, with foxes, raccoons, and skunks being primary carriers. Wolves and coyotes are dogs, and they also carry rabies. There are an estimated 55,000 human deaths annually from rabies worldwide—about 31,000 in Asia and 24,000 in Africa. The disease has been known since 3500 B.C.E. The first written record of rabies appears in the Codex of Eshnunna (ca. 1930 BC), which dictates that the owner of a dog showing symptoms of rabies should take preventive measure against bites. If another person was bitten by a rabid dog and later died, the owner was fined heavily.[42]

Japan once had wolves, but the Japanese exterminated their wolves by 1905, when the wolves contracted rabies.[43] Currently, Japan is considering reintroducing wolves to control deer.

Cases of rabies in wolves are numerous in the eastern Mediterranean, Middle East, and central Asia. Wolves commonly develop a "furious" phase of rabies, aggressively attacking whatever they encounter. This, coupled with their size and strength, make rabid wolves among the most dangerous of rabid animals. Heptner and Naumov state that bites from rabid wolves are 15 times more dangerous than those of rabid dogs.[44]

Linnell, et al., reported cases of rabid wolves biting humans in Italy, France, Finland, Germany, Poland, Slovakia, Spain, Baltic States, Russia, Iran, Kazakhstan, Afghanistan, China, India, and North America.[45]

With the disappearance of the buffalo, wolves had to look elsewhere for food or starve to death, for other species of wildlife were not nearly as common as some claim. As Dr. Charles Kay has pointed out, "...early explorers, who spent 765 days in the (Yellowstone) ecosystem on foot and on horseback between 1835 and 1876, reported seeing elk only once every 18 days, and bison were only seen three times, but not in the park itself."[46]

When wolves began killing livestock as a substitute for wild game, people began a campaign to exterminate wolves with poison, traps, and hunting.

When Young and Goldman looked for wolf attacks on humans before 1900 in North America, they found 30 accounts of attacks and six possible human kills.[47]

Wolf Attacks in Recent Times in North America

Today we tend to think of habituated or rabid wolves being the ones most likely to attack. However, the records show that healthy wolves have attacked people in wild settings in North America in the past and continue to do so. Let us look at some examples. Again, this is only a sample of what has occurred as formal records of wolf attacks have not been kept until very recently.

In February 1915 on the Coppermine River in the Canadian Arctic, a large white female wolf came into the camp of the Canadian Arctic Expedition of 1913-1918 as they were eating breakfast. The wolf attacked one man and then bit another man on the arm before one of the party shot the wolf, which was later mounted and is today in the collection of the National Museum in Ottawa.[48]

In 1922, an elderly trapper near Port Arthur, Ontario, disappeared in the woods. Two Indians were sent to find him. When they did not come back, a search party was organized, and the bodies of all three were found, killed by wolves.[49]

On January 26, 1950, a missionary driving a dog sled from Palmer, Alaska, to the Copper Basin, carrying supplies and gifts for orphans and needy families, was surrounded by wolves in deep snow near Sheep Mountain close to the headwaters of the Matanuska River. All the dogs were killed. The missionary survived by keeping the wolves away by building a fire.[50]

By the late 1930s, wolves had been virtually eliminated from the lower 48 states due to hunting, trapping, and poisoning. Wolves were still thriving in Alaska, however, and as more and more people moved to Alaska, encounters with wolves began to increase. In 1974-1975 in Fairbanks, wolves attacked and killed 165 dogs, a number of which were being walked by owners or were in the yards of their owners. In response, 13 wolves were shot in and around Fairbanks, and the incidents dropped dramatically.[51]

On April 18, 1996, five wolves attacked and killed 24-year-old Patricia Wyman in the Haliburton Forest and Park Reserve in Haliburton, Ontario. The wolves had been raised in captivity but had never been trained to socialize with people.[52]

In September 2006, a wolf attacked six people in Algonquin Provincial Park in Ontario. The wolf had a broken clavicle and tooth when it was shot by park rangers following the attacks, but it did not have rabies.[53]

Dogmatic teaching in colleges and universities pushed the belief that wolf attacks on humans occurred only abroad and surely not in North America! Two wolf attacks on humans that occurred in 2000 changed things forever.

In Icy Bay, Alaska, six-year-old John Stenglein and his nine-year-old friend were playing outside his family's trailer at a logging camp. When a wolf came out of the woods towards the boys, the boys ran. The wolf attacked young Stenglein, biting him on the back and buttocks. Adults, hearing the boy's screams, chased the wolf away. The wolf returned a few moments later and was shot. According to Alaska Department of Fish and Game, the wolf was a healthy wild wolf that apparently attacked without provocation.[54]

This incident moved the Alaska Fish and Game Commission to request that a comprehensive study of wolf attacks on people in recent times in North America be undertaken. In response to this request, in 2002 Alaskan wildlife biologist Mark McNay published a report, "A Case History of Wolf-Human Encounters in Alaska and Canada," about a two-year study documenting 80 aggressive encounters between wolves and people in North America in the 20th century—41 from Canada, 36 from Alaska, and three from northern Minnesota. Of these

80 encounters, 39 were aggressive encounters with healthy wolves, and in only 12 of the attacks were the wolves rabid. Since McNay's report came out, there have been two fatal attacks by healthy wolves—one in Canada and another in Alaska—and an unknown and increasing number of non-fatal aggressive encounters and attacks on people and their pets in the U.S. and Canada.[55]

Also in 2000, on Vargas Island, British Colombia, a 23-year-old university student, Scott Langevin, was on a kayaking trip with friends.

FOR IMMEDIATE RELEASE: March 25, 2011 No. 11-21
Contact: Mark Burch, 907-267-2387

Military Base Wolf Removal Effort Successful

(Juneau) – Alaska Department of Fish & Game (ADF&G) biologists, working in cooperation with military personnel, have removed nine wolves on the Joint Base Elmendorf Richardson (JBER). The wolves were considered a significant threat to public safety on the military installation and in surrounding residential areas.

"The effort was successful. We feel confident we have minimized public safety risks by removing specific wolves and significantly reducing wolf numbers in the area," said Mark Burch, Regional Supervisor, Division of Wildlife Conservation.

Over the winter, the nine wolves were taken by joint ADF&G/JBER trapping and ground shooting efforts. These actions are believed to have mitigated risks to public safety. However, ADF&G biologists and military personnel will continue to monitor the situation and take wolves if necessary.

The wolf removal effort was in response to habituated wolves in the area that were becoming increasingly aggressive towards humans and pets. Tissue, bone, and hair samples from the wolves will be analyzed to develop profiles of dietary habits. This information could prove useful for future comparative studies. In addition, samples will be used as part of a broader genetic study of Alaska wolves.

To prevent future problems, area residents must take precautions not to leave out garbage, pet food, or other attractants that might draw wolves near homes and into neighborhoods. In addition, homeowners should take precautions to secure pets and livestock. Negligent and intentional feeding of wolves is prohibited by state regulation (5AAC 92.230).

People should enjoy outdoor pursuits, but recognize risks and take precautions when recreating in wolf and bear country. Children should always be accompanied by an adult and dogs should be on a leash. For detailed information about coexisting with wolves, preventing habituation, and staying safe in wolf country visit: www.adfg.alaska.gov/static/species/livingwithwildlife/pdfs/living_in_wolf_country.pdf

ADF&G biologists and JBER officials would like people to continue to report wolf sightings or encounters on or near the military base. Please immediately report sightings to the base dispatcher at 522-3421, or call the Division of Wildlife Conservation Information Center at 267-2257 Monday-Friday from 8 a.m.-5 p.m.

They camped out on a beach in an area where other campers had stayed. In the middle of the night, Langevin awoke with something pulling on his sleeping bag. He suddenly came face to face with a wolf. Langevin yelled at the wolf, and it bit him on the hand. Langevin attempted to fight back, but the wolf jumped on his back and started biting him on the back of his head. Friends, hearing his yells, came to his aid and scared the wolf away. Fifty stitches were required to close Langevin's wounds. The B.C. Ministry of Environment speculated the reason for the attack was wolves being habituated at that location as a result of being fed by people who used that campground. Langevin's party did not feed the wolves.[56]

In Canada, wolves have been extensively hunted and trapped for many years with a bounty as an incentive. This has resulted in over 1,000 wolves a year taken in Alberta alone. Such trapping and hunting keeps the wolf population in check and drives remaining wolves back into remote wilderness areas. The attack on Langevin took place in a campground where wolves were protected, and it is likely they were fed. Farther east in Algonquin National Park in Ontario, which has a viable wolf population and special wolf howling trips for visitors, there is a history of wolf attacks on people, including five attacks over an 11-year period.[57]

Alaska has 7,000 to 11,000 wolves, which have also been hunted and trapped. However, as more and more people flock to Alaska and its urban areas increase in size, wolves appear to investigate these new enclaves where hunting is not permitted.[58] For example, in the winter of 2007, the Elmendorf wolf pack around Anchorage attacked a number of people walking dogs, killing some dogs.[59] This made the news in Anchorage and Fairbanks, but not much was heard farther south about wolf attacks until November 8, 2005. College student Kenton Carnegie was hiking on a road near Points North Landing in northern Saskatchewan when he was attacked and killed by wolves.[60] A thorough provincial inquest found that wolves were responsible. It was later determined that the wolves had become habituated to a garbage dump in that area and that, just a few days prior to Carnegie's death, some other people were confronted by habituated wolves in the same area.[61]

Another wolf attack in Saskatchewan occurred on New Year's Eve, 2005, when Fred Desjarlais was coming home from his job at Key Lake, about 550 kilometers north of Saskatoon. A wolf suddenly came at him from a ditch and bit him several times on the back, arm, leg, and groin.[62] These incidents caused Canadian media to do some research, which found more wolf attacks on people in Canada, including those in 1994, 1995, 1996, 2000, and 2004.[63] That research illustrated that many wolf attacks do occur but do not make the national news or the scientific literature.[64] Following are four examples.

58

Resort owner Ken Gangler in Manitoba reported that in the summer of 2008 one of his summer resort staff went out jogging on the landing strip for the private airport that serves Gangler's Resort. She was suddenly struck from the back by a large wolf that knocked her down. Before the wolf could attack any further, it was driven off by people nearby.[65]

In Idaho, in 2008, a father and son were out hunting coyotes using a predator call. The pair was separated by 100 yards. Suddenly a pack of wolves surrounded the son. The father joined his son, and both men fired shots to scare the wolves away. The wolves left reluctantly, the men said. This was the year before it became legal to hunt wolves in Idaho.[66]

On March 9, 2010, Candice Berner, a 32-year-old Special Education teacher working in Chignik Lake, Alaska, went jogging at dusk on a road near town and was attacked and killed by wolves. As this was on U.S. soil, finally the reality of wolf attacks on people in North America became front-page news in the U.S. The attacking wolves were not rabid.[67]

On March 8, 2013, Dawn Hepp pulled over on Highway 6 near Grand Rapids in Manitoba, Canada, to help another driver. When she walked over to the car, a wolf lunged at her, grabbing hold of her neck with its jaws. The wolf dug deeper into her neck with its jaws before letting go. Hepp was treated for puncture wounds and rabies.[68]

As this book was being prepared for publication, at least three wolf attacks were reported in North America in the summer of 2013. In June, a Montana resident was riding his motorcycle on Highway 93 in Kootenay National Park, one of 43 national parks in Canada, when he was chased by a wolf that came less than 10 feet from him before he outpaced it.[69]

In early July, an Idaho cyclist was on a 2,750-mile pedal fundraiser when his nightmare began with a gray wolf sprinting out of the woods 60 miles west of Watson Lake, a town on the Alaska Highway in southeastern Yukon, close to the British Columbia border. The gray wolf surprised the passing cyclist with an initial chomp that just missed his pedal. The panicked cyclist blasted the wolf with bursts of bear spray during the chase, which did not, however, deter the wolf. The driver of a passing RV came to the cyclist's rescue by quickly stopping, allowing the cyclist to jump off his bike and into the safety of the RV. The wolf attacked the bike and shredded the cyclist's backpack as though it were prey.[70]

In late August, in what's being called by the Department of Natural Resources "the first confirmed attack of a person by a gray wolf in Minnesota history," a 16-year-old Solway boy was attacked by a wolf at the West Winnie Campground on Lake Winnibigoshish in north-central Minnesota. Part of a church group, Noah Graham was preparing for bed when a 75-pound male grey wolf clamped its jaws onto the back of Graham's skull and wouldn't let go. The boy had to physically pry the jaws of the wolf open to free his head, and then the wolf stood there growling until chased away. The wolf's sharp teeth left a laceration on Graham's skull that required 17 staples to close. In addition, the boy had several puncture wounds behind his left ear. He also had to endure the painful regimen of shots in case the wolf was rabid. The 75-pound gray wolf, also known as a timber wolf, was captured by USDA trappers and killed to permit rabies testing. [71]

Dangerous Perspective

In spite of documented wolf attacks, Dr. Valerius Geist, whom we will hear at length from in Chapter 6, states that wolves are generally not dangerous when:

 they are well fed on abundant wild animals,
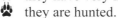 they have very little contact with people, and/or
 they are hunted.

Dr. Geist further says that, in general, the evidence indicates that wolves are very careful to choose the most nutritious food source that is most easily obtained without danger. They tackle dangerous prey only when they run out of non-dangerous prey, and they shift to new prey only very gradually, following a long period of gradual exploration.

Four Factors

In the Northern Rockies and northern Midwest, where wolves have not been hunted, big game herds of the wolves' favorite prey—elk, deer, and moose—have declined dramatically due to wolf predation. This then has set the stage for wolves to approach livestock, pets, and people as prey.[72]

In a study of wolf predation worldwide, Linnell, et al., found that there are four factors associated with wolf attacks on humans:

 rabies,
 habituation,
 provocation, as in people approaching dens with young wolves, and
environmental conditions including little or no natural prey, garbage dumps, children who are unattended, poverty, and limited availability of weapons, especially firearms.[73]

Unfortunately, there has not been a detailed follow-up study of wolf attacks on people like Mark McNay's, which concluded in 2002.[74] As wolf populations grow across North America, wolves are testing boundaries and moving into new territory. As wolves come into closer contact with people, the chance of attacks increases, especially if people approach wolves naively.

Wolf biologist Dr. David Mech advises people to never feed wolves, because it causes them to become habituated. He says that, if you meet a wolf, do not run away—yell, look as big as you can, and throw rocks. Pepper spray helps. The sound of a gun will let them know you mean business.[75] Wolves are smart, prolific, and adapt quickly. Mech says that, so long as there is adequate food and habitat, it is necessary to kill off between 28% and 53% in an area just to keep that wolf population stable, which is part of the necessary management program for wolves living in the 21st century in areas where so many people now live.[76]

Expanding wolf packs in North America are visiting the suburbs of Minneapolis, Minnesota, and walking the streets of Sun Valley, Idaho. There are numerous reports of four wolf packs living around Anchorage, Alaska, attacking dogs and people walking them.[77] And, in 2009 in Yellowstone National Park, where visitors line the roads to take pictures of wolves, rangers shot a wolf because it was chasing bicyclists.[78]

The evidence of history clearly shows that—when wolves and people do come together, especially when people are not armed and wolves are hungry—sooner or later, attacks will occur and someone is going to be hurt or killed. The historical record also shows that, on occasion, both healthy and rabid wolves do go on killing sprees that can include humans, especially children. This reality must be factored into any responsible wolf management program.

"Ivan, didn't the farmers around Gorky tell you
it is the wolf you do not see that you must fear?"
Tom Clancy, *The Hunt for Red October*[79]

Conclusion

Ultimately, by not accepting that wolves can and do attack people and have done so from time immemorial, people then believe that wolves are not dangerous to humans, an approach that leaves them vulnerable to wolf attacks.

It is assumed that an undisturbed animal community
lives in a certain harmony…
the balance of nature.
The picture has the advantage of being an
intelligible and apparently logical result
of natural selection in producing the best possible
world for each species.
It has the disadvantage of being untrue.
Charles Elton

🐾 🐾 🐾 🐾 🐾 🐾 🐾 🐾

[1] http://www.fws.gov/northdakotafieldoffice/endspecies/species/gray_wolf.htm (North Dakota Field Office, last updated March 14, 2011)

[2] http://www.nytimes.com/2011/08/16/science/16grizzly.html?_r=1&hpw

[3] http://latimesblogs.latimes.com/nationnow/2011/08/mountain-lions-flocking-to-greenwich-ct-them-too.html

[4] http://www.dogbitelaw.com/PAGES/statistics.html

[5] http://digitalcommons.unl.edu/vpc21/1/

[6] http://latimesblogs.latimes.com/outposts/2009/10/musician-taylor-mitchell-dies.html

[7] Vyhnak, Carola, "Meet the Coy-wolf," *Toronto Star*, August 15, 2009.

[8] http://en.wikipedia.org/wiki/Wolfdog&http://www.wolf.org/wolves/learn/intermed/inter_human/algonquin.asp

[9] http://en.wikipedia.org/wiki/Red_Wolf

[10] http://www.forwolves.org/wolves.html

[11] http://www.wolf.org/wolves/learn/scientific/challenge_mech.asp

[12] http://www.iucnredlist.org/apps/redlist/details/3746/0/full

[13] Moriceau, Jean-Marc, *Histoire du méchant loup: 3, 000 attaques sur l'homme en France*, 2007: 623.

[14] Thompson, Richard H., *Wolf-Hunting in France in the Reign of Louis XV: The Beast of the Gévaudan*, 1991: 367.

[15] *Ibid.*

[16] http://www.wolfsongalaska.org/wolves_humans_danger.html

[17] http://en.wikipedia.org/wiki/Wolf_attacks_on_humans

[18] http://www.physorg.com/news90260221.html

[19] Graves, Will N., *Wolves in Russia: Anxiety Through the Ages,* Detselig Enterprises, 2007.

[20] Pavlov, Mikhail P., Translated by Valentina and Leonid Baskin and Patrick Valkenburg, "The Danger of Wolves to Humans," *The Wolf in Game Management,* Moscow, 1982; http://westinstenv.org/wibio/2010/02/22/the-danger-of-wolves-to-humans/

[21] Graves, W., *Op. Cit.*

[22] http://westinstenv.org/wibio/2010/02/22/the-danger-of-wolves-to-humans/

[23] Knight, John, *Wildlife in Asia: Cultural Perspectives,* 2004, 280.

[24] http://hindu.com/2001/05/08/stories/1308017f.htm

[25] Maclean, Charles, *The Wolf Children*, Hill and Wang, 1980, 336.

[26] http://www.wolfsongalaska.org/Wolves_South_Asia_child.htm

[27] http://www.wolf.org/wolves/learn/intermed/inter_human/india_abstract_003.asp

[28] http://www.iranfocus.com/en/iran-general-/homeless-man-eaten-by-wolves-in-iran-01150.html

[28] http://risingkashmir.in/wild-animals-unleash-terror-in-baramulla-villages-2/

[29] http://en.wikipedia.org/wiki/Indian_Wolf#cite_note-19

[30] http://www.cnn.com/EARTH/9711/12/yellowstone.wolves/

[31] http://www.wolfsongalaska.org/wolves_in_american_culture.html

[32] Boitani, Luigi, *Wolves: Behavior, Ecology, and Conservation,* University of Chicago Press, 2003.

[33] Nelson, Richard, *Make Prayers To Raven: A Koyukon View of the Northern Forest,* University of Chicago Press, 1983.

[34] http://www.rangemagazine.com/archives/stories/summer03/ground-hog.htm

[35] Lopez, Barry, *Of Wolves and Men,* Charles Scribner's Sons, 1978.

[36] http://www.sinauer.com/groom/article.php?id=24

[37] http://fwp.mt.gov/mtoutdoors/HTML/articles/2006/LCmisadventures.htm

[38] Audubon, J.J., and J. Bachman, *The Quadrupeds of North America,* New York: Wellfleet Press, 1851–1854.

[39] Hampton, Bruce. *The Great American Wolf*, NY, NY: Henry Holt, 1977: 94.

[40] Grinnell, G.B.*, Trail and Campfire: Wolves and Wolf Nature,* New York, 1897; http://www.aws.vcn.com/wolf_attacks_on_humans.html

[41] http://en.wikipedia.org/wiki/Prevalence_of_rabieshttp://en.wikipedia.org/wiki/Prevalence_of_rabies

[42] http://bestjapanguide.com/483/

[43] Heptner, V.G., and N.P. Naumov, *Mammals of the Soviet Union,* Science Publishers, Inc., 1988, 267.

[44] http://www.lcie.org/Docs/Damage%20prevention/Linnell%20NINA%20OP%20731%20 Fear%20of%20wolves%20eng.pdf

[45] Kay, Charles E., "The Kaibab Deer Incident: Myths, Lies, and Scientific Fraud," *Muley Crazy*: January/February 2010.

[46] Young, S. P., and E. A. Goldman, *The Wolves of North America: Parts 1 and 2,* New York: Dover Publishing Inc., 1944, 636.

[47] Jenness, Stuart E., "Wolf Attacks Scientist: A Unique Canadian Incident," *Arctic* 38, June 1985, 129-132.

[48] http://www.rangemagazine.com/archives/stories/summer03/ground-hog.htm

[49] http://www.farnorthscience.com/2007/12/13/news-from-alaska/ravenous-wolves-attack-missionary/

[50] http://www.farnorthscience.com/2007/12/13/news-from-alaska/ravenous-wolves-attack-missionary/

[51] http://en.wikipedia.org/wiki/Wolf_attacks_on_humans

[52] http://www.propertyrightsresearch.org/2006/articles09/six_injuredin_rare_wolf_attack.htm

[53] http://juneauempire.com/stories/042700/Loc_wolf.html

[54] http://www.adfg.alaska.gov/static/home/library/pdfs/wildlife/research_pdfs/techb13_full.pdf

[55] http://www.cbc.ca/sask/features/wolves/attacks.html

[56] http://www.wolf.org/wolves/learn/intermed/inter_human/algonquin.asp

[57] http://alaska.wikia.com/wiki/Wolves

[58] http://www.alaskastar.com/stories/122007/new_20071220007.shtml

[59] http://en.wikipedia.org/wiki/Kenton_Joel_Carnegie_wolf_attack

[60] http://www.cbc.ca/canada/saskatchewan/story/2007/11/01/wolf-verdict.html

[61] http://www.cbc.ca/canada/story/2005/01/04/sask-timberwolf050104.html

[62] http://www.cbc.ca/sask/features/wolves/attacks.html

[63] http://missoulian.com/news/local/article_a43b34b2-5b16-11df-a302-001cc4c002e0.html

[64] Personal conversation, January 2009.

[65] Personal conversation, March 2009.

[66] http://www.msnbc.msn.com/id/35913715/ns/us_news-life/

[67] http://www.cbc.ca/news/canada/manitoba/story/2013/03/18/mb-wolf-attack-dawn-hepp-manitoba.html?cmp=rss

[68] http://www.montanaoutdoor.com/2013/06/man-on-motorcycle-takes-pictures-of-the-wolf-chasing-him/

[69] http://www.spokesman.com/blogs/outdoors/2013/jul/09/sandpoint-cyclist-survives-tense-wolf-encounter-alcan-highway/

[70] http://www.cbsnews.com/8301-201_162-57600272/teen-survives-first-confirmed-wolf-attack-in-minn/

[71] http://westinstenv.org/wp-content/Geist_when-do-wolves-become-dangerous-to-humans.pdf

[72] Linnell, et. al., 2002, *Op. Cit.*

[73] http://wolfclash.com/

[74] http://seattletimes.nwsource.com/html/localnews/2004087262_wolves22m.html?syndication=rss

[75] http://www.nationalparkstraveler.com/2009/05/yellowstone-national-park-rangers-kill habituated-wolf

[76] http://www.msnbc.msn.com/id/35913715/ns/us_news-life/

[77] http://www.npwrc.usgs.gov/resource/mammals/wpop/results.htm

[78] Clancy, Tom, *The Hunt for Red October,* New York: Berkeley Books, 1984, 6.

Chapter 4

RUSSIAN WOLVES AND AMERICAN WOLVES: NO DIFFERENCE

By Will N. Graves

With the outbreak of the Korean War, Will Graves volunteered for the U.S. Air Force and was trained as a Russian linguist. In order to accelerate and develop his skills in Russian, he started to read Russian wildlife magazines and books. Wolves were often discussed, and soon his interest became focused on wolves in Russia. He asked every native Russian he met if they had any knowledge of wolves and began to record data and resources. Graves' interest in wolves grew into a serious hobby that continued after the war.

Editors' Statement of the Issues

Stories about Russian wolves attacking humans are well-known and well-documented. What is not well-known or well-documented is that the wolves in North America are the same, attack the same, and will kill the same. There is no behavioral difference between them.

Editors' Statement of the Facts

Modern day attacks of wolves on people happen all too frequently, especially to children. These authenticated attacks happen for a variety of reasons: habituation, rabies, lack of wildlife, and just plain vicious predator behavior. As North American wolf populations increase, it is irresponsible to downplay the inherent risks associated with their growth and accessibility to communities.

> **Based on Russia's experience, the wolf is a different animal than we have been led to believe from North American books and movies, and the picture is not rosy.**
> Will N. Graves

Russian Experience

August 7, 2005. Authorities in Ukraine say a pack of wolves has attacked more than a dozen people in the country's eastern region.[1]

September 2, 2009. A rabid wolf has bitten six persons on the premises of the Chernobyl atomic power plant, four of them employees of SSE Chernobyl and two employees of the contractor organization UAB.[2]

January 31, 2012. A pack of wolves terrorized locals in the streets of a Karelian town, not returning to the woods until police opened fire, killing two. The wolves were reported by local residents from several locations in the city of Petrozavodsk, some 660 kilometres northwest of the capital Moscow.[3]

January 20, 2012. At least six people were attacked by wolves in southeastern Tajikistan in the last several weeks. A survivor of one of the wolf attacks, Ozodamoh Saidnurulloeva, told

66

RFE/RL: "Before, the wolves only attacked livestock; now they have started attacking humans." Saidnurulloeva, 89, said she walked into her yard early one morning this week to prepare for the Muslim morning prayer when a wolf attacked her, knocked her down, and dragged her for several meters. She said she thought the wolf was the "angel of death." A neighbor, who happened to be outside, saw the attack and started screaming for help. The wolf then left Saidnurulloeva and ran away. Locals in the Sijd and Rivak villages told RFE/RL they have pitchforks and shovels to fight wolves when necessary.[4]

Russians have lived with wolves for thousands of years. In contrast, contact between European settlers and wolves in North America for the entire U.S.-Canada region has only been slightly more than three centuries and much less than a century and a half for many regions. A good deal of the folk tales, sayings, fairy tales, and stories that warn us about wolves—that have been passed down through many generations and which are so often today discounted by pro-wolf advocates—originated from Russia. There is good reason for Russia to place so much interest in wolves. Russia has always had many wolves, but in the last century, the USSR and then the Russian Federation of States have had the largest population of wolves in the world, at least 45,000 on average.[5] That is almost three times the wolf population of North America. In addition, the Russians have never had the widespread ownership of firearms that citizens of North America have always enjoyed.

Average numbers, however, do not tell the whole story. As we have learned in North America, wolves are intelligent, hardy, skillful hunters and prolific, their populations quickly increasing when the opportunity presents itself. In Russia since the 1870s, there have been five periods when the wolf population swelled dramatically. During each population upswing, nature alone did not keep the wolf population in check. As their population increased and wild game became scarce, wolves came into more direct contact with people and livestock in their quest for food. Damage and danger quickly escalated. Each time, as the situation became critical, the government was forced to organize wolf control programs to intervene. Overall, approximately 1.5 million wolves have been killed by hunting, trapping, and poisoning during these population surges. Nonetheless, the current wolf population remains well above that which was present when the Bolsheviks overthrew the Czar in 1917 and established the USSR and its 15 republics, the largest of which was the Russian Soviet Federated

Socialist Republic (RSFSR), with its capital in Moscow.

Russia's long-time experience with wolves foreshadows what we can expect in North America as rapidly growing and expanding wolf populations come into contact for the first time with modern settlement patterns and big game populations that have been carefully built up since Teddy Roosevelt's time. Based on Russia's experiences, we know that the wolf is a different animal than we have been led to believe from North American books and movies, and the picture is not rosy.

First Period of Wolf Abundance

The first recorded period of wolf abundance in modern times in Russia was the 1870s in Czarist Russia. In 1873 in only 45 provinces, wolves killed 179,000 adult cattle and 562,000 sheep and goats. During the Czarist period, peasants were not allowed to own weapons. Not only was firearm ownership banned, but also farmers were not allowed to own metal pitchforks; they had to be made from wood. As the wolf population soared, the government instituted a number of management programs, and the wolf population declined, but these programs soon were curtailed as the country was moving toward revolution and civil war, and men and weapons were needed for national security.

Second Period of Wolf Abundance

There was a large increase in the Russian wolf population during World War I (1914-1918) and in the years immediately after. This was in part due to men of military service age being conscripted, leaving rural areas with little protection and few arms. In 1924-1925, wolves killed about one million head of cattle, or 0.5% of all the cattle in the entire USSR. Especially hard hit areas were the Volga area where 2.2% of all cattle were killed by wolves; Siberia where 1.6% of all cattle were killed by wolves; and Kazakhstan, which lost 1.5% of all its cattle to wolves. As is the case in North America, such predation was not spread evenly across an entire area. Some farms and areas suffered much more than others. As the damage reached crisis proportions, the government mounted a campaign of hunting, trapping, and poisoning, and the wolf population dropped back.

Third Period of Wolf Abundance

During World War II, which started in Europe in the early 1930s and ended in 1945, once again the Russian wolf population increased dramatically as men left communes to fight and few peasants in rural areas had firearms. Immediately following WWII, it was estimated that there were at least 200,000 wolves in every region in the USSR. In the

RSFSR in the late 1940s, the wolf population was estimated at 100,000, and these wolves were killing approximately 200,000 head of livestock a year. Not only were livestock and wild game severely reduced, but also wolf attacks on humans skyrocketed. In 1945, immediately after the war ended, wolf-hunting brigades using light aircraft, trucks, and jeep-like vehicles were formed, and bounties were offered. In 1946, 60,000 wolves were killed. From 1947 to 1951, the annual kill was 40,000 to 50,000.[6] Between 1947 and 1951, 240,000 wolves total were culled.[7] The wolf population did not significantly decline until the early 1960s when the annual kill dropped below 20,000. It continued to remain at this level through the mid-1970s. During this period, wolf damage to livestock dropped by a factor of 10.

As wolf numbers dropped, bounties were removed in many areas. The publication in the USSR of Farley Mowat's book *Never Cry Wolf* appealed to the growing urban Russian population, which had never had first-hand contact with wolves, and organized control programs ceased. From 1971-1975, the annual wolf cull throughout the entire USSR dropped to an average of 15,000 to 16,000 per year.

Fourth Period of Wolf Abundance

Between 1973 and 1976, the wolf population of the USSR increased by 50,000 to 60,000, setting the stage for an explosion of wolves that began about 1978. In just five years, the wolf populations in Ukraine, Belarus, and central areas of the RSFSR doubled. In the April 25, 1978, issue of *Kommunist Tadzhikistana*, A. Chutkov—Deputy Director of the Tadzik Republic State Forest Committee's Administration on Hunting, Game Preserves, and Nature Preserves—wrote:

> *Tadzhikistan's 10,000 wolves are harmful to agriculture, game animals, and privately owned livestock. ... The number of wolves is growing, as is the damage they do to the economy.... The bounties should be increased to 100 rubles for a mature wolf and 60 rubles for a cub, which are the going rates in the Russian and Ukraine Republics, and more airplanes and motor vehicles should be made available to hunters.*

One consequence of the growing wolf population in this era was that the incidence of rabies in wolves increased six-fold between 1977 and 1979.

In 1978, the popular Soviet outdoor sports magazine *Hunting and Game Management* undertook a campaign to increase the wolf kill by creating a coalition of professional hunters, who kill wild game and sell it for food; amateur hunters, of which there are about 3.5 million in Russia; and part-time professional hunters, who are largely farmers that hunt and trap in winter months. The total number of hunters in Russia has grown in recent years, but it still is small compared to the United States. A further limitation is that the amateur hunters, who must belong to a hunting collective like a club, may only possess and use shotguns.

In response to growing wolf populations, increased livestock predation, and numerous attacks on people, the number of wolves culled in the USSR doubled in the 1980s as compared to the 1970s. In January 1981, A. Sitsko wrote: "In 1976 in the RSFSR, there were 7,000 wolves culled; in 1980 there were more than 15,900 culled."[8] In 1985, 38,900 wolves were culled throughout the entire USSR. Again, when culling was increased, damage associated with wolves decreased.

Fifth Period of Wolf Abundance

When the USSR collapsed in 1992, the government began a process of reorganization. Wolves slipped into the background of governmental attention, and quietly they began to increase in numbers. From 1987 to 1997, the wolf population in the Russian Federation alone rose from 29,000 to 42,000. One consequence of this latest wolf population explosion is that, between 1988 and 1997, the moose population decreased 33%—from 904,000 to 604,000. In the same time period, the number of moose bagged by hunters dropped from 66,000 to 22,000.[9] During this era, the reindeer population in Anadyr in northeast Siberia dropped from 500,000 to 150,000, while in the Koryak Autonomous Area of the Kamchatka Peninsula, the wolf population increased from 150 to 800, and in 1998 wolves killed 10,000 reindeer.[10] Most reindeer herders are not authorized to have rifles, so they have little way to control hungry wolves. The native peoples in these areas were heavily impacted by the burgeoning wolf population of this period.

Meanwhile, in Kyrgyzstan during the same period, the wolf population increased three-fold, resulting in increased predation on livestock and attacks on people. In 2003 it was written, "Officials say attacks on people and domestic animals have increased overall, almost doubling in the last nine months. One man died after being bitten."[11]

Russia's long experience with wolves provides us with a wealth of information that is especially valuable as North American people experiment with reintroduction of wolves into areas with both an abundance of big game and human populations that are denser than at any other time in history.

A wolf is not afraid of a dog, but is afraid of a weapon.

Russian proverb

Let us briefly look at some of what the Russians have found from their lengthy experience of living with wolves and coping with the five population peaks.

Wolves as Disease Vectors

Russian scientists provide ample evidence that wolves carry many diseases including some that can be dangerous to humans and to livestock, such as distemper, brucellosis, hoof and mouth disease, mange, and of course rabies. The wolf is a major host of rabies in Russia, Iran, Afghanistan, Iraq, and India. It is well known that victims bitten by animals carrying rabies face a 100% chance of dying unless they receive treatment before symptoms appear. Rabies is primarily transferred by a rabid animal biting another. It causes a dramatic change in the behavior of infected animals. Rabid wolves become extremely agitated and aggressive, deserting their packs and traveling as much as 80 kilometers in a day. On such trips they lose all fear of humans and attack anything in their path. In 1957, a rabid wolf in Belorussia traveled 150 kilometers in one and a half days. During that time, it bit 25 people, about 25 livestock, and an unknown number of wild animals.

Rabid wolves tend to aim for the head and face with their attack. In 1976 to 1980, there were 30 registered cases of wolf rabies and 36 attacks on humans by rabid wolves in the Russian Kazakh and

Russian Wolves

IN the course of last Winter's campaign the wolves of the Polish and Baltic Russian stretches had amassed to such numbers in the Kovno-Wilna-Minsk district as to become a veritable plague to both Russian and German fighting forces. So persistent were the half-starved beasts in their attacks on small groups of soldiers that they became a serious menace even to fighting men in the trenches. Poison, rifle fire, hand grenades, and even machine guns were successively tried in attempts to eradicate the nuisance. But all to no avail. The wolves—nowhere to be found quite so large and powerful as in Russia—were desperate in their hunger and regardless of danger. Fresh packs would appear in place of those that were killed by the Russian and German troops.

As a last resort, the two adversaries, with the consent of their commanders, entered into negotiations for an armistice and joined forces to overcome the wolf plague. For a short time there was peace. And in no haphazard fashion was the task of vanquishing the mutual foe undertaken. The wolves were gradually rounded up, and eventually several hundred of them were killed. The others fled in all directions, making their escape from carnage the like of which they had never encountered. It is reported that the soldiers have not been molested again.

The New York Times
Published: July 29, 1917
Copyright © The New York Times

71

Georgian Soviet Republics; there were additional cases in the Russian Federation (RF). Between 1972 and 1978 in the Aktyubinsk region of Kazakhstan, wolves attacked 50 people, and 33 suffered bites from rabid wolves. More often than not, wolves attacked agricultural workers in the fields.

According to the World Health Organization, there were 30 cases of wolf rabies in Turkey from 1977 to 1985: four in Germany, five in Poland, one in Romania, and one in Poland. Foxes, raccoons, and stray dogs, as well as wolves, are the primary species for transmitting rabies in Eurasia. From 1990 to 1997, there were 76 cases of human deaths from rabies in 30 administrative territories of the RF.[12] In 1998, there were 2,868 cases of rabies registered in the RF.[13] In North America, we can also add bats and coyotes, as well as the various hybrid wild canids—wolf-dogs, coy-dogs, and coy-wolves that are increasingly being found in North America.

In addition to rabies, wolves carry over 50 types of parasites that cause dangerous diseases such as *echinococcus, cysticercosis, coenurosis, trichinosis, tularemia, listeriosis, and ascariasis.* They also can carry and spread diseases that affect livestock including anthrax, *neosporum caninum*, and hoof and mouth disease, although they themselves are immune to these diseases. Numerous studies have found that the majority of wolves carry at least one type and often several types of tapeworms. A Soviet scientist researching wolves in the extreme northeast of Siberia found that, in areas with high wolf populations, there was also a high incidence of brucellosis (which can be carried by wolves) in reindeer, which are a primary source of food and income for people in these areas.[14] (For more about wolves as vectors for zoonotic diseases, see Chapter 14: The Wolf as a Disease Carrier.)

Attacks on People

In his 1982 definitive book, *The Wolf*, M.P. Pavlov writes, "Facts about attacks of wolves on people are not myths but reality… This danger is increased by the appearance of wolf-dog hybrids."[15] There are countless documented cases of wolves attacking people in Russia, especially children. The attacks are by both rabid and non-rabid wolves. Following are some examples:

According to the Russian Police Department, from 1849 to 1851, wolves in the Russian Empire killed 376 people, including 266 children.

From 1896 to 1897, there were 205 attacks by wolves on

humans in the Vyatsjij Province of Russia, 10 attacks on people in Vologodskij, 18 attacks in Kostromskij, one in Arkhangelskij, and nine in Yaroslavskij.

Between 1886 and 1901, in Odessa, Ukraine, there were more than 150 cases of people being attacked and bitten by rabid wolves.[16]

In 1924, two rabid wolves attacked and bit 20 people in Kirov; 10 died.

In the Vladimirskij Region from 1945 to 1947, there were 10

Woman stabs wolf to death
to save elderly husband in Madinah village

By ARAB NEWS

Published: Dec 24, 2010 22:37

JEDDAH: A Saudi housewife stabbed a wolf several times with a kitchen knife after it attacked her husband in front of their home in a village in the Madinah province on Wednesday night.

The woman's husband, Saadi Al-Mitairi, who is in his 70s, had gone outside at midnight to tend to their goats when he came across the wolf in front of his house in the village of Ghidairah, Mahd Al-Dahab municipality. The wolf attacked Saadi who fought back and held on to its limbs for some 90 minutes.

Wondering where her husband was, Saadi's wife came outside and saw what was happening. She then went inside and returned with a kitchen knife, stabbing the wolf several times till it died.

"We only heard what my mom and dad did the next day. My dad's old but strong and was able to hold on to the wolf's legs for around an hour and a half," said Abdullah Al-Mitairi, Saadi's son.

"He didn't think he'd find a goat when he went out that night to check on the goats," he said, adding that his father was not hurt or injured, and that he and his brothers were out at the time.

The Saadi family has now hung the wolf's body from a tree in front of the house, as is the custom in rural Saudi villages, to scare off other wolves.

http://arabnews.com/saudiarabia/article222376.ece

---

attacks by wolves on humans. All wolves were shot, and they weighed at least 60 kilograms each and were healthy.

A 1947 report by Dr. P.A. Mantejfel that summarizes non-rabid wolf attacks on humans in the USSR finds that, in recent years, there were 12 documented cases of non-rabid wolves killing and eating 80 humans.

In 1957, it was reported that, in Czarist Russia from 1849 to 1851, there were reports of 266 adults and 110 children killed by wolves. In 1875, 160 people were reported killed by wolves.

In 1975-76, rabid wolves attacked 15 people in Ulyanovskij, seven in Kaluzhskij, six in Orenburgskij, and four in the Orlovskij Region.

On August 21, 1978, near the village of Styazhnoe, Bryansk Oblast, a wolf attacked and bit six people. After a hunter killed the wolf, its body was examined, and it was found to be healthy.

In August 2005 in a three-day period, wolves attacked 14 people in the eastern part of Ukraine in the Zaporizhia Region. One wolf that was killed by a car was found to be rabid.

Wolf Predation and Russian Damage

Russian wolves prey on three principle wild game species: moose in the forests, elk or wapiti in more open areas, and reindeer on the tundra. Moose and elk are wild game species, but a significant number of reindeer have been domesticated and represent a very important source of food for people living in the northern regions. A five-year study in the region north of Moscow found that, between 1970 and 1975, 82.4% of the food of wolves was moose. This research also found that bears and wolves were not in competition for food.[17]

Russian research on wolf predation also contradicts the widely touted claim by biologists in the U.S. and Canada that wolves are "nature's sanitarians." Numerous studies show that wolves do not just attack the weak and the old animals, thus making the herd healthier. Actually wolves seem to prefer healthy wild ungulates, especially the young and the female. As part of the five-year study, the bodies of 63 moose killed by wolves were examined. Only nine of the 63 total had any defects or deviations from the normal. Five had defects with their teeth; three had defects with their antlers; and one had defects with its hooves. During this same study in February 1971, it was noticed that wolves would approach some moose but would not attack them. The moose they did not attack were later found to have ringworm. Of the three moose that died from ringworm, wolves would not feed on any of them.

Numerous studies in Russia report that wolves engage in surplus killings. One of the most dramatic incidents was a three-week period in February 1978 when wolves embarked on a mass killing of young Caspian seals on the ice of the Caspian Sea near Astrakhan. It is estimated that wolves killed between 17% and 40% of all the young seals in the area and left their bodies strewn on the ice with few eaten. Large, adult seals were also killed and not eaten.[18]

It is more difficult to count the exact numbers of wild game killed by wolves, though reindeer herders in various regions have lost thousands of their semi-domesticated stock to wolves in many years. For example, between 1945 and 1954, about 75,000 reindeer were lost to wolves in northern regions. In a

WOLVES KILL BRIDAL PARTY.

Only Two Escape Out of 120 In Asiatic Russia.

Special Correspondence THE NEW YORK TIMES.

ST. PETERSBURG, March 8.—Tragic details of the fate of a wedding party attacked by wolves in Asiatic Russia while driving on sledges to the bride's house, where a banquet was to have taken place, are now at hand, and in their ghastly reality surpass almost anything ever imagined by a fiction writer.

The exceptionally severe weather has been the cause of many minor tragedies in which the wolves have played a part, but perhaps none has ever been known so terrible as that now reported, since in this instance no fewer than 118 persons are said to have perished.

A wedding party numbering 120 persons set out in thirty sledges to drive twenty miles from the village of Obstipoff to Tashkend.

The ground was thickly covered with snow, and the progress was necessarily delayed, but the greater part of the journey was accomplished in safety.

At a distance of a few miles from Tashkend the horses suddenly became restive, and the speculation of the travelers changed to horror when they discerned a black cloud moving rapidly toward them across the snowfield.

Its nearer approach showed it to be composed of hundreds of wolves, yelping furiously, and evidently frantic with hunger, and within a few seconds the hindmost sledges were surrounded.

Panic seized the party, and those in the van whipped up their horses and made desperate attempts to escape, regardless of their companions, but the terrified horses seemed almost incapable of movement.

A scene frightful almost beyond description was now enacted. Men, women, and children, shrieking with fear, defended themselves with whatever weapons they could, but to no avail, and one after another fell amidst the snarling beasts.

The wolves, roused still further by the taste of blood, rushed toward the leading sledges, and though the first dozen conveyances managed to stave them off for a time, it was only at a terrible cost, since it is asserted that the women occupants were thrown out to be devoured by the animals.

The pursuit, however, never slackened, and the carnage went on until only the foremost sledge—that containing the bride, and bridegroom—remained beyond the wolves' reach.

A nightmare race was kept up for a few hundred yards, and it seemed as though the danger was being evaded, when suddenly a fresh pack of wolves appeared.

The two men accompanying the bridal couple demanded that the bride should be sacrificed, but the bridegroom indignantly rejected the cowardly proposition, whereupon the two men seized and overpowered the pair and threw them out to meet a horrible fate.

Then they succeeded in rousing their horses to a last effort, and, though attacked in turn, beat off the wolves and eventually reached Tashkend, the only two survivors of the happy party which had set out from Obstipoff.

Both men were in a semi-demented state from their experience.

The New York Times

Published: March 19, 1911

number of documented cases, wolves killed as many as 30 reindeer in one attack, leaving most of them uneaten. V. Glushkov reports that, from 1967-1972 in the Vyatskij forests, every year wolves killed about 28.6% of all moose, while grizzly and black bears only accounted for 4.7%.

When wild species decline, then wolves turn to livestock, which occurs when either a food shortage or severe weather causes wildlife population declines or the wolf population increases. Wolf predation on domestic animals in Russia has at times, especially during the five peak periods, been extreme. Following are some examples:

In 1944 in the Buryatskij region ASSR, wolves killed over 5,300 cattle; in the Penzenskij region, 8,700 cattle; in the Kujbyshevskij region, 4,200 cattle; and in the Tambovskij region, 8,000 cattle.[19]

In 1987, there were approximately 62,000 wolves in Kazakhstan which were causing damage to both livestock and game animals. In the first half of 1988, wolves killed 38,719 domestic animals. The Central Administration of Hunting organized 300 teams of professional hunters to cull wolf numbers. This was in addition to other private and professional hunters, who were also culling wolves. However, their combined effort was not sufficient to hold back increases in the number of wolves. In 1992, there were approximately 90,000 wolves, and there was a tendency for the numbers of wolves to increase. In October 1994, at the International Symposium in Poland titled "Large Predators and Man," a Russian scientist reported that presently there were probably 100,000 wolves in Kazakhstan.[20]

By 1995, the damage by wolves in just one region of the Russian Federation, the Altai Region, was estimated to be about 10,000,000 rubles per year. It was also reported: "There are sounds of a calamity coming."[21]

In the Sakha Republic, whose name was formerly the Yakutsk Republic, it was reported in 2013 that the population of wolves was about 3,500. The Veterinarian Department of the Ministry of Agriculture reported that wolves killed 9,995 reindeer in 2000, 8,869 in 2001, and 9,540 in 2002. In 2012, wolves killed 313 horses and 16,111 reindeer—the wolves make no selection if the reindeer are sick or healthy. The *estimated value of one reindeer is $328.00, which adds up to a loss of about $5 million to reindeer herders; 750 wolves have been recently culled. The governor of Sakha, Mr. Borisov, said in January 2012 that the goal of Sakha is to reduce the wolf population from 3,500 to 500. As an incentive for hunters and herders to cull wolves, he has placed a bounty of about $630.00 per each wolf pelt. Vladimir Krever, the Russian World Wildlife Federation's Head of the State Committee Biodiversity Program, said: "When wolves start attacking deer and livestock, they have to be killed and the population controlled. This is the right policy." He added: "Even if they were able to kill 3,000 wolves, the population would recover quickly."[22]*

Conclusion

To summarize, since 1870 in Russia, the area of the world with the most wolves, during five historical periods when governments reduced support for wolf culling, the wolf population expanded rapidly. During these population peaks, wolves significantly reduced wild ungulates, and when wild game became scarce, the wolves turned their predatory attention to domestic animals. The wolves grew bolder and bolder, culminating in wolves entering villages and preying on dogs, cats, sheep, goats, geese, and large livestock. In many cases, wolves killed far more animals than they needed for food. The wolf predation overall caused considerable economic damage, as well as loss of food. Attacks of rabid and non-rabid wolves on people also escalated, as did incidence of diseases carried by wolves.

The wolf population was not controlled by some "balance of nature." Population explosions could only be controlled by significant human intervention. As ownership of firearms was and still is primarily limited to the government troops, law enforcement, and a relatively small number of the upper class, when wolves did enter villages and communes, they met with little resistance. The interventions had to be supported by the government, as the general populace did not have the resources to undertake large-scale culling.

The lessons from the Russian experience with wolves are many for North America. What Russia tells us about wolves is that, while wolf behavior is always unpredictable, the overall patterns of wolf behavior adapting to modern settlements follow Dr. Geist's model of habituation and have consequences that are not pleasant for man or animals, wild or domestic.

If wolves are howling near a village in winter,
it portends hunger and high prices for food.
Russian saying

1 http://www.freerepublic.com/focus/f-news/1459944/posts
2 http://www.rense.com/general87/rabid.htm
3 http://en.rian.ru/strange/20120131/171045959.html
4 http://www.rferl.org/content/tajikistan_wolf_attacks/24457676.html
5 http://www.bbc.co.uk/nature/animals/conservation/wolves/intro.
shtml?oo=15732
6 Biblikov, D., "The Wolf," *Science*, 1985; Biblikov, D., Hunting and Game Management, July 1978.
7 *Hunting Management in the USSR*. Moscow: Forest Industry, 1973: 196.
8 Sitsko, A., Hunting and Game Management, January 1981: 1-2.
9 Utilin, Dr. A., President of the Central Board of the All Russian Association of Hunters and Fishermen, Conservation Force Web Page, 2001.
10 Andrei Ivanov, IPS World News, March 26, 1998, Internet.
11 Davis, Catherine, Sky Wolf News, BBC Central Asia Correspondent, January 27, 2003.
12 Bectimirov, T., and A. Movsesyants, "Epidemic and Epizootic Rabies Situation in the Russian Federation," Moscow: L.A. Tarasevich State Institute, Moscow. February 9, 2003.
13 Vedernikov, V.A., V.A. Sedov, et. al., Moscow: Institute of Experimental Veterinary Medicine. Internet, October 8, 2003.
14 Zheleznov, N.K., *Wild Ungulates of the USSR North-East*, Vladivostok: Academy of Science 9f of the USSR, 1990: 411.
15 Pavlov, M.P., *The Wolf*, Moscow: Forest Industry, 1982: 109, 129.
16 Milenushkin, Y.I., *Wolves and Their Destruction*. Moscow: Military Press, 1957: 103.
17 Moskvin, N., "The Wolf of Mologo-Sheksninskij Mezhdurech," *Hunting and Game Management*, February 1978: 12-13.
18 Rumyantsev, V., and L. Khuraskin, "Wolves and Caspian Seals," *Hunting and Game Management*, February 1978: 22-24.
19 Makridin, V., "Do Not Cultivate Wolves, But Develop the Economy," *Hunting and Game Management*, August 1968, p. 18-19.
20 International Symposium on "Large Predators and Man," Poland, October 1994, attended personally by Will Graves. Dr. Mech was also there.
21 Internet, Pravda, March 12, 2003.
22 Internet, "Wolves: An Emergency Situation," January 23, 2013; also recent email exchange between Will Graves and a resident of Sakha.

Chapter 5

A Conservation Tragedy

By Cat Urbigkit

Cat Urbigkit is an award-winning writer and photographer, who lives on a Wyoming sheep ranch where she finds daily inspiration for her work. In addition to eight nonfiction books for children, Cat has written two adult nonfiction titles, including *Yellowstone Wolves: A Chronology of the Animal, the People, and the Politics,* of which an adaptation appears in this chapter. *Shepherds of Coyote Rocks: Public Lands, Private Herds, and the Natural World* is an action-packed true story of her journey alone to spend a season on Wyoming's open range, tending to a flock of domestic sheep as they give birth amid the challenges of nature—from severe weather to a wealth of predators. Her only companions are the guardian animals that repeatedly prove their worth in their devotion to protecting the flock. The book reveals the broad spectrum of the human relationship with nature, from harmony to rugged adventure. Cat's newest title, *When Man Becomes Prey*, is slated for release in the fall of 2014 and examines fatal encounters between humans and five North American predator species.

Editors' Statement of the Issues

The Canadian wolf was introduced under the guise of protecting an endangered species, while, in fact, the program deliberately introduced animals that would eliminate the integrity of the native wolf species, alive and well in the northern United States.

Editors' Statement of the Facts

Wolves were *not* extinct in the northern United States; however, the unwelcome fact that there was hard evidence of existing wolf populations was both denied and suppressed. The USFWS chose instead to ignore its obligation to protect our native wolf, thereby signing its death warrant. The federal government should not supersede local and state authority for wildlife decisions.

> *The judicial reversal—declaring that the Canadian wolves*
> *could stay and that the protection of a subspecies was*
> *not mandated by the Endangered Species Act—was a*
> *tragedy. Rather than a victory for wildlife, we viewed it with*
> *heartbreak—an action that would cause the extinction of*
> *populations of a truly distinct animal, our native wolf.*
> Cat Urbigkit

Jerry Kysar and the wolf he shot in the Teton Wilderness late in September 1992. Photo courtesy Jerry Kysar. Reprinted from *Yellowstone Wolves: A Chronicle of the Animal, the People, and the Politics.*

Like Shooting a Dinosaur

In 1992, an elk hunter shot a black wolf, one of a pack of five, as the animals ran through the Teton Wilderness south of Yellowstone National Park. The hunter said it was like shooting a dinosaur—a member of a species long thought to be extinct from the region. The timing of the event gave it national significance.

When the black wolf was shot, the U.S. Fish and Wildlife Service (USFWS)—the federal agency charged with managing the recovery of endangered species—was in the process of finalizing plans to reintroduce gray wolves, *Canis lupus*, into Yellowstone National Park, an area the agency claimed had not heard the howls of wolves for more than 60 years. USFWS immediately launched a publicity blitz to head off the obvious conclusion that there was at least one pack of wolves inhabiting the Yellowstone area and that reintroducing wolves was, therefore, unnecessary.

Federal biologists with USFWS claimed that the four animals that were with the wolf when it was killed were, amazingly, coyotes. The agency did not acknowledge the numerous reports that had accumulated over the years that multiple wolves had been sighted in the Teton Wilderness, wolves that had been seen by federal and state resource agency employees. The agency also did not say that the legality of the reintroduction proposal hinged on there being *no* wolves in the area.

Wolf Sightings

For nearly a decade before the black wolf was shot, however, a Wyoming couple—namely, my husband Jim and I—had been collecting reports of wolf sightings in western Wyoming. The wolves were already there, and although elusive, they had left diverse physical evidence of their presence:

> *The wolves left carcasses of elk they had fed on near the Green River Lakes entrance to the Bridger Wilderness of the Wind River Mountain Range, which we photographed.*

> *They left piles of dung on Straight Creek in the Wyoming Range Mountains, which we collected.*

> *They left tracks in the foothills of the mule deer winter range near Half Moon Lake, from which we made plaster casts.*

They left hair on a branch in an overhang used to wait out a
winter storm, which we gathered.

And our howls were returned from wild wolves in the Mount
Leidy Highlands and Johnson Ridge.

Local residents of the mountain country—men who grazed cattle
on high during the summer months, outfitted and hunted in the fall,
and cut wood in the spring—would see the wolves, their kills, and
their sign and send us back on the trail of the elusive predator. The
evidence we collected led us to know that the wolves were remnant
populations of the native wolf—the supposedly extirpated subspecies
Canis lupus irremotus.

The wolf biologists who were pushing the reintroduction were
based in offices in Helena, Montana, hundreds of miles away from
Yellowstone. The reintroduction would be the biggest experimental
relocation of a large carnivore population ever undertaken in the
history of wildlife management and the basis of a huge fundraising
opportunity for national environmental interests.

Grand Gesture

USFWS pushed the concept of wolf reintroduction into Yellowstone strongly and developed support for the project from across the
nation. Returning wolves to Yellowstone, the nation's first national
park, would, according to USFWS, be a grand gesture that would
correct an injustice done to nature many years prior. USFWS claimed
that the subspecies of wolf native to Yellowstone had been extirpated
early in the 20th century through predator control efforts, so wolves
belonging to a different subspecies, *Canis lupus occidentalis*, would
be brought from Canada and released into the park. The Canadian
wolves were breathtaking to behold—one of the largest subspecies of
wolf in the world and the most numerous.

With the existing wolves making their presence known, causing
pressure on the agency, USFWS Director and Wyoming native John
Turner advocated wolf reintroduction over natural recovery, arguing:

If, on the other hand, we sit back and wait for wolves to
recolonize the Yellowstone area on their own, the opportunity
to design locally responsible and flexible management
strategies will be lost. Once they reach the area on their
own, the experimental population option is foregone since

the Endangered Species Act stipulates that an experimental population must be "wholly separate geographically" from non-experimental populations of the same species.

Lawsuits Pursued and Lost

As support for the USFWS wolf relocation proposal grew and as evidence of the wolves already present in the Yellowstone area was ignored or dismissed, Jim and I came to feel that no one was willing to fight to save the native wolf. To us, it was apparent that they were facing deliberate extinction, so we took steps to prevent it. We fought their extinction all the way to the federal Tenth Circuit Court of Appeals.

On December 12, 1997, five years after the black wolf was killed in the Teton Wilderness, federal Judge William Downes ruled that the reintroduction of wolves was illegal because wolves already existed in the area. By that time, however, the reintroduction effort was already underway, and some 160 Canadian wolves were present in the Yellowstone area. Soon a higher court overruled Judge Downes' decision, declaring that the Canadian wolves could stay and that the protection of a subspecies, the basis of our litigation, was not mandated by the Endangered Species Act. This judicial reversal was a tragedy. Rather than a victory for wildlife, we viewed it with heartbreak—an action that would cause the extinction of a truly distinct animal, our native wolf.

Continuity of Wolf Sightings

The reports of wolf presence inside and adjacent to Yellowstone Park had demonstrated the continuity of wolf sightings in the area from the last days of the predator control era to just prior to the release of Canadian wolves into the Park. The available records indicated that a few wolves had survived in the area and that both the National Park Service and USFWS confirmed the presence of wolves during this time.

Documenting wolf occurrence in an area where most people viewed wolves as extinct should have been of great scientific value. Zoologist and wolf taxonomy expert Ronald Nowak, when speaking of the possibility that the wolf killed in the Teton Wilderness may have been a survivor of a remnant population, stated, "It would be one of the major zoological finds of the century."

It would have given us a chance to learn about an animal of which we knew relatively nothing. However, instead of working to preserve wolves native to the area, the federal government undertook a deliberate program due to political expediency. The wolf reintroduction program was soon touted as one of the nation's top achievements of endangered species conservation.

Increasing Wolf Population and Recovery Goals

The wolf population in the tri-state region around Yellowstone continues to increase. Recovery goals of the reintroduction program were first achieved in 2002, more than a decade ago. Yellowstone National Park's wolf population soon saturated the Park, bringing with it social strife and disease as the new challenges facing the booming population.

With the original recovery goal of 30 breeding pairs and more than 300 wolves well distributed among Idaho, Montana, and Wyoming for at least three consecutive years, the Northern Rocky Mountain wolf population has exceeded that goal every year for the last decade. For example, the 2010 count exceeded 111 breeding pairs and a minimum of 1,650 wolves in Idaho, Montana, and Wyoming.

Delisting Efforts

The states of Idaho and Montana assumed responsibility for managing the wolf population in their states in 2005 and 2006, respectively, under agreements with USFWS. Continued litigation from animal and wolf advocates eventually led to Congressional action protecting delisting efforts in those states from further court challenges. Wyoming reached agreement with USFWS late in 2007, a situation that led to the Yellowstone wolf population being briefly removed from the list of protected species early in 2008. The legal morass of on-again/off-again federal protection for wolves in Wyoming continues.

Under Wyoming's wolf management plan, wolves are managed under a dual classification system in which wolves in northwest Wyoming are designated and managed as "trophy game animals" and wolves in the rest of Wyoming are designated as "predatory animals." The management plan also includes a flex area known as the "Seasonal Wolf Trophy Game Management Area," where wolves are classified as trophy game animals from October 15 to the last day of February of the subsequent year and as predatory animals for the remainder of the year. Wolf hunting is prohibited in Grand Teton National Park and the National Elk Refuge.

The Wyoming Game and Fish Department has established wolf hunting seasons in 12 separate hunt areas in the Wyoming Trophy

Game Management Area (WTGMA) and Seasonal Wolf Trophy Game Management Area (SWTGMA). Hunting seasons in each hunt area begin October 1 and end December 31, except in Area 12 (the SWTGMA), which opens October 15 and closes December 31. The hunt season for each area remains open until the wolf quota for the area is reached or until December 31, whichever occurs first. Hunters must call a wolf hotline prior to hunting to ensure that the quota for wolves has not been reached in a particular hunt area. In addition, all hunters who successfully harvest a wolf in the designated hunt areas are required by law to report the kill within 24 hours to the wolf hotline and to present the skull and pelt to a game warden or WGFD regional office for registration.

In areas where wolves are designated as predatory animals, there are no closed seasons or bag limits, and a license is not required to kill a wolf. However, hunters who take a wolf in these areas are still required to report the kill to a district game warden, district WGFD biologist, or other WGFD personnel within 10 days.

Throughout the remainder of the state, however, wolves will be considered predators, and WGFD will not manage nuisance activities in those areas; nor will it compensate livestock producers for livestock that are killed by wolves where they are designated as predatory animals. Those having conflicts with wolves in these areas are simply on their own, and local predator management boards will feel the pressure to provide wolf control for those in need.

While humans continue howling in court and in the press, the wolves will continue on their own path. Whether it is the native subspecies of wolf, the Canadian transplants, or a cross between, we know that humans have little control over these intelligent beasts.

We also know that environmental groups and their governmental counterparts will pay for this mess for years to come. One result has been the creation of a groundswell of animosity toward wildlife management among many that live in the West. There were new organizations created solely from the backlash against the wolf program.

But more importantly, the standard for protection of unique ecological units has been lowered to the ground. This will be hard to overcome once the environmental groups finally pick the life form they actually wish to protect, since the current standard is that anything less than a species is entirely "discretionary" when it comes to deserving or qualifying for protection.

Taxonomic Truths and Homogenized Species

The quest for taxonomic truth in wildlife seems to have been abandoned, as wildlife management has turned to pursuit of homogenized species. Cases abound as wildlife managers move animals around without regard to maintaining the taxonomic integrity of ecological forms, affecting everything from numerous fish species and orangutans to pronghorn antelope and bighorn sheep.

The determination of what constitutes a species, a subspecies, and a distinct population is a critical factor in future implementation of the Endangered Species Act. The words of the Act are powerful, including the provision calling for conservation of threatened and endangered species, including "any subspecies of fish or wildlife or plants and any distinct population segment of any species." Taxonomic units lower than subspecies are to be protected, according to the law, including those "distinct population segments," which should be reason to rejoice. But it is not.

Conclusion

Those who were, and are, determined to have the reintroduction of wolves into Yellowstone, at any price, have helped to create a collusion of sorts between government and environmentalists to sacrifice taxonomic units below the species level. Today, USFWS is busy delisting large carnivore populations that it designates "distinct population segments"—not because of any real ecological distinction, but because of the distinctions of jurisdictional lines, including state boundaries. The legacy left by the wolf reintroduction program includes harm to the conservation of biological diversity far into the future.

Torsten Lorenz / Shutterstock.com

Chapter 6

SEVEN STEPS OF WOLF HABITUATION

By Valerius Geist, Ph.D.

Valerius Geist was born in Ukraine in 1938 and immigrated to Canada in 1953. Geist earned a bachelor's degree in zoology from the University of British Columbia in 1960. His wildlife studies began with feral goats and continued with moose, Stone's sheep, bighorn sheep, and Dall's sheep. In 1966, Geist earned a doctorate in ethology, the study of animal behavior. He returned to Germany for a year of research and the writing of his first award-winning book. Geist joined the faculty of the University of Calgary in 1968, where he was a professor of environmental science and biology until 1995. Geist has shared his research, knowledge, and ideas through 15 books; 7 wildlife policy reports; more than 120 scholarly papers, book chapters, and commentaries; and numerous magazine articles and contributions to documentary films covering an array of wildlife topics and species. Particularly relevant to this book, Geist has served as expert witness in wildlife law enforcement, environmental policy, native treaties, and animal behavior cases in the U.S. and Canada.

Editors' Statement of the Issues
Reassessing the danger posed by the introduced wolf packs with current data is critical to the safety of the American public. Since modern day people have no history of wolf attacks, they assume that wolves pose no danger to them. This is a dangerous belief to hold.

Editors' Statement of the Facts
As introduced wolf populations grow, wolf/human encounters will increase. Successive encounters tend to cause habituation in wolves, multiplying our danger as their fear dissipates. As populations grow, packs become fearless, as they have no predators, save man. Until detailed reporting of wolf attacks become available to all, the idea that the wolf is relatively harmless will prevail—to our danger.

There is another very important point made by native people and routinely ignored by officialdom—namely, wolves eat the evidence as well as distribute it widely or bury it.
Valerius Geist, Ph.D.

Four Decades of Field Research
Nothing convinces like personal experience! As an academic, I confess to this with some distress because—by training, experience, and attitude—I should be above it. That I am not alone in this habit is of little comfort. And so it is with wolves.

In my four decades of field research on ungulates in North America—mountain sheep, goats, elk, deer, moose, etc.—I also observed wolves, and my experience with continental wolves matches that of my colleagues; namely, wolves are intelligent and wary and keep their distance from people. Consequently, during my academic career and four years into retirement, I thought of wolves as harmless, echoing the words of more experienced North American colleagues, while considering the reports to the contrary from Russia, Europe, and Scandinavia as interesting, but not relevant to an understanding of North American wolves. I was wrong!

Wolves Up Close and Personal
I saw my first North American wolf in the wild early one morning in May 1959 on Pyramid Mountain in Wells Gray Provincial Park, British Columbia. In the spotting scope, his image was crisp and clear. It was an ash-gray wolf with a motley coat, sitting and watching me

from a quarter mile away with an eager, attentive look about his dark face. His red tongue was protruding, while golden-yellow morning light played on his fur. I do not know if my heart skipped a beat upon sighting that wolf, but it well might have. Whose wouldn't?

About five months prior, in early January, I had had an informative brush with a wolf pack just a few miles from that spot. A friend and I were observing moose. We were in the midst of a migration, and some two dozen, mostly bulls who had shed antlers, were dispersed over the huge burn. A few were feeding on the tall willows, but most were resting in the knee-deep snow. Suddenly we heard a low, drawn-out moan. When I glanced at the moose, I saw that all were standing alert, facing down the valley. We were green then and perplexed about this unearthly sound. As if to answer us, a high-pitched voice broke in and then another and another, and we realized that we were hearing wolves. Within minutes, a chorus was underway, and so were the moose. All were hastily moving up the valley, and 10 minutes later, the moose had vanished. I opted to stay at our lookout while my friend, taking my rifle, decided to search for the wolves. He saw them at dusk as they walked across a small lake, a pack of seven. Try as he may to fire the rifle, it would not fire as it had frozen in the severe cold. This may have been kind fortune, for the first wolf I shot with that rifle instantly attacked me, although it collapsed before reaching me. The second screamed, which has triggered pack attacks in the past. Had the pack attacked, I would have been minus a friend in minutes. While a large man can subdue an attacking wolf, even strangle it, there is no defense against an attacking pack.

Two years later during my study of Stone's sheep in northern British Columbia, I had exceptional opportunities to observe wolves in pristine wilderness. My closest neighbors, a trapper family, lived some 40 miles to the west, and the closest settlement, Telegraph Creek, was about 80 miles to the north. Timberlines were low, and the wolves spent much time in the open, plainly visible. I watched them for hours on end. These large, painfully shy wolves on occasion even panicked over my scent. Though they killed a few sheep, their hunts were largely unsuccessful. However, I began to appreciate their strategies and tenacity as hunters. In traversing the valley, one crossed a wolf track about every 50 paces. They were scouring the valley for moose.

On rare occasions, a wolf would follow my tracks, sit, and listen to what I was doing in my cabin at night. (Grizzly bears did that, too.) One evening three wolves began to surround me on a frozen lake. One raced towards me, but scrambled madly to get away once he got downwind of me. Another cut my fresh track, then jumped straight up, and raced back. Thus, my early experiences with mainland wolves indicated that they were inquisitive, but shy and cautious. Moreover, my contacts with wolves were few compared to the huge number of Osborn's caribou. I then thought that this was normal. Years later, a first doubt arose when a student of mine could hardly find a caribou where I had seen hundreds. However, a wolf pack of 43 individuals was recorded where I had observed for years a pack of seven.

Evidently, my experiences with wolves were anomalous for, a decade earlier, there had been massive broadcast poisonings of wolves to control rabies. The "pristine wilderness" had been tampered with; it had experienced a "rebound" of ungulate populations after they had been freed from severe predation. When my wife and I tell of forests of antlers as caribou bulls gathered on the Spazisi Plateau for the rut, colleagues look at us as if we came from another age.

Threatening Wolves

Nothing in my previous studies had prepared me for what I was to experience with wolves on Vancouver Island following my retirement in 1995 from the University of Calgary, where I had served as Program Director and Professor of Environmental Sciences for many years. In my student days, in the late 1950s, wolves on Vancouver Island were so scarce that some thought they were extinct. In the early 1970s, they reappeared and swept the island. The annual hunter kill of black-tailed deer dropped swiftly from about 25,000 then to less than 3,000 today. There were reported incidents of wolves threatening people, and a colleague was treed by a pack. When he spoke of this, nobody believed it, so he clamped up. Wolves threatening people? Ridiculous!

According to my colleagues, massive clear-cutting of old-growth forests and the rapid spread and growth of the wolf population caused the carnage to ungulates. Those who witnessed it tell of deer carcasses everywhere—and then no more deer. The loggers left standing small patches of mature timber as deer winter range. However, wolves, cougars, and black bears discovered those patches and cleared out the remaining deer. The clear-cuts also led to a population explosion of black bears, some of whom became experts in killing elk calves and deer fawns.

Deer are still so rare in the mountains of the island that I see about three dozen bears for every deer. However, deer are common in towns and suburbs and about farms, where they are relatively safe, at least from wolves. The elk population is holding its own, but at a low level compared to the vast amounts of food on the clear cuts. The bulls are huge, with massive antlers, but with a predator-induced silence during the rut. Enough calves perish so that there is little recruitment, and hunters are held to one permit per 40-150 applicants.

I retired to an agricultural area on Vancouver Island. During walks near our home, I explored through all seasons a meadow system associated with dairy, beef, and sheep farming. These meadows and adjacent forests contained, year-round, about 120 black-tailed deer and half a dozen large male black bears. In winter came some 60-80 trumpeter swans, as well as large flocks of Canada geese, widgeons, mallards, and green-winged teal. Pheasants and ruffed grouse were not uncommon. In the fall of 1995, I saw one track of a lone wolf. I cannot recall seeing any wolf tracks in the four years following.

Empty Landscape

Then in January 1999, my oldest son Karl and I tracked a pair of wolves in the snow, suggesting a breeding pair and thus pack formation. A pack did, indeed, arrive that summer. Within three months, not a deer was to be seen, or tracked, in these meadows—even during the rut. Using powerful lights, we saw deer at night huddling against barns and houses, where deer had not been seen previously. For the first time, deer moved into our garden and around our house, and the damage to our fruit trees and roses skyrocketed. The trumpeter swans left. The geese and ducks avoided the outer meadows and lived only close to the barns. Pheasants and ruffed grouse vanished. The landscape looked empty, as if vacuumed of wildlife.

Wolves Become Aggressive

Wolves attacked and killed or injured dogs, at times right beside their shouting, gesticulating owners. Wolves began following our neighbors when they rode out on horseback. A duck hunter shot one wolf and wounded fatally another as three wolves attacked his dog. Wolves ventured into gardens and under verandas trying to get at dogs and ran after quads, tractors, and motorcycles to attack the accompanying farm dogs. My neighbor warded off three such attacks on his dogs with his boots, and his hired man ran back with a tractor in panic with the wolves in pursuit of the two dogs that tried to stay under the running tractor. One wolf approached within about 15 paces of my wife and a group of 11 visitors that were taking an evening stroll about half a mile from our house. The wolf howled and barked at the people. Our neighbor then went out armed with his dogs, and the wolf, a small female, promptly attacked the dogs and was shot at 50 feet. Nine days later, my neighbor killed a second wolf that was openly following and barking at him—this wolf may have been defending a sheep it had killed and dragged half a mile. These wolves weighed between 60 and 70 pounds, small for wolves, a sign of poor nutrition.

Wolves of the "misbehaving" pack began to be killed as they persisted in attacking livestock and pets and aggressively approaching people. Wolves had been seen in the neighborhood sitting and observing people; we know from captivity studies that wolves are observation learners. One wolf, a male, approached my wife, my brother-in-law, and me across a quarter mile of open meadow and stood looking us over for a very long minute about 10 paces away, before moving on into the forest. Neighbors, myself included, repeatedly saw wolves showing interest in us.

A neighbor raising sheep lost many to wolves, so he acquired five large, sheep-guarding dogs. These dogs and the wolf pack engaged in a "dear enemy" ritual of frequent, nightlong barking and howling duels at the forest edge. I observed subsequently, on the evening of October 19, 2002, how the last of the pack, a male, fraternized successfully with the sheep dogs. He kept it up and was eventually shot March 12, 2003, while sitting among these dogs. However, before that, he visited us when our German longhair pointer female, Susu, was in heat and barked at my wife in our doorway; that is, he acted like other male dogs that were attracted to Susu in heat, only bolder.

There was a quiet spell of two years after the first wolf pack had been killed off. Migratory waterfowl began to reappear, but only half of the swans returned. Then on March 27, 2007, a second "misbehaving" pack appeared and took up where the first pack had left off.

The worst incident happened about 350 yards from our house. Our neighbors went in the morning to inspect their dairy cattle and pastures. Their old dog ran ahead of them. Just as they entered the forest, five wolves attacked the dog. My neighbor grabbed a cedar branch and advanced on the wolves, which turned towards him, snarling. His wife jumped into the caboose of their excavator that just happened to be standing there. My neighbor's energetic counter-attack freed the dog and intimidated all but one wolf, who advanced on him, snarling. However, he too withdrew, albeit reluctantly. While my neighbor ran home to get a gun, his wife ran to us, shouting for me to get a rifle. We did not see the wolves, though they were sighted briefly in the evening, and a neighbor walking his dog had an encounter with two wolves about a mile away; he was able to chase them away. The following morning, our neighbors took a rifle along during their inspection trip of their property. The wolf pack promptly went for them again, and my neighbor shot the most aggressive one, a male weighing 74 pounds.

From Misbehaving to Murderous

I saw the neighbors' cattle, spooked by a wolf, crash through fences while fleeing for the security of their barn. I found two of the three cattle killed and eaten by wolves; the third was severely injured about the genitals, udder, and haunches and had to be put down. I saw the docked tails, slit ears, and wounded hocks on the dairy cows. Our neighbor's hired man saw from a barn a wolf attacking a heifer with a newborn calf. He raced out and put the calf on his quad. As he ran to the barn, the wolf ran alongside, lunging at the calf, right into the barn! A predator control officer was called, and 13 wolves were removed within a mile of our house from the first "misbehaving" pack and four from the second "misbehaving" pack.

The "tameness," "hanging around," and increasing boldness and inquisitiveness are the wolf's way of exploring its potential prey and assessing the strength of its potential enemies. These incidents with the two "misbehaving" packs moved me to look for other situations where patterns of habituation might be found. I did not have to look far. Two wolves in June 2000 severely injured a camper on Vargas Island just off the coast of Vancouver Island. These wolves became tame before the attack—they nipped at the clothing of campers, licked their exposed skin, and ate wieners from their hands.

Our observations here suggested that wolves, attracted to habitations by the scarcity of prey, shift to dogs and livestock, but also increasingly, though cautiously, explore humans, before mounting a first, clumsy attack. I reported such at a Wildlife Society conference on September 27, 2005, in Madison, Wisconsin, in an invited paper on habituation of wildlife. That was about six weeks before wolves in northern Saskatchewan killed Kenton Carnegie.

On November 8, 2005, four wolves at Points North Landing, Wollaston Lake area in northern Saskatchewan, killed a 22-year-old, third-year geological engineering student at the University of Waterloo by the name of Kenton Joel Carnegie. This case is unique in that it is the first direct human fatality from a wolf attack in North America in recent times to receive a thorough investigation.[1] There have been people bitten by rabid wolves and killed, but such kills "do not count" as it is the rabies virus, not the wolf-bite, that killed.

I was asked by Kenton's family to investigate matters as were—independently—Marc McNay from Alaska, who ultimately served as an expert witness in the coroner's investigation, and Brent Patterson from Ontario. Our findings—independently conducted and assisted by other colleagues from Alaska and Finland—came to the same conclusions as did the earlier investigations by a native coroner, Mrs.

Rosalie Tsannie-Burseth, and a native Royal Canadian Mounted Police, Constable Alfonse Noey, who attended the scene within hours of Kenton's death. Both had been raised in the Saskatchewan Wilderness and were highly experienced with native wildlife. Mrs. Burseth was

Todd Svarckopf confronting a wolf in the same area where, four days later, Kenton Joel Carnegie was killed by wolves.

not only the coroner, but also Chief of the Hatchet Lake Band and Director of Education; she held two university degrees and was working on a doctorate. However, her and Constable Noey's thorough investigation was discounted by the Saskatchewan coroner. The two scientists called in by the Saskatchewan coroner, who examined the photography of Constable Noey, mistook the wolf tracks across an overflow on the frozen lake for bear tracks and proclaimed a bear as a killer. The subsequent scandal led to a coroner's hearing, which concluded that wolves had killed Kenton Carnegie.

Unfortunately, this coroner's inquiry had a narrow focus and did not deal with policy. Consequently, it never became public that Saskatchewan's legislation was, in good part, to blame for Kenton's death and that, in British Columbia, Kenton would have lived. In Saskatchewan, wolves can be killed by trappers or wardens or by special permit only. In British Columbia, any licensed hunter can shoot garbage-habituating wolves virtually throughout the year.

Fresh snow allowed accurate track reading that Mrs. Burseth was eminently qualified in, raised as she was by her father as a hunter well before her formal schooling in her teens. Constable Noey is also an experienced Northern hunter. Kenton was by himself when the wolves, both from behind and from the front, approached him. He fell three times before failing to rise. Constable Noey fired three shots to spook the wolves from Kenton's body and posted an armed guard— the husband of Mrs. Burseth and a work colleague, both experienced hunters and track readers—while he and Mrs. Burseth examined and photographed the kill and feeding sites.

We are aware that the four wolves in question, long observed by others, were garbage fed and were photographed four days earlier while attacking two employees of the camp, who beat back the wolves. Kenton was aware of this encounter. Unfortunately, neither he nor those who discussed the matter with him, as reported by Constable Noey's report and by reporters of the *Saskatoon Star Phoenix* of November 14, 2005, were aware that tame and inquisitive wolves are a signal of danger. Consequently, the first requirement is that the general public and especially outdoorsmen know that—when they see tame, inquisitive wolves—they should get out of the area quickly, but without undue haste, while being prepared to defend themselves. Running away invites an attack.

Additional Documented Wolf Attacks

As Mark McNay and others have established, there have been other attacks in Canada, historical and recent.[2] Five-year-old Marc Leblond was killed by wolves on September 24, 1963, north of Baie-Comeau, Quebec, as indicated by tracks and signs in the snow and the sighting close by of a wolf reluctant to leave, but his case drew little media attention. On April 18, 1996, 24-year-old Patricia Wyman was attacked and killed by five adult North American grey wolves (*Canis lupus ssp.*) at the Haliburton Forest and Wildlife Reserve near Haliburton, Ontario. And on December 31, 2004, Fred Desjarlais was attacked and wounded by a wolf in northern Saskatchewan.[3]

There are also plenty of unreported recent attacks by wolves like this one in Saskatchewan, not all fatal and not reported in any central formal wolf attack research center. A local rancher was attacked by three wolves while deer hunting; he killed two. Another attack happened on July 5, 2007, on Anderson Island off Aristazabal Island, north of Bella Bella, British Columbia. The man was a fit 31-year-old. He was put into the hospital by an emaciated, old she-wolf with broken teeth, even though the man stabbed her nine times. She was killed by a shotgun blast two hours later; there was no sign of rabies. Candice Berner, a 32-year-old schoolteacher, was killed on March 8, 2010, by wolves in the village of Chignik Lake on the Alaska Peninsula.

The Problem of Evidence

There is another very important point made by native people and routinely ignored by officialdom—namely, wolves eat the evidence as well as distribute it widely or bury it. Had the search parties looking for Kenton Carnegie not disrupted twice the wolves feeding on his body, very little of the body would have been left by the next morning. Within five hours, two or three wolves had consumed a very large portion of Kenton's body. Little wonder that only about one in seven calves killed by ranch-visiting wolves are ever attributed to wolves!

It is important to note that wolves learn differently than dogs. Wolves learn by observing, and they are insight learners; that is, they can solve problems by observing, such as how to unlock a gate. In some studies of captive wolves, researchers have found that wolves, and coyotes for that matter, not only learn to open their own cages, but also those of others. With these intelligence traits, wolves also develop an ability to assess the vulnerability of prey. For example, the sight of a human—walking boldly and carrying a firearm—will give them enough information to know that the potential prey is not vulnerable.

Seven Stages of Habituation

By studying the two "misbehaving" packs of Vancouver Island wolves and comparing this data with data obtained from scientists in Russia, Scandinavia, Europe, and the Middle East, a seven-stage model of habituation has emerged that progressively leads wolves from shy, wild animals to those that begin targeting people as prey. What we know is that, when wolves run out of wild prey, they first begin targeting livestock and then pets and finally exploring people as prey. The presence of garbage also helps draw wolves closer to people. It should be noted here that the considerable number of attacks by wolves in Europe and Asia is at least partially due to the absence of firearms in the hands of the citizenry. Distilled from my own observations and reviews of research in North America and abroad, these are the seven stages of habituation leading to an attack on people by wolves.

 Wolves move in closer to people. Within the pack's territory, prey is becoming scarce—not only due to increased predation on prey animals, but also by the prey evacuating home ranges en mass, leading to a virtual absence of prey. Alternatively, wolves increasingly visit garbage dumps at night. We observed the former in the summer and fall of 1999. Deer left the meadow systems occupied by wolves and entered boldly into suburbs and farm, causing—for the first time—much damage to gardens and sleeping at night close to barns and houses, which they had not done in the previous four years. The wintering grounds of trumpeter swans, Canada geese, and several species of ducks were vacated. The virtual absence of wildlife in the landscape was striking.

🐾 Wolves begin to approach human habitations at night in search of food. In our experience, their presence was announced by frequent and loud barking of farm dogs. A pack of sheep-guarding dogs raced out each evening to confront the wolf pack, resulting in extended barking duels at night. The wolves were heard howling even during the day.

🐾 Wolves appear in daylight and, at some distance, observe people doing their daily chores—wolves excel at learning by close, steady observation. Wolves approach buildings during daylight.

🐾 Wolves act distinctly bolder in their actions. Small-bodied livestock and pets are attacked close to buildings even during the day. Wolves preferentially pick on dogs and follow them right up to the verandas. People out with dogs find themselves defending their dogs against a wolf or several wolves. Such attacks are hesitant, and people save some dogs. At this stage, wolves do not focus on humans, but attack pets and smaller livestock with determination. However, they may threaten humans with teeth exposed and growl when humans are defending dogs or show up close to a female dog in heat or close to a kill or carrion defended by wolves. The wolves are still establishing territory at this stage.

🐾 Wolves explore large livestock, leading to docked tails, slit ears, and damaged hocks. Livestock may bolt through fences, running for the safety of barns. The first seriously wounded cattle are found; they tend to have severe injuries to the udders, groin, and sexual organs and need to be put down. The actions of wolves become more brazen, and cattle or horses may be killed close to houses and barns where the cattle or horses were trying to find refuge. Wolves may follow riders and surround them. They may mount verandas and look into windows.

 Wolves turn their attention to people and approach closely, initially merely examining them carefully for several minutes on end. This is a switch from establishing territory to targeting people as prey. The wolves may make hesitant, almost playful attacks—biting and tearing clothing and nipping at limbs and torso. They withdraw when confronted. They defend kills by moving towards people, growling and barking at them from 10 to 20 paces away.

 Wolves attack people. Initial attacks are clumsy, as the wolves have not yet learned how to take down efficiently the new prey. Persons attacked can often escape because of the clumsiness of the attacks. A mature, courageous man may beat off or strangulate an attacking wolf. However, against a wolf pack, there is no defense. Even two able and armed men may be killed. Wolves as a pack are hunters so capable a predator that they may take down black bears and even grizzly bears. Wolves may defend kills.[4] Attacks on people may not be motivated by predation, but rather may be a matter of more detailed exploration unmotivated by hunger. This explains why wolves on occasion carry away living, resisting children; why they do not invariably feed on the humans they killed, but may abandon such, just as they may kill foxes and just leave them; and why injuries to an attacked person may at times be surprisingly light, granted the strength of a wolf's jaw and its potential shearing power.

Upon presenting this model at professional meetings and sharing it with other wildlife biologists, I met Dr. Robert Timm at the University of California at Davis, who has been studying coyote attacks. It turned out that coyotes targeting children in urban parks act in virtually the same manner.[5]

Peer Review

Peer review is essential to science. My work has been translated into Swedish, Finnish, and German, and the progression became known in Finland as *the seven steps to heaven*! Then a review of the Russian wolf experiences by Professor Christian Stubbe in Germany and Will Graves' book *Wolves in Russia* added more support for the seven-stage model. Italian and French historians have also published papers and books detailing how thousands of people had died in earlier centuries from wolf attacks.

Some historians rightly asked the question, "How did North American scientists ever conclude that wolves were harmless and no threat to people?" We now know the answer: In the absence of personal experience or sound language competence, they chose to disregard, even ridicule, the accumulated experience of others from Russia, France, Italy, Germany, Finland, Sweden, Iran, Kazakhstan, India, Afghanistan, Korea, and Japan. They were also unaware that, during most of the 20th century, tens of thousands of trappers in the heartland of wolf distribution in Canada and Alaska were killing every wolf they could, legally and illegally. Moreover they were encouraged to do so by bounties, while concomitantly predator control officers removed wolves, aerial poisoning and shooting campaigns were carried out, and wolves were free to be killed by anybody wishing to do so. Little wonder that wolves were scarce and very shy, attacks on people unheard of, livestock losses minimal, and wolf-borne diseases virtually unnoticed.

Conclusion

The argument—that there is little danger from wolves because they have rarely attacked humans in North America—is fallacious. There are very good reasons why wolves in North America, as opposed to Europe, have attacked people rarely. In North America in the 1700s and 1800s, wolves were killed off and driven into wild places. In their absence in the past decades, we have experienced in North America a unique situation: we had a recovery of wildlife. Few North Americans are aware today that a century ago North America's wildlife was largely decimated and that it took a lot of effort to bring wildlife back. This restoration of North America's wildlife, and thus this continent's biodiversity, is probably the greatest environmental success story of the 20th century. Such a recovery begins with an increase in herbivores. It is followed after a lag-time by an increase in predators. While predators are scarce and herbivores are abundant, wolves are well fed. Consequently, they are very large, but also very shy of people.

We expect to see then no tame or inquisitive wolves. Wolves are seen rarely under such conditions, fostering the romantic image of wolves so prevalent in North America today. However, when herbivore numbers decline while wolf numbers rise, we expect wolves to disperse and begin exploring for new prey. That is when trouble begins.

Historical and current evidence indicates that one can co-exist with wolves where such are severely limited in numbers on an ongoing basis so that there is continually a buffer of wild prey and livestock between wolves and humans, with an ongoing removal of all wolves habituating to people and domestic animals. The current notion that wolves can be made to co-exist with people in settled landscapes—in multi-use landscapes surrounding houses, farms, villages, and cities—is not tenable. Under such conditions, wolves—becoming territorial—will confront people when they walk dogs or approach wolf-killed livestock. In addition, even well fed habituated wolves will test people by approaching such, initially nipping at their clothing and licking exposed skin, before mounting a clumsy first attack that may leave victims alive but injured, followed by serious attacks. While a healthy man can fight off a lone wolf with some chances of success, a lone person cannot defeat a pack.

One cannot defend the current romantic notions about harmless, friendly, cuddly wolves. It is necessary that the public be informed that there exists a large amount of experience and information to the contrary. And the public should know the signs of danger before heading into the wilds. Tame, inquisitive wolves are one such sign.

[1] Geist, V., "Death by Wolves and the Power of Myths: The Kenton Carnegie Tragedy," *Fair Chase*, Winter 2008, 23.
[2] McNay, Mark, "A Case History of Wolf-Human Encounters in Alaska and Canada," Alaska Department of Fish and Game, 2002.
[3] http://www.cbc.ca/news/canada/story/2005/01/04/sask-timberwolf050104.html
[4] Geist, V., "Let's Get Real: Beyond Wolf Advocacy, Towards Realistic Policies for Carnivore Conservation," *Fair Chase*, 2009, 24.
[5] Baker, R.O. and R.M. Timm, "Management of Conflict between Urban Coyotes and Humans in Southern California," 18th Vertebrate Pest Conference, University of California-Davis, 1998.

Chapter 7

MATHEMATICAL ERROR OR DELIBERATE MISREPRESENTATION?

By Don Peay

Don Peay earned his Bachelor of Science in Chemical Engineering and his Master's degree from BYU. After working at an aerospace firm, Don started his own engineering consulting firm, Petroleum Environmental Management Inc. With deer herd populations crashing, ranchers wanting to greatly reduce Utah's elk herds, and a fish and game agency saying the future was non-hunting programs, Don founded Sportsmen for Fish and Wildlife in 1994. Since then, Don has received numerous conservation awards in Utah and throughout the West, including *Outdoor Life* naming him one of the Top 25 Conservationists in North America in 2008. In 2009, the Utah Legislature passed a unanimous resolution thanking Don for his work to help military veterans, to protect wildlife and their habitats, and to increase jobs, tourism, and quality hunting opportunities.

Editors' Statement of the Issues

The proponents of wolf introduction set forth a series of statements indicating unreasonably optimistic outcomes regarding the type of prey wolves seek, expected growth of wolf populations, their expected behaviors, and their impact on existing ungulate herds.

Editors' Statement of the Facts

When attempting to reconcile the discrepancy between promised outcomes and actual data, there arises an almost insurmountable gulf between the proponents' presentation and the undeniable facts. This gulf gives credence to the concern that there was a deliberate obfuscation of the truth. The outlandish travesty is that many of these statements and the program itself were funded by taxpayers.

I knew the answer to the problem was not biological.
It was not legal. I knew the answer was a political answer.
Don Peay

Red Flags and Failed Experiments

The Foundation for North American Wild Sheep (FNAWS) strongly opposed the introduction of wolves from the beginning, one of the few sportsmen-based conservation groups that did. The rest of the groups drank the Kool Aid—that wolves would only kill the sick and the weak, thereby helping the health of the herds. The FNAWS members, most of whom have hunted in Canada, along with many Canadian outfitter members knew the wolf restoration effort in the Northern Rockies would be a disaster for abundant ungulate populations and the hunting and ranching industries. They should know, as they live with wolves.

I started the fledgling Utah Chapter of FNAWS in 1991. That group has raised more than $5 million to support an amazing wild sheep restoration effort in Utah, going from some 500 bighorns in just a few isolated herds in the 1990s to about 5,000 desert and rocky mountain bighorn sheep in over 28 different places in Utah today.

From the beginning, I was personally opposed to the wolf restoration and went on record to that effect with both the Utah Chapter of FNAWS and Sportsmen for Fish and Wildlife, which I founded in 1994.[1] The red flag went up when I saw that the mathematical models and projections being proposed by the "professional biologists" were completely flawed—as this failed experiment has proven. I wasn't a professional biologist; I had a degree in Chemical Engineering, an MBA, and some successful business experience, but I did have another credential. I was a "rocket scientist" and had worked on some very complex mathematical equations while working at an aerospace firm that manufactured nitroglycerin-based propellants for strategic rocket motors. One very critical part of our jobs was to find out when processes went from linear to exponential transitions—from smooth flight to catastrophic explosions. Similarly, in my MBA training, I spent a fair amount of time looking at financial modeling and the growth of investment portfolios. The rate of return, with small variations over a 10- to 15-year period of time, had tremendously different outcomes.

Either Wrong or Lying

Many of my college professors told us that math is a universal language and that, if you understood math well, you could work in any profession. My experience has proved those thoughts to be true. So, from the very beginning, when some "professional biologists" told us that we didn't have a biology degree and that we had no idea what we were talking about, I was very leery of what the wolf reintroduction experts were telling us. I will give two specific examples, early on, that proved to me that the biologists pushing for wolf introduction and others were either wrong or they had an agenda and were not telling the truth.

Besides the mathematical modeling, engineering, and business training, I had spent my life in the outdoors of the western United States, and I have a pretty good understanding of basic biology. When I looked at the models that were being used to forecast wolf populations back in the 1990s, I could see a similar mathematical process with wolves and elk about to happen—a catastrophic failure in 10 to 20 years.

Biggest Myth

The biggest myth perpetrated on the public with wolf reintroduction is that nature perfectly balances itself and that wolves only eat the sick and the weak, thus actually helping the herds become more healthy. The federal and state biologists stated in their own documents that wolves would only have a 7-13% impact on elk and very little, if any, impact, on moose. Well, as we now know, there has been an 80% reduction in the greater Yellowstone elk herds, moose are for all practical purposes gone from Yellowstone, and now the bison are the final prey... and they are declining as well.

Second Major Myth

The second major myth was that the wolf populations in each state would only grow at a 3-5% growth rate, based upon wolf growth rates in Canada and Alaska where a "steady state" has been reached. It doesn't take a rocket scientist to understand that, when you dump a few predators in the middle of the "Garden of Eden" of wildlife, the wolves are going to multiply at very high rates—25-35% a year.

Mathematical Implications

Let's take a look at the mathematical implications of these errors. Let's assume that you have a $2 million portfolio and that your goal is to never touch the principal. You are, on average, getting an 8% return, so you develop your plan to spend $160,000 a year, and things are going along fine.

Before the wolf introduction, excess big game animals were taken by hunters, with hunter harvest very carefully regulated through permits for both the male and female species, and for years, the populations of elk (or your $2 million investment portfolio) remained fairly constant. Your base principal is there, and you can spend $160,000 every year for a long time to come.

Now, let's assume two things happen. Number one, your interest rate of return drops to 4% (the beginning effect of wolves eating calf elk). Your investment manager tells you that you have to cut back your spending to $80,000 a year or you will be eating into your principal. Translating this to the elk situation, state agencies cut back the number of hunting permits, but they knew that wolves were *not* cutting back but were growing exponentially. The net effect of wolves killing calves and then adults is like this effect on your investment portfolio: in just five short years, your principal is now $1 million, not $2 million, as the growing pack of wolves ate half of it, and compounding the problem, with less cows (your principal), your interest rate is now dropped to a 2% return. Without dipping into your principal, the most you can spend is $20,000 a year. In just a short time, say five to seven years, you have gone from being able to spend $160,000 a year and never touching your principal to only being able to spend $20,000 a year, and even with reduced spending, you know your principal is dropping dramatically. Again translating this to wolves and elk, the wolves are rapidly gobbling up the $1 million portion.

The Yellowstone elk herd dropped from over 19,000 head in 1995 to just over 4,000 in 2012. During this time period, wolf numbers were growing exponentially. Anyone who understands math at all could see this biological train wreck coming. Your consumption is growing exponentially, your interest rate (calf survival) is dropping exponentially, and your principal (total elk population) is dropping rapidly. Even though hunting permits were ended—we told biologists to end them sooner—elk populations still dropped precipitously.

The second point that concerned me is that these biologists had a political agenda. Ed Bangs, Wolf Recovery Coordinator for the U.S. Fish and Wildlife Service, presented a talk about the wolf recovery effort at the University of Utah three or so years after wolf reintroduction began in 1995-96. In his analysis, he presented some data that painted the Disney World picture—wolves were eating the old, the sick, and the weak, and hunters were killing the healthy elk. Bangs' data showed that, at that time, wolves were killing elk that averaged eight years old and that hunters took elk that averaged four years old—i.e., wolves were good, and hunters were bad. I raised my hand and asked one question: "Did you count in your average age of harvest the baby calf elk component into the wolf take?" Bangs replied, "Uh, well, no."

A similar analogy would be like talking to 500 people that make $50,000 a year, then throwing in a few billionaires, and then telling everyone that, on average, they each make $250,000 a year so everyone should be happy! WRONG! If the biologists had included the total number of calf elk—ranging from two days to six months—killed by wolves with the adult elk killed, the average age of elk killed by wolves would have been dramatically younger than the age of elk taken by hunters. This discussion proved to me beyond a shadow of a doubt that these wolf biologists had an agenda to let wolves run free and destroy game herds and that they were using phony math to buy time.

Out of Hand in a Hurry

Let me put one other issue in mathematical context. Assume you have 100 wolves in 1995, and you predict that they are going to grow at 5% per year. Compounding annually, by 2010 you will have 208 wolves. However, if that same population of wolves grew at 25% compounding annually, by 2010, there would be 2,842 wolves—an ERROR almost 14 times larger than the promised number. And unfortunately, this is exactly what has been happening. We were promised the goal of 10 breeding pairs and a total population of 100 wolves in Montana, Idaho, and Wyoming. However, in 2012, we had at least 3,000 wolves, and some experts believe twice that many in the three states. Don't you wish your $100,000 investment grew not to $208,000 from 1995 to 2010, but to $2.84 million? This is the magnitude of the error, and with some very quick and easy mathematical modeling, I knew this wolf situation was a complete fiasco that would unfold in a period of 10 to 15 years.

Let me give you one last example of putting numbers in terms of what they really mean. With 500 wolves in the woods (200 more than the agreed upon minimum population for the three states) and with an average kill per wolf of 23 elk per year, the wolves would be killing around 11,500 elk a year. With just 2,500 wolves, they would be killing 57,500 elk a year. Considering that states like Utah only have 65,000 elk in the entire state and hearing that the Animal Rights groups claim they want 5,000 to 8,000 wolves, it doesn't take a calculator to see that wolves would dramatically reduce game herds and destroy the hunting and ranching industries—and then, after the game herds are gone, the wolves would kill each other and focus on livestock and pets, while considering people. We knew that, even though wolves were years away from Utah, we had better solve this problem before it got to our doorstep. As an engineer and businessman, you clearly know that an exponential function can get out of hand in a hurry.

Even though the math is pretty clear and the models are easy to predict, it still amazed me how many biologists and others told us that wolves would NOT greatly reduce our flourishing game herds or the tens of millions of dollars in economic activity that abundant herds sustain. These biologists had been taught in school that, with good habitat, predators would have minimal impacts on game herds and the rest of the wonder wolf fable.

The Real Fairy Tales

It is easy to see that a huge mistake has been made by the mismanagement of wolves in the Northern Rockies. The original predictions in the Environmental Impact Statement of wolf population goals and what impact the wolves would have on the wildlife and the economics of the states where wolves take up residence were the real fairy tales about wolves. When wildlife biologists in Wyoming looked at the difference between what the original USFWS Environmental Impact Statement for the wolf recovery program said and what actually happened, my predictions were more than confirmed.

In 2005, 10 years after the relocation took place, the Wyoming Game and Fish Department did a review of the predictions made by the USFWS in that Environmental Impact Statement (EIS). This is what they found:

🐾 The wolf population in the Greater Yellowstone Area (GYA) in 2005 was at least 3.3 times the original Environmental Impact Statement (EIS) prediction for a recovered population.

🐾 The number of breeding pairs of wolves in the GYA in 2005 was at least twice as high as the original EIS prediction, and the number of breeding pairs in 2004 was at least 3.1 times the original EIS prediction.

🐾 In 2005, the wolf population in Wyoming outside Yellowstone National Park exceeded the recovery criteria for the entire region and continues to increase rapidly.

🐾 The estimated annual predation rate (23 ungulates per wolf) is 1.8 times the annual predation rate (12 ungulates per wolf) predicted in the EIS.

🐾 The estimated number of ungulates taken by 325 wolves in a year (7,150) is six times higher than the original EIS prediction.

🐾 The percent of the northern Yellowstone elk harvest during the 1980s currently taken by wolves (50%) is 6.3 times the original estimate of 8% projected in the EIS.

🐾 The actual decline in the northern Yellowstone elk herd (more than 50%) is 1.7 times the maximum decline originally forecast in the EIS.

🐾 The actual decline in cow harvest in the northern Yellowstone elk herd (89%) is 3.3 times the decline originally forecast in the EIS.

🐾 The actual decline in bull harvest in the northern Yellowstone elk herd is 75%, whereas the 1994 EIS predicted bull harvests would be "unaffected."

🐾 Since wolf introduction, average ratios of calf elk to cow elk have been greatly depressed in the northern

Yellowstone elk herd and in the Wyoming elk herds impacted by wolves. In the northern Yellowstone elk herd and in the Sunlight unit of the Clarks Fork herd, calf to cow ratios have been suppressed to unprecedented levels below 15 calves per 100. The impact of wolves on calf recruitment was not addressed by the 1994 EIS.

Conclusion

To me, knowing that hunters and ranchers have invested hundreds of millions of dollars to restore and sustain abundant game herds and then seeing a group of people—willing to make reckless representations in management and models, change agreements to have 100 wolves in each of three states, then allow for state management, and go to court and have judges use ancillary and irrelevant biological arguments to achieve their agenda of no wolf management—was very disturbing. The people pushing for no wolf management have contributed little if any money to growing and sustaining the very food sources of "their" wolves. To sportsmen and women, who know that wildlife belong to the people and that we have fought and worked so hard to have abundant herds, it was very offensive.

I grew up in a single-wide trailer with no indoor plumbing, and I worked hard for 50 years. Many of my sportsmen friends are of the same mold. We have worked hard and sacrificed a lot, and we do not intend to let a handful of activists destroy our abundant game herds and all the intrinsic and economic values that come from having abundant game herds.

I knew the answer to the problem was not biological. It was not legal. I knew the answer was a political answer. Working with Ted Lyon and many others, we have set a precedent that will honor the work of sportsmen for many years, and we have made headway on updating the Endangered Species Act to respect the rights of people as well as wildlife.

🐾 🐾 🐾 🐾 🐾 🐾 🐾 🐾

[1] Sportsmen for Fish and Wildlife (SFW) has played a major role in the establishment and obtaining of the funding for the 750,000-acre Watershed Restoration Act, and it has helped increase funding for wildlife and land conservation by more than $200 million. SFW gave a voice to sportsmen in Utah, a state that had been dominated by non-wildlife-friendly interests. SFW also worked to restore world-class trophy bull elk on Utah's public lands; has led the aggressive effort to transplant new herds of bison, antelope, wild turkey, mountain goats, and bighorn sheep; and has been instrumental in the Mule Deer Recovery Act.

Chapter 8

THE CARIBOU CONSERVATION CONUNDRUM

By Arthur Bergerud, Ph.D.

Dr. Arthur T. Bergerud has been a population ecologist involved in research on caribou populations in North America since 1955 and is considered the world's foremost authority on woodland caribou. His 30-year study (1974 to 2004) of two caribou populations, one in Pukaskwa National Park (PNP) and the other on the Slate Islands in Ontario, is considered the most comprehensive study of caribou ever done. Along with Stuart Luttich and Lodewijk Camps, Bergerud authored *The Return of the Caribou to Ungava* (2008), which is the story of the George River caribou herd that increased from 15,000 animals in 1958 to 700,000 in 1988, becoming the largest herd in the world at the time. For over 20 years, Bergerud, the former chief biologist with Newfoundland and Labrador's Wildlife Division, and his associates studied the George River herd all across Canada's tundra and taiga.

Editors' Statement of the Issues
Many current environmental concerns incorrectly center around mankind's impact on wildlife populations, ignoring specific biological studies to the contrary and raising the question why these facts are not in their dialogue.

Editors' Statement of the Facts
Wolf activists have chosen to ignore irrefutable data showing the real impact of predators on existing wildlife populations. When wolves are reduced in number, wildlife populations recover—a key fact not published in the wolf activists' position statements. Environmental activist groups should not be permitted to legislate at the federal level to protect their current favorites with taxpayer money.

> *It is assumed that an undisturbed animal community lives in a certain harmony... the balance of nature. The picture has the advantage of being an intelligible and apparently logical result of natural selection in producing the best possible world for each species. It has the disadvantage of being untrue.*
> Charles Elton

Fifty Years of Research
Caribou or reindeer (*Rangifer tarandus*) are an Arctic and Subarctic species of deer found all around the world in northernmost regions. There are two general groups: the tundra caribou that live in the far north in comparatively barren tundra habitat, and the woodland caribou that reside in northern forests.

Woodland caribou (*Rangifer tarandus caribou*)—a subspecies that is found primarily today in Canada but once was found in the North American boreal forest from Alaska to Newfoundland and Labrador and as far south as New England, Idaho, and Washington—are now listed as "threatened" with some herds listed as "endangered."[1]

Based on well over 50 years of research on woodland caribou, I can only conclude that the primary reason for their decline is increased predation from expanding wolf numbers. To reduce wolves to lower densities, we need the support of the general public, especially the environmentalists that are most concerned about the caribou. However, there are two huge problems in gaining public support: for decades, biologists and naturalists have argued that habitat determines animal numbers and that caribou are wilderness animals that cannot coexist with logging and industrialization. Secondarily, environmentalists and some biologists and naturalists have accepted the pseudoscientific myth that there is a balance of nature, a stable equilibrium between prey and predators that prevents extinctions in the absence of anthropogenic factors (of, relating to, or resulting from the influence of human beings on nature). They argue that, if caribou are declining, it must be anthropogenic factors such as logging and human disturbance that are causing the declines.[2]

Predator-Prey Relationship Misconceptions

Fundamental to understanding predator-prey interactions is the debate whether ecological systems are structured from top down (predator-driven) or bottom-up (food-limited). This question has been debated since at least 1960[3] and has been rekindled with the introduction of wolves to Yellowstone National Park and Idaho in 1995-96 and with the rapid decline in recent years of woodland caribou across Canada.

The decline in woodland caribou, however, has been in progress since the early 1900s.[4] Common early explanations were that the decline was the result of the loss of boreal forests from logging and the loss of the lichen's ranges from forest fires and early settlements. That argument continues today. Based on years of research that began in the 1950s in Newfoundland and Labrador, in 1974 I published that the decline of both woodland and barren-ground caribou resulted from increased predation, primarily from wolves, and *not* from a decrease in lichen pastures by fire and overgrazing of terrestrial lichen.[5] My 1974 paper is now generally quoted by young caribou biologists as showing that the proximate cause (an act from which an injury results as a natural, direct, uninterrupted consequence and without which the injury would not have occurred) for the decline of caribou is wolf predation with the implied implication that the *real* reasons are anthropogenic factors such as logging that have upset the "balance of nature." Many older wolf biologists still believe in the balance of nature concept, have helped indoctrinate the general public to that view, and still do not support management intrusion into natural systems.

Research Findings

To resolve the question of top-down or ground-up, I selected two study areas following all the controversy of the 1974 paper. For the experimental area, we chose an insular caribou herd on the Slate

Islands in Lake Superior near Pukaskwa National Park, Ontario. Those islands had no terrestrial ground lichens and also no natural predators of caribou—i.e., no wolves, bears, lynx, or wolverines.

As the control population, we selected the caribou in Pukaskwa National Park (PNP), 50 kilometers distant. In the Park, there was a normal boreal fauna of wolves, bears, lynx, moose, and caribou and little anthropogenic disturbance. We compared the demography to these two populations over 30 years (1974-2004). On the Slates and in the absence of wolves, the population persisted for 30 years with the highest density of caribou in North America, up to 10 caribou/km^2 with numbers varying from 150 to 600 animals. Total numbers were regulated by starvation from a shortage of summer foods, not lichens.[6] The animals went into the winter too weak to survive, regardless of the abundance of winter forage. This density was 100 times greater than the control herd in PNP where the density was 0.06 caribou per km^2—100 times less than the Slate Islands. This original low density was due to the predation of wolves existing at 11 wolves per 1,000 km^2. As the study continued, the PNP herd declined, and by 2009, only four caribou were left. The herd may now be extinct. It was clearly regulated top-down by predation and was favorably critiqued in the prestigious textbook *Ecology 6th Edition*, 2009, by Emeritus Professor Charles Krebs.

Near the end of our Slate Island study, a natural experiment occurred in 1994 when Lake Superior froze and two wolves crossed to the Slate Islands. We lost nearly all the calves born in 1994 and 1995, and in those two years, the wolves caused sufficient mortality of adult females to change the sex ratio from the nominal sex ratio of one male to two females (approximately 35% males) to 55% males, a decline of 100 animals, mostly females. This mortality was replicated in 2003 and 2004 when another duo of wolves reached the islands and, again, the wolves killed nearly all the calves on the island.

Caribou Decline

In the 1970s, I researched the demography of mountain caribou in northern British Columbia and published three peer-reviewed papers in 1984, 1986, and 1987.[7] The 1984 paper documented that 90% of the calves were killed each spring before the age of six months mostly by wolves and bears. The 1986 paper documented that the decline of caribou in British Columbia occurred when the moose population increased, which brought higher wolf populations to northern British Columbia. Additionally, we compared the recruitment and adult mortality of caribou for all herds in North America where wolf abundance had been measured (a sample size of 750,000 caribou). The wolf density where caribou mortality and recruitment were balanced for caribou [the stabilizing density (r^S)] was 6.5 wolves/1,000 km^2 [finite rate of increase $(\lambda) = (1\text{-}\% \text{ adult mortality})/(1\text{-}\% \text{ recruitment})$]. When wolf densities were higher, caribou declined. With caribou, the

Jon Nickles, USFWS
www.weforanimals.com

stabilizing recruitment of calves to balance adult mortality should be 15% of the population, or 25 calves per 100 females measured, when calves are 10 to 12 months of age. The pregnancy rate in caribou is normally greater than 80% (80+ calves per 100 females).[8] To maintain caribou populations, wolf numbers need to be below the densities of eight or less wolves/1,000 km[2].[9] In multi-ungulate systems, wolf densities are commonly 15-25 wolves/1,000 km[2].[10]

In the 1980s, I joined Dr. John Elliott, regional fish and wildlife biologist in British Columbia, in studying the demography of caribou, moose, elk, and Stone's sheep in northeastern British Columbia. At that time, all four species were in decline. As in the case of the Ontario studies, we had control and experimental areas replicated in two separate regions: the Kechika region (18,400 km[2]) and the Muskwa region (19,000 km[2]). The two control replications were left undisturbed, and in the two experimental areas, we removed wolves—Kechika: 492 wolves in four years, and Muskwa: 505 wolves in three years. The grizzly bear and wolverine populations were left undisturbed. The study was 10 years in duration.[11]

In the paper, we showed that, in the areas where we reduced wolves, recruitment was greatly increased based on calves/100 females at six months to 12 months. For moose, recruitment averaged 40 calves/100 females, elk 49/100 females, Stone's sheep 41/100 females, and caribou 39/100 females. For those herds with no wolves removed, the recruitment was moose 11 calves/100 females, sheep 20/100, elk 26/100, and caribou 7/100. The elk and moose populations in this study that had wolf management increased from 18,000 to 33,000. In all removals, wolves were harvested in the spring before wolf denning, and in all years, young wolves were already dispersing into the removal area but insufficiently organized for denning so that the four prey species secured positive recruitment. At the end of the 10 years, the wolf population was a healthy 20+ wolves/1000 km[2].

In the other areas in the North where wolves have been managed, the percentage of calves before and after wolf reductions were as follows: Nelchina herd, Alaska, 1960s, 24% vs 40%; Forty Mile herd, Alaska, 1970s, 5% vs 31%; Delta Herd, Alaska, 1-9% vs 25%; and Beverly Herd, Northwest Territory, 1960s, 7-8% vs 20-25%.[12]

The Predator Pit

The last study I wish to mention is the George River caribou herd in northern Quebec and Labrador. I counted the herd in 1958 at 15,000 animals. In 1958, there were no wolves; hunters that had been going in the country for decades had seen practically none on the land. The wolverine had also gone extinct. Starting in 1958, the herd increased yearly, reaching 300,000 by 1980.

In the 1970s, wolves reappeared and increased as the herd grew. By 1980, the wolves finally reached sufficient numbers that their predation halted the growth of the herd (recruitment equaled adult mortality). Then in 1981 and 1982, the wolves got rabies and declined by 80%. The caribou herd exploded with numbers reaching 537,000 by 1985 and 650,000 by 1987, the largest herd in the world at that time. The herd started to decline in 1988 when densities reached 10/km², similar to the density that resulted in decline on the Slate Islands. However, unlike the Slate Islands, there was no winter starvation. In the George River herd, the females died in the summer from malnourishment with lactation problems and insect attacks with the result that pregnancy rates declined.[13] In addition, wolf predation continued as did hunting, which was unabated. The herd reached 74,000 in 2010, well within the carrying capacity of forage, but it is continuing to decline from heavy predation of calves and adults. The herd is now in what some biologists call "the predator pit," which means that each time the herd starts to recover, the predation intensifies, and the herd remains limited.

The herd could be turned around now and started back up if wolves were managed, but in Canada, there has been little or no wolf management in recent decades. The herd will continue down until the wolves go elsewhere; that was what happened in the decline in the 1890s, from a high of 700,000 to the 15,000 I counted in 1958. If the herd was managed now at 74,000, it could increase rapidly as did the Western Arctic herd in the 1970s. That herd had declined from 242,000 in 1970 down to 75,000 by 1976. Then Alaska Fish and Game took over, and the harvest was drastically curtailed with the natives fully cooperating, and predators were managed. With wolves managed, the herd was back to 113,000 by 1979 and continued higher.[14]

Support of the Population

The only wildlife agency in North America that manages wolf numbers as a standard practice is in Alaska. This management has the support of the majority of the population in a system where the populace commonly depends on moose and caribou as subsistence food.[15] This dependence is protected under both state (state subsistence statue) and federal (Alaska National Interest Lands Conservation Act— ANILCA) and in agreement with the National Research Council 1997 standards.[16] There has been some sporadic management of wolves in Alberta and of coyotes and bears on the Gaspe Peninsula—150 animals are left at this time. Recently British Columbia had planned to manage wolves for their remaining endangered arboreal caribou (less than 1,400), but environmental groups thought the high abundance of wolves was more valuable than the last remaining arboreal caribou and forced the government to cancel the management.[17]

Going Extinct

Species around the world are going extinct. In Africa, many endemic species are declining from predation,[18] and predators themselves are going extinct, including the large cats (tigers, cheetahs, and lions). But in North America, wolves are prospering, spreading across the western United States from the Yellowstone and Idaho introductions and increasing in Canada as their prey base of moose and deer expand with climate change.

In 1992 or 1993, I received a questionnaire from a polling company apparently under contract to the U.S. Fish and Wildlife Service (USFWS), asking how wolf introduction to Yellowstone Park would impact the other species and the vegetation in the Park. The participants of this poll became known as the Delphi Committee, which was made up of ungulate and wolf biologists. It was an anonymous committee. I still do not know who else took part, but I believe it was weighted to pro-wolf biologists.

Based on our research in northern British Columbia, I predicted a major decline in elk and moose, if wolves were introduced. My major comment was that, if the introduction went ahead, the wolves would have to be managed in Yellowstone Park and prevented from spreading beyond the Park. Well of course, there was no intention to manage wolves; they have now reached California and Colorado, and Utah is trying to hold the line. The Northern Yellowstone elk herd in the Park is mostly gone—from 20,000 in 1994 prior to the wolf relocation to 4,635 in 2011—and elk hunters may have seen the end of their hunting in that area. Animal rights activists and anti-hunting organizations have won.

I could not believe USFWS would introduce wolves to Idaho where the last caribou herd was classified as endangered (the Selkirk herd), but they did, and the wolves are now hunting these last caribou. An even more unethical act, the USFWS is adding an additional area of 152,000 hectares (61,538 acres) to the preserve surrounding the herd that is off-limits to the local residents. The caribou do not need this additional land grant. They need predator control of the mountain lions and now the introduced wolves, but there will be none by the USFWS. Lion predation has been a problem in the past. Patrick Valkenburg, a caribou biologist, e-mailed me stating: "With 300 million people in the United States now, is it realistic to just let 'wolves do their thing'? Wolves belong in the boreal forest, not the Great Plains and the Great Basin where there are no longer any buffalo. Perhaps the USFWS did not know."

Some older wolf biologists—i.e., Dr. Victor Van Ballenberghe, Dr. James Peek, Dr. John Theberge, and Dr. Paul Paquet, etc.—are emotionally involved with wolves and do not want to see wolves managed. I can relate to an emotional involvement relative to caribou. There is nothing more heart rending than watching a female caribou that has had her calf so completely eaten by a wolf or bear that there is not even any scent left. She does not understand that her calf is dead and will stand and look and look and then go back to the last undisturbed location where they were together and then finally back to the birth site. I have seen a female travel up a long line of cows and calves, scenting each calf, seeking recognition of her lost progeny. Some cows will stay near the remains of their dead calves for many days, waiting for them to get up and guarding the carcass from the lynx that often monitor attack and kill sites.[19] Such cows can be called close by imitating calf bleating.

However, such emotional views will not help us manage the moose-caribou-wolf system. Those wolf biologists that have elevated the wolf to icon status have done a disservice to wildlife management in fostering the balance of nature myth. In contrast, coyote biologists have remained unattached and objective, relating the damage to livestock and the dangers of these animals in cities to small children and pets.

Balance of Nature

Charles Elton—the father of ecology, who discovered the 3-4 year cycle and the 10-year cycle of mammals and who wrote the first ecology book[20]—told us at the beginning:

> *It is assumed that an undisturbed animal community lives in a certain harmony... the balance of nature. The picture has the advantage of being an intelligible and apparently logical result of natural selection in producing the best possible world for each species. It has the disadvantage of being untrue.*

117

In recent years, three widely respected biologists in Canada have concluded that woodland caribou are endangered from the increasing wolf population triggered by the increase in the moose and deer prey base with climate warming: Professor Emeritus Charles J. Krebs[21], Professor Emeritus A. R. E. Sinclair[22], and Professor Emeritus Valerius Geist[23]. These men are the leading ecologists in Canada.

The most respected wolf biologist in the world is Dr. L. David Mech. In his 1996 monograph "The Wolves of Isle Royale," he used the words "stable equilibrium" to describe moose-wolf interactions ("stable equilibrium" is a synonym for the "balance of nature"). In later years, the balance completely disappeared.[24] Mech accepted this disproof, stating in 1998: "The Isle Royale moose and wolf populations have fluctuated greatly over time, and there is little correlation between wolf and moose numbers in any given year."[25] From 1986 to 1994, Mech and his students studied the Denali caribou herd in Alaska.[26] On page 159 of *The Wolves of Denali*, Mech states: "The Denali wolf-caribou relationship is a good illustration of the dynamics of populations and why stability or balance at a variety of levels should not be considered inherent in natural systems."[27] Good scientists try to disprove their own hypotheses, and this is what he has done. This is significant because Mech, with his decades of studies of and publications about wolves, unintentionally has contributed probably more than any other scientist to furthering the pseudoscientific myth—the balance of nature.

Woodland caribou survived the Pleistocene in the Appalachian Mountains, spacing themselves away from the large predators of the megafauna living at lower elevations. These caribou lived with a completely different set of species than they do today; i.e., sloths, peccaries, tapers, and austral faunal elements. In *Science*, Graham et al.[28] stated that the fossil mammal fauna of the Late Quaternary[29] at 2,945 fossil sites supported the Gleasonian community model, which assumes that species respond to environmental changes in accordance to their individual tolerances with varying rates of range shift. They stated, "Modern community patterns emerged only in the last few thousand years, and many late Pleistocene communities do not have modern analogs." Those who say caribou and wolves evolved together for hundreds of eons have not checked the fossil record.

Each species, through individual selection, evolves its own distinct behavior-habitat strategies to persist, but that does not guarantee continual survival. Each species walks its own road down through time—there is no magical balance of nature. An article by Elisa Beninca et al.[30] in the prestigious journal *Nature* stated: "Advanced mathematical techniques proved indisputable presence of chaos in this food web; short-term prediction is possible, but long-term prediction is not."

Critical Habitat

In Canada, when animal species are classified as "endangered," the federal government requests that the provinces involved develop a management plan for the species in danger of extinction, and they require that critical habitat be identified. For barren ground caribou, it is generally accepted that the calving grounds are the critical habitat. However, some still do not recognize the key element of that habitat is reduced predation risk and not forage.[31] The critical habitat for montane and boreal woodland[32] is the areas where the cows calve and where there is reduced risk for newborns. The montane caribou calve on alpine peaks spaced away from moose and wolves at lower elevations, and the boreal caribou calve on the islands, shoreline, and muskegs of the boreal forest.[33] The critical habitat is not the winter ranges where boreal woodland caribou seek lichens. They adapted long ago to rotating their winter ranges in response to lichen destructions by forest fires and over-utilization.[34] However, for the montane caribou, the old growth high alpine forests with their lichen loads and protective deep snow that serves as a barrier to wolves are the critical habitats. In British Columbia, 2.2 million hectares (890,688 acres) of old-growth forest are protected from industrial forestry activity, sufficient for the present low population of the remaining 1,400 animals.

The Canadian Parks and Wilderness Society (CPAWS) in February 2012 delivered a petition of 32,000 signatures to the Environmental Minister to save woodland caribou by protecting their boreal habitat from industrialization. The World Wildlife Fund's solution to the caribou conservation conundrum is to create more parks.[35] None of these efforts will save the caribou. The boreal ecosystem is structured from the top down by predation. The boreal forest can be logged if the calving habitat is left undisturbed and if the wolf population is managed to the same levels as in Alaska, fewer than 6.5 wolves/1,000 km^2.[36]

Mistaken Beliefs

Some believe that caribou are wilderness animals, cannot tolerate anthropogenic[37] disturbance and cannot tolerate the presence of man.[38] These are mistaken beliefs. The two biologists who hold this belief—Schaeffer and Vors—are modelers with very little experience studying the behavior of caribou on the ground. I have been on the calving grounds studying calving behavior for 33 years in five provinces, as well as the Northwest Territory and Alaska. In addition, I have established 13 herds and maintained a captive herd for several years. There are numerous examples of caribou adapting to the presence of man. For example, caribou have wintered at the Armstrong Airport in Ontario for the past 37 years so they can avoid the presence of wolves. In Alaska, the caribou have calved for decades in the Prudhoe Bay oil field, the herd increasing from 5,000 to 30,000 animals in the absence of wolves and bears. Skeptics should Google "Slate Island caribou" and see pictures of the caribou, visiting occupied campgrounds and seeking handouts and the ashes in the fire pits. Caribou adapt well to the presence of benign humans and have been domesticated by the reindeer herders in Eurasia.

In contrast, the caribou are now extinct because of predation in the Canadian National Parks of Banff, Glacier, and Revelstoke, where there is no logging, and anthropogenic disturbance are minimal.[39] The arboreal caribou in British Columbia are facing extinction from predation in their protected habitat set aside for them. Wittmer showed that old-growth forest with arboreal lichens was adequate for the herd.[40] There are possibly only three or four caribou left in Pukaskwa National Park, and wolf predation was the cause.[41] The woodland caribou in Newfoundland from 1970 to 1996 increased from 7,000 to 96,000 animals.[42] These were years of intense anthropogenic disturbance: logging near calving grounds, roads built across the caribou habitat, lakes and rivers dammed, and increased mining[43] —yet the caribou were able to increase, despite all this disturbance, because wolves had been extinct for several decades in Newfoundland.

The Caribou-Moose-Wolf System

Woodland caribou spent the Pleistocene[44] south of the Laurentide ice sheet[45] in the Appalachian Mountains spaced away from the megafauna at lower elevations. They left the mountains 13,000-12,000 BC and reached Ontario at 10,000 BC, passing through deciduous forests. Fossils have been found near Udora, Ontario, 10,500 BC[46] and near Atikokan, Ontario, 9,940 BC.[47] The caribou were blocked going further north by the ice sheet and, through time, evolved the anti-predator tactics of using islands and shorelines to calve where they could escape wolves by swimming.

A second segment of the Appalachian caribou gene pool moved though Maine and New Brunswick following the shore around the ice at about 9,000 BC and turned north in tundra where forests were delayed by the cold temperature of the Labrador current. They reached

Indian House Lake at 7,000 BC, as did the Paleo Indians.[48] This population became the "barren ground" race, the George River Herd. Wolves must have followed these two separate lines, but fossils have not been found (the acid soils of the Canadian Shield erodes fossils).

Moose did not reach the boreal forest until the end of Little Ice Age, 1,850–10,000 years *later* than the caribou. The caribou had persisted in the boreal forest with the single wolf prey system for 10,000 years. However, the arrival of moose signaled the decline of caribou.[49] Moose provided the biomass for a much higher population of wolves. Biologists like Fuller[50] calculate wolf numbers based on the total ungulate prey biomass weighted as follows: moose at six, elk at five, caribou at three, and deer at two. Moose provided the biomass for many more wolves than were provided previously with the single prey system of caribou. Switchover by predators between prey species is well recognized and researched in biology. In the boreal forest with a two prey system of moose and caribou, wolves commonly switch over to the caribou that are easier to kill than moose, which may stand and fight,[51] but the large biomass of the moose is what maintains the higher wolf population. This results in inverse density dependence mortality—the Allee Effect[52]—and this generally leads to extinction unless safe refuges exist, like islands.[53]

Now we have a warming climate and more moose and deer than ever that are expanding farther north.[54] The increased predation from wolves is causing the extinction of caribou along the southern edge of their range. Moose have now reached the arctic coast. Caribou can only cope with wolves in densities of 6-8/1,000 km^2—this then is a species diversity problem for caribou, which generally do not have a shortage of forage. If wolf populations were managed, we could have caribou densities of greater than 2/km^2, but because of the Allee Effect, caribou can only persist by being rare—densities commonly of only 0.06/km^2.[55] The only reason caribou cannot exist in logged areas with their abundance of deciduous forage is the greater abundance of moose supporting more wolves. The sequence is settlement + climate change = more deciduous forage = more moose = more wolves = more predation on caribou = more extinctions. In Ontario, the extinction line north for caribou has only halted because of the safety of islands in several large lakes in two provincial parks and Lake Nipigon.

Conclusion

Environmentalists are contributing more to the extinction of caribou than anyone else is. Fish and game departments are not going to manage wolves without the support of the public, especially the vocal environmentalists. Environmentalists, as well as some biologists, are always blaming anthropogenic factors for declines in caribou like roads, seismic lines, pipelines, logging, etc. They say that such disruptions actually assist wolves in locating caribou. To the contrary, caribou have gone extinct in national parks, large parks that have not been logged, and where roads are minimal, due to predation. The supposed anthropogenic disturbances are absent, and the supposed effect persistence is also absent—the cause is not necessary.[56] Caribou have persisted since the Ice Age in a simple one prey-one primary predator system and continue to persist with climate change and increased species diversity. The bottom line is that wolves or the alternate prey species (moose and deer) will have to be managed if caribou are to survive.

🐾 🐾 🐾 🐾 🐾 🐾 🐾 🐾

[1] COSEWIC (Committee on the Status of Endangered Wildlife in Canada) Report, 2002. [COSEWIC is a committee of experts that assesses and designates which wildlife species are in some danger of disappearing from Canada.]
[2] Schaefer, J.A., "Long Term Recession and Persistence of Caribou in the Taiga," *Conservation Biology,* 2003, 17: 1435-1439; Vors, L.S., J.A. Schaefer, B.A. Pond, A.R. Rodgers, and B.R. Patterson, "Woodland Caribou Extirpation and Anthropogenic Landscape Disturbance in Ontario," *Journal of Wildlife Management,* 2007, 71:1249-1256.
[3] Hairston, N.G., F.E. Smith, and L.B. Slobodkin, "Community Structure, Population Control and Competition," *American Naturalist,* 1960, 194:421-425.
[4] Cringan, A.T., "Some Aspects of the Biology of Caribou and a Study of the Woodland Caribou Range on the Slate Islands, Lake Superior, Ontario," Master's Thesis, 1957, University of Toronto, Toronto, Ontario, Canada.
[5] Bergerud, A.T, "The Decline of Caribou in North America Following Settlement," *Journal of Wildlife Management,* 1974, 38: 757-770.
[6] Bergerud, A.T., W.J. Dalton, H.E. Butler, L. Camps, and R. Ferguson, "Woodland Caribou Persistence and Extirpation in Relic Populations on Lake Superior," *Rangifer, Special Issue,* 2007, 17: 57-78.
[7] Bergerud, A.T., H.E. Butler, and D.R. Miller, "Anti-Predator Tactics of Calving Caribou: Dispersion in Mountains," *Canadian Journal of Zoology,* 1984, 62:566-575; Bergerud, A.T. and J.P. Elliott, "Dynamics of Caribou and Wolves in Northern British Columbia," *Canadian Journal of Zoology,* 1986, 64:1515–1529; Bergerud, A.T. and R.E. Page, "Displacement and Dispersion of Parturient Caribou as Calving Tactics," *Canadian Journal of Zoology,* 1987, 65: 1597-1606.

8 Bergerud, A.T., "A Review of the Population Dynamics of Caribou and Wild Reindeer in North America," 1980, pp. 556-581 (in "Proceedings of 22nd International Reindeer/Caribou Symposium," Edited by E. Reimers, E. Gaare, and S. Skjenneberg, Roros, Norway, 17-21 September, 1980).

9 Bergerud, A.T. and J.P. Elliott, "Dynamics of Caribou and Wolves in Northern British Columbia," *Canadian Journal of Zoology*, 1986, 64:1515–1529; Thomas, D.C., 1995. "A Review of Wolf-Caribou Relationships and Conservation Implications in Canada," pp. 261-273 (in L.N. Carbyn, S.H. Fritts, and D.R. Seip, editors, *Ecology and Conservation of Wolves in a Changing World*, Canadian Circumpolar Institute, Edmonton, Alberta, Canada); Lessard, R.B., "Conservation of Woodland Caribou in West-Central Alberta: A Simulation Analysis of Multi-Species Predator-Prey Systems," Ph.D. Thesis, 2005, University of Alberta, Edmonton, Alberta, Canada.

10 Messier, F., "Ungulate Population Models with Predation: A Case Study with the North American Moose," *Ecology,* 1994, 75:478-488.

11 Bergerud, A.T. and J.P. Elliott, "Wolf Predation in a Multiple-Ungulate System in Northern British Columbia," *Canadian Journal of Zoology,* 1998, 76:1,51-1,569.

12 Bergerud, A.T. and J.P. Elliott, "Dynamics of Caribou and Wolves in Northern British Columbia," *Canadian Journal of Zoology*, 1986, 64:1515–1529.

13 Bergerud, A.T., S.N. Luttich, and L. Camps, *The Return of Caribou to Ungava,* 2008, McGill and Queens University Press, Montreal, Quebec, Canada.

14 Davis, J.L., P. Valenburg, and H.V. Reynolds, "Population Dynamics of Alaska's Western Arctic Caribou Herd," in "Proceedings of 22nd International Reindeer/Caribou Symposium," Edited by E. Reimers, E. Gaare, and S. Skjenneberg, Roros, Norway, 17-21 September, 1980.

15 Titus, K., "Intensive Management of Wolves and Ungulates in Alaska," 72nd North American Wildlife and Natural Resources Conference, 2007, pp. 366-377.

16 Regelin, W.L., P. Valkenburg, and R.D. Boertje, "Management of Large Predators in Alaska," *Wildlife Biology in Practice*, 2005, 1:77-85.

17 Wittmer, H.U., A.R.E. Sinclair, and B.N. McLellan, "The Role of Predation in the Decline and Extirpation of Woodland Caribou,*"* *Oecologia,* 2005, 144:257-267.

18 Sinclair, A.R., E. Pech, R.P. Dickman, C.R. Hik, S. Hik, P. Mahon, and A.E. Newsome, "Predicting Effects of Predation on Conservation of Endangered Prey," *Conservation Biology,* 1998, 12: 564-574.

19 Bergerud, A.T., "The Population Dynamics of Newfoundland Caribou," *Wildlife Monograph,* 1971, 25.

20 Elton, C., *Animal Ecology*, Macmillan, London, Sidgwick, and Jackson, 1927.

[21] Krebs, Charles J., Professor Emeritus, *Ecology: 6th edition,* 2009, pp. 197-198.

[22] Sinclair, A.R.E., Professor Emeritus, co-author (with H.U. Wittmer and B.N. McLellan), "The Role of Predation in the Decline and Extirpation of Woodland Caribou*," Oecologia,* 2005, 144:257-267.

[23] Geist, Valerius, Professor Emeritus, *The Deer of the World: Their Evolution, Behavior and Ecology*, Mechanicsburg, Pennsylvania: Stackpole Books, 1998.

[24] Peterson, R.O., "Wolf Ecology and Prey Relationship on Isle Royale," *Science Monograph Series* No. 11, U.S. National Park Service, 1977; McLaren, B.E., and R.O. Peterson, "Wolves, Moose, and Tree Rings on Isle Royale," *Science* 266, 1994, 1,555-1,558.

[25] Mech, L.D., L.C. Adams, T.J. Meier, J.W. Burch, and D. Dale, *The Wolves of Denali*, Minneapolis, Minnesota: University of Minnesota Press, 1998.

[26] *Ibid.*

[27] Pimm, S.L., *The Balance of Nature: Ecological Issues in the Conservation of Species and Communities*, Chicago: University Chicago Press, 1991.

[28] Graham, R.W. et al., "Spatial Response of Mammals to Late Quaternary Environmental Fluctuations." *Science* 272, 1996, 601-606.

[29] Past 0.5 to 1.0 million years.

[30] Benincà, Elisa, et al., "Chaos in a Long-Term Experiment with a Plankton Community," *Nature*, February 14, 2008.

[31] Bergerud, A.T., S.N. Luttich, and L. Camps, *The Return of Caribou to Ungava,* McGill and Queens University Press, Montreal, Quebec, Canada, 2008.

[32] Montane means of mountains and other high elevation regions; Boreal means of or relating to the forest areas of the Northern Temperate Zone.

[33] Simkin, D.W., *"A Preliminary Report of the Woodland Caribou Study in Ontario*, Department of Lands and Forest, Section Report No. 59, 1965; Shoesmith, M.W. and D.R. Storey, "Movements and Associated Behavior of Woodland Caribou in Central Manitoba," Proceedings of International Congress of Game Biologists, 1977, 13:51-65; Hatler, D.F., "Studies of Radio-Collared Caribou in the Spatsizi Wilderness Park Area," Smithers: British Columbia Spatsizi Association for Biological Research, 1986; Edmonds, E, J., "Population Status, Distribution, and Movements of Woodland Caribou in West-Central Alberta," *Canadian Journal of Zoology,* 1988, 66:817-826.

[34] Bergerud, A.T, "The Decline of Caribou in North America Following Settlement," *Journal of Wildlife Management,* 1974, 38: 757-770.

[35] Petersen, B., A. Iaconbelli, and E.E. Kushny, *"The Caribou Conundrum: Conservation of Woodland Caribou and Designing Protected Areas,"* World Wildlife Fund, Toronto Poster Presentation, 1998, 8th North American Conference, Whitehorse, Yukon.

[36] Bergerud, A.T. and J.P. Elliott, "Dynamics of Caribou and Wolves in Northern British Columbia," *Canadian Journal of Zoology*, 1986, 64:1515–1529; Bergerud, A.T., "The Need for the Management of Wolves," an open letter, *Rangifer*, Special Issue, 2007, 17:39-50.

[37] Of, relating to, or resulting from the influence of human beings on nature.

[38] Schaefer, J.A, "Long-Term Recession and Persistence of Caribou in the Taiga," *Conservation Biology,* 2003, 17: 1435-1439; Vors, L.S., J.A. Schaefer, B.A. Pond, A.R. Rodgers, and B.R. Patterson, "Woodland Caribou Extirpation and Anthropogenic Landscape Disturbance in Ontario." *Journal of Wildlife Management*, 2007, 71:1249-1256.

[39] Serrouya, R. and H.U. Wittmar, "Imminent Local Extinctions of Woodland Caribou from National Parks," *Conservation Biology,* 2010, 24: 363-364; Hebblewhite, M., C. White, and M. Musiani, "Revisiting Extinction in National Parks: Mountain Caribou in Banff," *Conservation Biology,* 2010, 24:341-344.

[40] Wittmer, H.U., A.R.E. Sinclair, and B.N. McLellan, "The Role of Predation in the Decline and Extirpation of Woodland Caribou," *Oecologia*, 2005, 144:257-267; Wittmer, H.U., R.N.M. Ahrens, and B.N. McLellan, "Viability of Mountain Caribou in British Columbia, Canada: Effects of Habitat Change and Population Density," *Biological Conservation*, 2010, 143:86-93

[41] Bergerud, A.T., W.J. Dalton, H.E. Butler, L. Camps, and R. Ferguson, "Woodland Caribou Persistence and Extirpation in Relic Populations on Lake Superior," *Rangifer*, Special Issue, 2007, 17: 57-78.

[42] Wildlife Division Newsletter, June 6, 2009.

[43] Bergerud, A.T., H.E. Butler, and D.R. Miller, "Anti-Predator Tactics of Calving Caribou: Dispersion in Mountains," *Canadian Journal of Zoology,* 1984, 62:566-575.

[44] Geological epoch that lasted from about 2,588,000 to 11,700 years ago.

[45] This ice sheet was the primary feature of the Pleistocene epoch in North America, commonly referred to as the Ice Age. The ice sheet was up to two miles thick in Quebec, Canada, but much thinner at its edges.

[46] Storck, P.L. and A.E. Spiess, "The Significance of a New Faunal Identification Attributed to an Early PaleoIndian Occupation at the Udora Site, Ontario, Canada," *American Antiquity,* 1994, 59:121-142.

[47] Jackson, L.J., "First Ontario C-14 Date for Late Pleistocene Caribou," *Archeology,* 1989, 89:4-5.

[48] Bergerud, A.T., S.N. Luttich, and L. Camps, *The Return of Caribou to Ungava,* McGill and Queens University Press, Montreal, Quebec, Canada, 2008.

[49] deVos, A., and R.L. Peterson, "A Review of the Status of Woodland Caribou in Ontario," *Journal of Mammalogy*, 1951, 322:337.

[50] Fuller, T.K., "Population Dynamics of Wolves in North-Central Minnesota," *Wildlife Monograph* 105, 1989.

51 Simkin, D.W., "*A Preliminary Report of the Woodland Caribou Study in Ontario*," Department of Lands and Forest, Section Report No. 59, 1965.
52 Allee Effect: Mortality accelerates as prey numbers decline.
53 Wittmer, H.U., A.R.E. Sinclair, and B.N. McLellan, "The Role of Predation in the Decline and Extirpation of Woodland Caribou*,*" *Oecologia*, 2005, 144:257-267; Bergerud, A.T., W.J. Dalton, H.E. Butler, L. Camps, and R. Ferguson, "Woodland Caribou Persistence and Extirpation in Relic Populations on Lake Superior," *Rangifer*, *Special Issue*, 2007, 17: 57-78.
54 Latham, A.D.M., M.C. Latham, N.N.A. McCutchen, and S. Boutin, "Invading White-Tailed Deer Change Wolf-Caribou Dynamics in Northeastern Alberta," Journal of Wildlife Management, 2010, 75:204-212.
55 Bergerud, A.T., "Rareness as an Anti-Predator Risk for Moose and Caribou," *Wildlife 2001 Populations*, editors D.R. McCullough and R. H. Barrett, New York: *Elsevier Applied Science*, 1992, pp. 1008-1021.
56 Hempel, C.G., *Philosophy of Natural Science*, Englewood Cliffs, N.J., 1966.

tarasov / Shutterstock.com

Chapter 9

WOLVES, A SERIOUS THREAT TO LIVESTOCK PRODUCERS

By Heather Smith-Thomas

Heather Smith-Thomas grew up on a cattle ranch near Salmon, Idaho, and started writing about horses and cattle in high school and college. She graduated from the University of Puget Sound in 1966 with a BA in English and history. Heather, who writes regularly for more than 25 farm and livestock magazines and 30 horse publications, has written 20 books on such topics as horse care and cattle raising. She and her husband, Lynn Thomas, have been raising beef cattle and horses since 1967.

Editors' Statement of the Issues

Lofty ideas promoted without regard to either historical data or current documentation have no rightful place in the management of wildlife or the environment. Stakeholders must have a strong voice before any legislation is passed regarding wolf management.

Editors' Statement of the Facts

It has become the norm for wolf activists to disregard the voice of those directly affected by wolves when, in fact, reason insists that those whose lives are impacted by wolves should have the primary voice for wolf management. Coupled with any reasonable management policy is the foundation of factual documentation that will limit current and future predator populations.

> *Cattle and wolves are never going to be compatible. Some people keep saying that wildlife people and cattlemen are going to have to learn to manage for wolves, but that's impossible. It's hard to "manage" when you are always on the losing end.*
> Len McIrvin

Imaginary Scenarios

Wolves have become the noble, beautiful creatures of myths and movies in the minds of most Americans; it's easy to believe imaginary scenarios when a person has no personal experience with the reality of the beast. The average American has had no reason not to believe the myths about wolves—since most published materials perpetuate the myths. Some Western ranchers, however, remember their parents and grandparents telling about bloody carnage when wolves attacked their herds and flocks, often killing for sport or leaving half-dead animals in their wake after a night's spree. These stories and accounts of wolf attacks have been discredited by modern "wisdom" about wolves. Now ranchers are paying the price.

Despite predator control campaigns, wolves were never eliminated, never endangered; there were, in fact, thriving populations in Canada, Alaska, and Minnesota. Yet most U.S. ranchers from the 1950s through the mid-1990s could sleep easy at nights, not having to worry about wolves wreaking havoc with their livestock. This all changed when government agencies transplanted an "experimental" population of wolves in the northern Rockies in 1995.

Controversy over "Reintroduction" of Wolves

Rural communities in the West fought against wolf "reintro-duction," knowing that these animals would have a significantly adverse impact on wildlife and livestock and possibly human safety. Their views and arguments were overshadowed by the broader voice of an Eastern public as well as Western cities (where the average person is completely removed from the realities of where his/her food comes from and wolves live in zoos). Both of these large segments of population thought it would be nice to have wolves on the Western landscape again. Reintroducing a "native" animal, considered to be an important part of this ecosystem, sounded like a great idea, but this reasoning was flawed for two reasons.

First, wolves were not historically present full-time in the Intermountain West, as there was no good food source. Wolves were primarily plains animals, following bison, elk, and antelope herds. From the Lewis and Clark expedition that came through in 1806, we know there were very few bison, deer, or elk in central Idaho, for instance, in pre-settlement times. The plains wolves in pre-settlement times came into central Idaho only when the bison did, on a sporadic basis. The Native Americans of that era nearly starved for lack of meat and lived on fish. They had to travel annually across the mountains onto the plains of what is now Montana to hunt bison to obtain meat to help them get through winter. Wolves moved into Idaho after the settlers came, and the early livestock industry provided a prey base. The wolves that harassed livestock in the central Rockies during the late 1800s and early 1900s were diligently hunted and eliminated. Big game herds expanded with less hunting pressure and with the introduction of elk in the 1930s by the Idaho Fish and Game Department.

Second, the wolves that were dumped into the central Rockies were big Canadian wolves that could weigh 130 pounds or more, not the smaller native wolves of the American plains. Wolves are killing machines, and big wolves have big appetites.

Additional Issues

When the federal government made public its plan to introduce wolves, many Westerners feared that wolves were being used as part of a larger agenda to wrest control and management of private land away from landowners in the West. Those who raised livestock were also fearful that the introduced wolves would not stay in their "designated areas" like Yellowstone Park. The elk herds in the Park generally come down out of the high country in winter and often mingle with ranchers' cattle on feeding grounds. Most of the game herds in the West spend a lot of time on private land, sharing pastures and hayfields with domestic animals. Wolves don't just stick to wild prey; livestock are easy targets; and ranchers felt that wolves would simply follow the prey source.

The wolf reintroduction plan not only infringed on ranchers' property rights and ability to protect their livestock (especially if wolves denned on a ranch, that area would be off-limits for traditional uses), but it also boded economic losses due to depredation. Ranchers feared and resented this infringement and were afraid of the personal tragedies that would result. Those fears have become reality.

For people who raise livestock as a way of life, their animals are not just a paycheck. After spending a lifetime delivering baby calves or lambs, doctoring the sick ones, and fine-tuning every ability and instinct to become competent care-takers of livestock, the rancher won't forsake his animals. The rancher is responsible for their well-being and is morally committed to taking care of them. To find a favorite old cow lying hamstrung and bleeding or young calves totally missing—with their mothers upset and bawling—is unacceptable.

The wolf advocates promised there would be ways to reimburse ranchers for their losses, but such compensation would never cover all the economic losses or any of the emotional losses. If a baby calf is killed, would the rancher be reimbursed for what it would have been worth in the fall as a big steer or grown up to be a herd sire or replacement heifer in the herd? How do you put a value on genetic potential lost forever? And how do you alleviate the personal loss and anguish the rancher suffers when watching his animals being maimed and killed? Or how do you compensate for the fear of wolves attacking ranchers and/or their children?

Ranchers fight bad weather, livestock diseases, and other threats. In good conscience, they cannot stand meekly by and watch wolves slaughter their livestock. To demand that we couldn't shoot a wolf to protect our livestock or wildlife created an inevitable clash that would make lawbreakers out of most ranchers and many sportsmen. Wolf protection according to the Endangered Species Act was an unworkable law. If the majority of people can't abide by or live with a law, it's a poor law, and this should be reason to re-think it. Little is said in popular press about the viewpoint of the people who live with wolves in the northern Rockies. This article lets people hear the voices of those of us who live with wolves.

Wolf Depredations

Within days after the first wolves were turned loose in the backcountry of central Idaho, a wolf killed a young calf belonging to Gene Hussey, an 83-year-old rancher on Iron Creek near Salmon, Idaho. The wolf was shot, perhaps by someone driving by the pasture on the road, since in ranch country it is common practice to protect livestock from predators. When the dead wolf and dead calf were discovered side by side, Hussey was afraid of what the feds might think and called a local veterinarian, Dr. Robert Cope, to perform a post-mortem examination on both the calf and the collared wolf.

Later, when the federal experts arrived, they began harassing Hussey, assuming he shot the wolf. They took both bodies for their own post-mortem examinations and claimed that the calf was stillborn and that the wolf did not kill it but was simply scavenging. What they didn't know was that Dr. Cope videotaped the original autopsy as he performed it, which showed that the calf had milk in its stomach and fully inflated lungs. According to Cope, this was indisputable evidence that the calf was born alive. When the feds found out about the tape, they realized they'd gone too far, twisting the facts to fit their own purposes and to keep up the façade about wolves.

The swat team mentality of the government agents, who threatened to persecute anyone who dared shoot a wolf, was emphatically resented by rural communities. It didn't make sense that the penalty for killing a wolf to protect one's animals or in self-defense was worse than for killing a human for the same reason.

Sheep Depredation

The Soulen family near Weiser, Idaho, owns one of the sheep outfits that early on was significantly affected by wolf depredation. Margaret Soulen Hinson ranches with her father and brother in a family-owned corporation. "My grandfather started our business in the 1920s, and I'm the third generation," she says proudly.

Soulens have about 10,000 ewes and 1,000 cows that graze on a mix of public and private land. Margaret explained:

We own about 50,000 acres and also run on BLM and Forest Service, as well as some state grazing leases and private leases. Our range operation covers about 450,000 acres of rangeland located in eight Idaho counties. We had our first loss to wolves in 1996, shortly after they were introduced. Most of our wolf depredations have been on the Payette

131

National Forest near McCall, Idaho. We're only about 80 miles (as the crow flies) from where wolves were turned loose in the Frank Church Wilderness. It didn't take the wolves long to find our sheep. They love lamb!

The first year, Soulens lost 30 head.

We didn't know what to expect when we started losing sheep. We didn't know if the killing would continue or not. The first night they tore up 13 head. We then did everything we could to keep them out of our sheep, since it was illegal to kill the wolves. We took extra people and slept on the bedgrounds with the sheep, but we still lost 30 head that summer. It was a learning experience, because wolves have been gone from our area for a long time. Even my grandfather didn't have any experience with wolves.

The next year they lost 28 head. The loss numbers bounced around: 37 head in 1999; 5 in 2000; none in 2001; 2 in 2002; 55 in 2003; and 332 in 2004.

In 2004, the Cook pack, about 13 wolves, hit the sheep before we even made it to the National Forest. They killed 78 head in one night and continued to hit the sheep. This pack was living in an area where all our bands had to pass through, and they hit almost every band. We also had many other wolves in the area that year. Every band had wolves around them.

Finally the Soulens received help from U.S. Fish and Wildlife Services, which came in and removed the Cook pack, but by that time, the wolves had done a lot of damage. "It was almost August by the time they took out those wolves," she says. In 2004, it was still controversial about whether the government agency would allow removal of any wolves, and many people still didn't think wolves would ever be much problem for livestock.

Soulens also worked with Defenders of Wildlife and other wolf advocate groups to try to do whatever they could to co-exist with wolves.

We bought extra guard dogs and had extra people sleeping out with the sheep. We tried rag boxes, cracker shells, gun training for the herders, etc. Our herders became more experienced at recognizing when wolves are a danger. When the guys know wolves are around, they stay with the sheep 24/7. But the reality is that sometimes you have to remove the wolves that are causing too much damage. They become habituated to preventative measures and won't let up. By taking out some of the wolves, you can break that cycle.

In 2005, we lost about 220 head; in 2006, we lost 175; in 2007, we lost 125; and in 2008, we lost 75. We did a graph to look at what the wolf population numbers have been, in comparison with our losses. There was an exponential increase in wolf population—and our losses reflected that increase. One good thing about all this was that the Idaho Conservation League actually sent out a press release in support of removing that pack, which helped people recognize what some of these issues were. I think people are starting to recognize that wolves kill livestock. The stories early on were that they only kill the weak and the lame or that they only eat deer and elk and leave the livestock alone. In reality, wolves are predators and opportunists. The idea that they kill only what they will eat is also false. With sheep, especially, they'll just tear them up. I've seen instances where they rip them open and string the guts out. They may eat only a small portion, or they may eat the entire animal to where there's only a skull left that looks like it's been boiled clean. Sometimes you never find anything; the animal is completely gone. Hopefully, the delisting and hunting season will put some fear into the wolves, if they are shot at more often.

It may be difficult for hunters to find them, however, because wolves are often nocturnal in their depredations and are very good at hiding during daylight hours.

Our losses typically occur at night. Where sheep are herded, at least someone is there, and you can get the problem identified right away. It's more difficult for cattle people to protect their animals than it is for sheep producers, because cattle are scattered and not as closely watched like the sheep with their herders. You can't be out with cattle that much; often you never find the animals that are killed and eaten.

In order to have any compensation for losses or to document a wolf problem, livestock owners must prove that it was actually wolves that did the killing.

This can be an issue in our remote range areas. The herders have a challenge to get the word out and to get Wildlife Service people up there to document and verify the kill before the carcass has deteriorated or something else has eaten on it. Wildlife Services and the U.S. Fish and Wildlife have been very responsive when we had problems. They have been very good to work with.

Soulens were compensated for most of their early losses. State funding—federal money that came through the state—paid for the probable wolf kills.

Defenders of Wildlife paid for every documented wolf kill and were fair on prices. We were one of the first ranchers to experience losses. They've been fair, also, in what they've paid. One issue is the challenge of getting it documented in a timely manner.

There are significant losses when dealing with wolves, however, that are hard to quantify and document and for which the producer will never be compensated.

The compensation doesn't make up for all the depredation losses, because there are some things you never get compensated for—like all the extra labor, the stress on the sheep, and the lower weight gains when wolves are harassing them.

Wolves create a lot of extra work for the herders, the foreman, and others who are dealing with this situation. We've had to herd the bands a lot tighter. They're not as spread out, and it's a little harder on the land. The lambs are not gaining as well, as a result. The herders have to sleep out with the sheep or bring them back closer to camp at night. This is not optimum management. Typically we'd herd them a little looser, and they would graze more.

134

The wolves have added many additional expenses to their sheep operation.

We've also needed more guard dogs. Our herders know what to expect when they hear warnings from the guard dogs. The dogs make a different sound when it's wolves than when it's bear or coyotes. Our foreman said that, when wolves are around, the dogs cry. The dogs are afraid of them. We've lost guard dogs to wolves. We run four guard dogs per band. If it's just a single wolf, the guard dogs gang up, and the wolf will back down, but if there's more than one wolf, they kill the guard dogs. We used to use Pyrenees guard dogs, but now we use Akbash and Akbash-cross dogs. They are a little more athletic and aggressive. The dogs are expensive; the last ones we bought were $650 apiece.

During that first year, when there was a lot of effort trying to trap the first pair of wolves on Pearl Creek to relocate them, Wildlife Services didn't have manpower to deal with other problems. They did finally relocate the wolves, but in some of the other bands we were experiencing bear and cougar damage, and nobody was available to deal with those issues. The wolves have taken a lot of time and a lot of money.

When wolves were introduced, the USFWS didn't realize how well they would do or how fast their numbers would grow. With total protection for more than a dozen years, wolves became bold.

If you look at it from the wolf advocates' point of view, this is a tremendous success story. Moving forward with delisting and a more pro-active management system for wolves must be the next step. With our operation, we were granted "take permits" early on, so we could shoot a wolf if it was in our sheep. But in all those years, we never managed to shoot a wolf. Wolves can be very difficult to actually shoot.

Ranchers hope there won't be too many roadblocks (lawsuits by wolf advocates to stop the wolf hunting seasons) in progressing toward workable management of these predators.

I was disappointed that Defenders of Wildlife and some of the other groups filed for the injunction to halt hunting.

They're not being fair. The wolves' numbers had exceeded their goal, so it's time to move forward with the delisting and start managing them. This blocking effort is not productive. It makes the environmental community look bad.

Ranchers want something logical and workable, and when the environmental and wolf advocate groups refuse to work toward a management solution, it's viewed in a negative light, and cooperation becomes more difficult.

Learning Curve

Nearly 20 years after the wolf project was launched, almost every rancher in central-eastern Idaho has seen wolves or their tracks, and livestock losses are all too common. In 1999, Ralph McCrea, a rancher near Leadore, Idaho, lost 20 lambs to a wolf that came into his flock two different nights.

This wolf was collared and had been turned loose in Yellowstone Park at noon on a Wednesday. The next day he went through Dillon, Montana (70 miles away), and by Friday night, he'd come over the mountains and was at my place killing sheep.

McCrea has also been losing calves; for several years wolves killed some of his big calves on the range, and then wolves began coming into his calving pens at night, killing baby calves right next to his house.

James Whittaker runs 1,000 cows near Leadore, Idaho, and experienced his first confirmed losses (three bred heifers) in 2007. In 2008, Whittaker had two confirmed kills within a mile of his house.

In 2009, we had four confirmed kills on our range. It's hard to find them soon enough to get the losses confirmed; our range is 35 miles long. All told, we lost 45 calves that summer. Our annual loss went from 2% to nearly 5% that year.

When you lose a calf from other causes, you usually find the carcass, but with a wolf kill, you may not find anything.

It's like the animal evaporated. We get paid for confirmed kills but not for the ones that disappear or that we find too late to determine the cause of death.

136

Bruce Mulkey, who ranches near Baker, Idaho, recalls wolf advocates on a panel discussion at Idaho State University saying that only 1% of livestock are killed by wolves and that ranchers lose more to disease and other problems. "But if you are part of that 1% and happen to lose 15% to 20% of your calves, that's a huge loss."

Jay Smith, on a ranch near Carmen, Idaho, did some research of his own, looking at ranchers' BLM use reports for several years. He checked as far back as these records go to assess range losses on the Diamond-Moose allotment.

> *I documented how much loss ranchers experienced before wolf introduction, compared with losses after the wolves came. Very soon after wolves were brought in, John Aldous and some of the other ranchers on that allotment had huge losses.*

Aldous lost 34 calves in one small area near Leesburg the first year.

> *Then the wolves scattered out and hit other ranchers besides me. They had dens on the border between our range and Edwards'. Our elk population dropped from 300 in Moose Creek Meadows 15 years ago to where now you're lucky to see 20 elk. The USFWS eliminated the Gerano wolf pack three times, and they just kept coming back. I think some of the offspring that go off on their own come back and regroup.*

Fay and Eron Coiner have a grazing ranch high in the mountains above Salmon, Idaho, and spend summers there with their cattle. Fay says they had more than 100 elk on their ranch, but by the third summer of wolves, they were down to 13 elk.

> *Now we are lucky to see any elk. But we lose at least one calf a year to wolves, even with Eron riding out there every day to try to chase them off. He watched one of our calves being killed, and that was gruesome. Eron was given a permit to kill any wolves that were with our cattle, and he's shot a few, but they still came back.*

Coiners worked closely with the USFWS and Fish & Game to help them manage wolves rather than have to kill them. Fay explained:

We did our best to make the wolves understand that this is our territory and that our place should be off limits. When the government agencies got together to do a flagging experiment around our ranch to try to deter the wolves, it didn't work very well because wolves are curious. They came to see what was going on, and that summer we lost three calves. We've used other deterrents, too, such as rubber bullets and firecrackers to scare them away, but the wolves came right back.

Often wolves don't kill an animal outright, but bite at it numerous times (leaving characteristic scratch marks and bruising the underlying tissues), wearing the animal down, and they often start eating on it before it dies. Coiners have had numerous maimed and injured cattle that don't survive, even with intensive medical care, which costs a lot of money.

Ross Goddard, a rancher near Tendoy, Idaho, says his cattle on summer range are very unsettled—constantly nervous and on the move.

They don't graze as much or gain as much. And I can't use my range allotments properly. If we put cows in a certain area and they're supposed to be there for a month, two days later, they may all be out of there and in the next unit, and you can't get them to go back. It disrupts your whole season's grazing management.

His grazing use is under strict observation because of fish habitat.

We have to keep cattle off riparian areas because of endangered fish, so we shove the cattle into higher country right into wolf domain.

Bruce Mulkey says that, after wolves harass cattle, you can't gather or move them with dogs anymore. "The cows are all stirred up, and all they want to do is fight the dogs." Even the cattle that have been trained to work with dogs get on the fight and are overly protective of their calves. Coiners saw wolves harassing a cow-calf pair until the cow was utterly exhausted trying to defend her calf. "That cow was bellowing and frothing at the mouth and on her last legs," says Fay.

Whittaker says cattle are much more flighty when wolves are around. "One group of cows ran and hit the fence and tore the gate

138

down, getting into the adjacent pasture and mixing with another group of cattle." McRea says that, when wolves go through his private pastures, the cows go crazy, bellowing and running around the field trying to find their calves. Many ranchers have lost young calves—trampled and killed by stampeding cows.

> *I had a calf run over; he got rolled around and ended up with a crooked neck, but survived. I had a month-old calf chewed up by wolves, with his intestines hanging out. The vet came, and we sewed him back up and gave him antibiotics, but he only lived two days.*

Cattle that survive a wolf attack or calves crippled by stampeding cows can't be sold for top dollar with the rest of the group at sale time. The rancher has to butcher them or sell them at a loss. Ranchers are not compensated for the costs of doctoring injured animals or for cows that must be sold because they lost a year of production due to loss of their calves, which eats into the sustainability of the cow herd and makes it necessary to raise or buy more replacement heifers. Ranchers are also not compensated for lower weights on market calves. McCrea said:

> *In the winter, wolves have been coming in to kill at night, and my calves lose weight. I have to corral them every night so they're close enough that I can get to them when the wolves come in. By spring the cattle don't want to go back into the corral at night, but I don't dare leave them a mile and a half away from the house with no chance to protect them. When you get up in the night to check cows, it's horrible that part of your calving equipment has to be a rifle on your back. I was within 25 yards of one wolf, right in the corral with the cows, with just a flashlight.*

Now he takes a gun with him. McCrea says he realizes that we can never completely control wolves to keep them from killing livestock, but he points out that it would help if ranchers could get full compensation for missing animals. "If the public insists we have wolves, they should pay the price and help us resolve the problems wolves have created." Many ranchers feel that wolf introduction—together with the impositions it has caused—is a taking of private property by the federal government.

Working with the Agencies

Every state's laws are different in how problem wolves can be handled, and ranchers must work with the state as well as federal agencies. When the wolves were still listed as endangered, however, in all instances, a wolf kill of livestock had to be confirmed before any action was taken to remove the wolves. Rick Williamson, Idaho branch of USFWS, said:

> When I'm called to a rancher's place to investigate a suspected depredation, we skin out the dead animal(s) and look for evidence such as bite marks and tracks. We have to determine whether the animal was actually killed—not just fed on—by wolves. There are four categories in our evaluation: confirmed kill, probable kill, possible kill, and other. For a while, the Defenders of Wildlife would pay a rancher 100% for the loss if it was confirmed and 50% for a probable kill.

> Once we confirmed a kill, we conferred with state Fish & Game because they have the final authority to tell us whether we can remove a wolf or if we can trap and collar one. For instance, we collared a wolf on Dale Edwards' place, and he had a telemetry receiver so he could check for presence of that wolf.

> We have some other tactics, like radio-activated scare boxes that we can put out near a den site or rendezvous site to deter wolves. But we have to be careful because, in some situations, we may be just moving them from one producer to his neighbor. I think the best thing we can do, if we have depredation, is to remove the problem wolves.

Now things have changed with the wolves delisted and the state involved in managing wolves.

A few years ago, a graduate student, John Oakley, did a study for his master's degree, looking at numbers of livestock killed by wolves. Williamson explained:

> Oakley was trying to determine how many kills you don't find. He and his helpers radio-tagged more than 350 calves on the Diamond-Moose allotment. What Oakley discovered was that, for every kill found, there could be up to eight head you don't find.

Williamson said:

In our state, most of the wolf problems with livestock are occurring on private property. People are seeing wolves near their homes. A few years ago, the White Hawk pack was causing problems on Carmen Creek, and over a three-year period, there were more than 50 livestock kills—and all but two of those were on private land. When we later removed that pack, it stirred up a hornets' nest of protest around the country. But people need to understand that wolves have no business being on private property.

Wolves, Smart and Cagey

Control of problem wolves has proven difficult and expensive. Problem wolves are generally relocated or removed by USFWS, but this, too, can be labor intensive and frustrating. Whittaker says one year, when the wolves hit him hard, there was a den two miles from his house, and the government trappers worked all summer trying to get the wolves and only killed two.

They trapped one and shot one from a plane. It's hard to get one from a plane, because if you don't get them the first time, they hide when they hear it coming.

Allen Bodenhamer, a rancher near Baker, Idaho, says wolves were killing calves in his barnyard at night and hiding up on the mountain in the timber during the day. "The government hunters flew our area six times trying to find the wolves and never got any of them." During calving season, Bodenhamer and his wife were out checking the cows every hour but didn't see the wolves that came in to kill their calves. "For those wolves to come in between checks with our spotlights, they must have figured out our pattern."

Jay Smith runs cattle on 75,000 acres of public land.

At least one of us (myself, my wife, my mom and dad, or the hired help) is out there horseback almost every day. We've only had close enough encounters for shooting opportunity about 10 times since 1995. It's a lucky accident if a person is able to shoot a wolf. The only two wolves I've seen that were close enough to shoot, I wasn't carrying a gun at the time. Also, if they feel pressured by humans, they go nocturnal and kill livestock at night.

Aldous says you have to be lucky to shoot a wolf. "In our area there's so much timber that we may only get a glimpse of one when we're going down a trail with a herd of cows, and then it's gone."

Wolf Hunts Not Enough

Most ranchers and sportsmen feel the wolf hunts won't be enough to control the expanding wolf population. They also feel that wolf numbers are much higher than federal and state game departments estimate. Only a small percent of the population is actually radio-collared, and there's no way to see and count them all. Officially estimated numbers have increased enough, however, that wolves in the central Rockies were finally delisted and no longer protected by the Endangered Species Act. Some states have been allowed to manage them as a game animal and initiate a hunting season.

In Idaho, an unlimited number of wolf tags can be sold by Fish and Game, but the hunt in each unit closes as soon as the allotted number of wolves is harvested. For the first hunt in 2009, there were 26,384 tags sold in Idaho—for a chance to kill 220 wolves! Due to lack of hunting success, however, several units remained open after the late December deadline and stayed open into spring. Some quotas were still not filled by summer 2010.

Jay Smith says that having a hunting season is a step in the right direction, but "You could open the season year round, and we wouldn't put a dent in the wolf population."

Whittaker feels the hunting season won't make much difference. "Amateur hunters just make them more elusive. If you shoot at a wolf and miss, he won't give you that opportunity again."

With the increasing impact on elk herds in certain areas of the state, the Idaho Department of Fish and Game (IDFG) now realizes there must be more control of wolves than the hunting seasons are providing. As of January 2013, the IDFG began considering two more ways to work toward its goal of reducing current wolf numbers to a manageable level without risking coming under federal protection again. The first proposal is to take $50,000 from their coyote control program in eastern Idaho and use it to supplement wolf control efforts in areas where elk herds are in decline. The second idea is to develop relationships with successful wolf trappers and possibly give them financial assistance to offset some of their costs, as long as the trappers are working in those same areas where elk are being significantly impacted.

Wolf-Cattle Interactions Study

The increasing wolf population in the West is having a significant impact on ranchers and livestock, but the extent of this impact had not been well documented. The Oregon Beef Council (with help from USDA) funded a 10-year study project that began in 2008 to look at how wolf activity affects cattle behavior. GPS collars are being used on some of the cattle on ranches in the study area and on a few wolves to monitor movements of these animals and when and where they interact.

Dr. Pat Clark, Range Scientist, Northwest Watershed Research Center, USDA-ARS at Boise, Idaho, has been involved with this study from its beginning and was instrumental in helping with the GPS collars. He had used collars on range cattle for an earlier study looking at how cattle changed their use patterns after a fire. Those first commercial GPS collars cost about $5,000 apiece, so Clark decided to build his own. With the help of some Boise State University engineering students, he created a more functional, less expensive GPS collar. By 2008, they had one that could log GPS data every five minutes for more than a year and cost less than $500 each.

Clark began using the collars, in a pilot wolf-cattle interaction study in central Idaho in 2005, to determine the effects of wolf presence on cattle range-use patterns. Part of the challenge of GPS collaring a wolf was getting the collar into a wolf-sized package. A cow can carry a fair amount of bulk and weight, but wolf collars had to be smaller and lighter.

The Oregon Beef Council (OBC) heard that Clark was using GPS collars to collect data and to address some research questions about the wolf-livestock issue in central Idaho. The OBC realized wolves would soon be coming into Oregon from Idaho and wanted to evaluate how this would impact rangeland cattle production in their state. The OBC asked Clark to extend the wolf-cattle interactions research he was doing in central Idaho to the border area of northeastern Oregon (where there were no wolves yet) and to western Idaho (where wolves were common). The proposal for the study was developed and submitted to OBC in 2007. The Oregon-Idaho Wolf-Cattle Interactions project began in 2008 with three study areas in western Idaho and three in eastern Oregon. Clark said:

> *We selected our Idaho study areas from the western part of the state where range cattle grazing is the dominant land use and where wolves were common. We searched for study areas in Oregon that had similar topography, vegetation*

143

types, soil types, cattle breeding, calf age at turn-out, grazing schedules, and so on.

This type of experimental design is called a BACIP (Before-After, Control-Impact, Paired) design. The Idaho study areas served as the Control since wolves were already present. What would eventually change (and create impact) was wolf presence in Oregon. Wolves were expected to travel from Idaho into Oregon and establish packs in the Oregon study areas.

In Oregon, we wanted to collect at least a couple years of "Before" period data before wolves arrived in enough numbers to have a true impact, and then we could contrast these "Before" data to "After" period data collected after wolves set up shop in Oregon.

We put GPS collars on at least 10 cows in each study area. Cow-calf herd sizes in these areas were 300 to 450 cows. The 60-collar total sample size was limited by available funds from OBC.

The study was later expanded to include two more paired ranches, providing enough sample size for adequate data to detect differences in cattle range use and other behaviors. GPS data tracks the location of collared cows and determines how fast they are traveling, enabling the researchers to classify cattle activity into resting, foraging, or traveling. Clark explains:

From this, we can calculate how much time a cow spends foraging during the day versus standing looking around or bedding. A hyper-vigilant animal will spend a lot of time standing and watching, rather than foraging, which could impact nutritional status. With GPS data collected every five minutes, we can detect changes such as increase in vigilant behavior and decreases in foraging time.

One of the Idaho study sites is the OX Ranch near Council, Idaho, managed by Casey Anderson, where data has been collected since 2008. This ranch experienced wolf depredation before the study started, but didn't have documentation.

In 2009, we were able to confirm more of the wolf kills—18 animals—but were also missing five cows, two yearlings, one bull, and about 70 calves that were unaccounted for.

There were not many wolves in Oregon when the study began— just a few passing through and returning to Idaho. Within two years, however, the wolves coming from Idaho established packs in northeastern Oregon, and the study moved into the "After" design— comparing high wolf pressure with low wolf pressure (rather than no wolf pressure). "Having wolves and cows collared enables us to match up their location and time and to determine where and when wolf-cattle interactions are occurring," says Clark. Data on wolf presence and activities is augmented by wolf scat sampling along forest roads and use of trail cameras.

Interesting Study Results
Even though only one wolf was collared in 2009, this gave some interesting results, said Clark:

We found interactions between all 10 of the collared cows and that single GPS-collared wolf within a very extensive study area (over 50,000 acres). The 10 collared cows were dispersed throughout the landscape, part of the larger herd of 450 cow-calf pairs, yet the data showed this wolf was within 500 yards of every one of the 10 collared cows many times throughout the summer (for a total of 783 contacts).

The collared wolf's area was 210 square miles with a 55-mile perimeter. The least distance he traveled was about six miles per day, and the most, 29 miles per day. "On a day of confirmed depredation, we could look back to the data and see that the wolf was in the immediate area at that same time," he says. This wolf was a member of an 11-wolf pack. There were three different packs, totaling 34 wolves, roaming that study area during 2009. The rancher on that study area suffered more than 40 confirmed or probable wolf depredations.

The collared cow with the least number of interactions with the collared wolf had 23 interactions within 500 yards during that grazing season. The cow with the most encounters with that wolf interacted 140 times over the course of 137 days (mid-June through early November 2009—when the collar was removed from the wolf to retrieve the data. All 10 collared cows had interactions within 250 yards, and only one of the 10 cows didn't have interaction at 100 yards or less. Envision 450 cows with more than 30 wolves moving among them. If one of those wolves had 783 encounters with just 10 of those cows, think how many encounters there must have been by 30 wolves among the 450 cows; in other words, the cattle were constantly impacted by the presence of wolves. Two of the collared cows' calves disappeared that summer. Clark said,

> *If the behavior of this wolf was representative of the larger local wolf population, it is easy to imagine that this cattle herd, and likely other neighboring herds, was exposed to an intensive wolf predation threat.*

> *We wondered how many of these GPS-detected wolf-cattle interactions resulted in depredations, but we will never know. We did go back and look at the wolf GPS data and compare the tracking logs with locations of known 2009 depredation sites. We could see tight, spiral patterns in the wolf movements that occurred at those sites. These circling activities may illustrate prey-appraisal or pursuit events.*

> *The following spring of 2010, we hiked into the sites where spiral patterns had occurred in the 2009 wolf log and inspected the sites, looking for signs of depredation, and found fresh cattle bones at one of those sites.*

Rendezvous Sites

Certain areas are used by wolves to park their pups with a babysitter wolf while other members of the pack go back and forth on hunting excursions. Dave Ausband, a researcher with the Montana Cooperative Wildlife Research Unit at the University of Montana, developed a wolf rendezvous prediction site model and, from that, created a predicted wolf rendezvous site map for Idaho. Clark said:

> *These rendezvous sites anchor the distribution patterns of the entire wolf pack to specific points. Out of curiosity, we plotted both our wolf GPS data and our cattle GPS data from one of our Idaho study areas onto Ausband's map and found the clusters of concentrated wolf location data lined up with areas predicted on the map as being high quality wolf rendezvous site habitat. We wondered if the map and*

model could be used to predict where wolf-use patterns and cattle-use patterns frequently overlap—increasing the potential for wolf depredation.

Wolf rendezvous sites tend to be in grassy areas close to water—not necessarily riparian areas, but grassy meadows without a lot of overhead cover. These areas are also good foraging areas for cattle. Rendezvous site information could enable ranchers to be aware of certain danger areas and to check for signs of depredation in those areas. They might choose to not put cattle into those areas until later in the season when the pups are grown and the pack is not tied so closely to those sites.

Indirect Effects

Clark explained about the indirect effects of wolves upon cattle. "There was some earlier research done on death and injury losses due to wolf predation on cattle, indicating that these impacts are underestimated," says Clark. Some cattle just disappear; a pack of wolves can consume an entire carcass overnight.

However, no one has researched the indirect effects of wolf presence on rangeland livestock production. These indirect effects may have much larger impacts on livestock producers than livestock death and injury losses to wolves. Indirect effects of wolves on livestock range-use patterns include impact on foraging efficiency, disposition, and stress levels.

These effects could cascade down to affect cattle diet quality, nutritional status, and disease susceptibility. Cascading further down, wolf presence may indirectly affect (and reduce) calf weaning weights and cow body condition in the fall and result in increased winter feeding costs, along with conception and pregnancy failures, increased veterinary care and supply costs, and death loss to disease. This complex cascade of effects could substantially impact a ranch's bottom line.

Casey Anderson at the OX Ranch was weighing and assessing his cows when they were collared in the spring before turnout and again when they came off the range to be hauled down to winter pasture and the collars removed. "What we were seeing across the board in our herd was that the cows were coming home a full body score less than they had been in the past. With a mature cow, each score is about 100 pounds," he explains. This means extra feed costs to get them through the winter. "We were also seeing the conception rate plummet. With our herd health program and mineral and protein supplement, there was no reason for this other than the wolves. What normally would be 90 to 95% conception rate in our herd went down to as low as 82%," says Anderson.

No Single Effect, But Significant Losses

Some of the direct impact went beyond actual kills; additional animals were severely maimed. "Although we spend a lot of time doctoring these and trying to save them, we end up with a bunch we can't sell because they are crippled," Anderson says. During the summer of 2012, he had nine calves with wolf bites in just one week. "The alpha female in the pack is using the calves to teach pups how to hunt. These injured calves may live, but you have to doctor them." In addition, four young cows in the first-calf heifer group had big abscesses behind the shoulder and/or above the flank area, due to infected wolf bites.

Wolf depredation is generally not a single effect; it usually stimulates a change in cattle use patterns and different areas for grazing. "We are also seeing changes in the way cattle use the range. They are bunching up more against fences, and wolves are using the fences to corner them," says Anderson.

This may be a shift from a high-quality foraging area to a low-quality foraging area just because cattle are worried about wolves as Clark explains:

> Cattle may get up on an open hillside where they can see farther to detect if wolves are approaching, but the time of year they are doing this might be when the grass is so dried out it doesn't have much protein—so cattle diet quality declines. Cattle under threat bunch up, standing in open areas where they feel safer, but these areas soon become dry and dusty from concentrated trampling, and the rising dust can lead to respiratory disorders—especially in stressed cattle that have become more vulnerable to disease.

If cattle are being bred on the range, as is the case for the majority of outfits raising commercial cattle, all these factors work toward decreasing the number of cows that get bred on time and increasing the number that end up breeding late or coming up open in the fall, according to Clark. Late calves change the uniformity of the calf crop, which adversely affects the price received for those calves, which is another financial impact. This may also affect the number and quality of replacement heifers a rancher is able to keep, and the future productivity of the herd is adversely affected.

Cattle Temperament Changes

There are also changes in cattle temperament. Clark explains:

In one of the central Idaho herds, a wolf followed cattle down to the calving areas, and the cows were very aggressive in trying to protect their newborn calves. The rancher couldn't tag a calf without being attacked by the mother cow; he had to get the calf into the back of a pickup to safely do anything with it.

Anderson says that, before wolves came, their cattle were all dog-broke and easy to work with dogs.

Now our cattle chase dogs; we can't use dogs anymore to herd them. In this steep, rugged country, dogs have always been a good tool. Now it is extremely difficult to move cattle.

The OX Ranch calves in late May through June. The calves are branded when the cattle are gathered in the fall. "By that time, the calves are 350 to 400 pounds, and when they come into the corral, they size you up and take you," he says. They are totally focused on defending themselves, attacking a dog or a person. "This is not normal behavior. These cattle were all nice to handle before the wolves came. Now they attack anything that comes close to them," says Anderson. He had a horse that had to be sold after it was run through a fence by wolves. Even though the horse's injuries healed, it was no longer safe to have dogs near the horse as it would strike and kick at the dogs. Threatened, stressed cattle are unpredictable and often become more aggressive, especially when people are working on foot, but also in rangeland situations where ranchers are trying to herd them on horseback. If cows are prone to fight or take off and go through fences, this affects ranchers' costs, and injury risk for ranch personnel also becomes an issue, as Clark explains:

There are more handling-related injuries to cattle that have been exposed to depredation threat, along with increased frequency of bunching and flight events. Handling flighty, nervous cattle may lead to more trampling deaths of small calves and broken legs in larger calves. Nervous cattle are more prone to chute-crashing, ramming fences, crawling up corral walls, and other flight-type behavior, which can result in spinal injuries, rendering the injured animals unmarketable.

Dr. John Williams (OSU Extension) says one of the purposes of this 10-year study is to try to understand how cattle react differently due to presence of wolves. "We are finding that cattle temperament changes drastically when they have to live with wolves," says Williams.

This winter (2013) we hope to put collars on cattle, do blood tests on the ones that have been living among wolves, and compare their stress level (measuring cortisol levels in the blood) with cattle that have not been living with wolves. We know that cows are individuals, and some will be stressed more than others.

We want to find out how long this effect lasts. If we were able to measure the stress on a cow that is attacked, her cortisol level the next day would be very high. Our ranchers have been telling us that many of these cows' changed behavior lasts for a while. In some cows, it becomes less pronounced, in some cows it disappears, and in others it does not.

Neil Rimbey, a range economist with University of Idaho, has been evaluating economic impact on ranches in the study, using economic models to assess the economic impact of management alternatives. The depredation loss (killed animals) is what gets attention, but the first years of this study have shown that indirect losses have more impact on ranch profitability and sustainability. Rimbey says:

A recent study in Wyoming indicated that ranchers are able to find only about one of every seven animals killed by wolves, so in Wyoming they are compensating ranchers on a one to seven basis. This research in Wyoming also suggests that some of the indirect losses—more open cows, reduction in weight gain on calves, more veterinary treatments for injured calves or for stressed calves that got pneumonia— should raise the compensation rate (from one to seven) up to one to 13 or 14.

The change in cattle behavior in the presence of wolves impacts management. Cattle may not stay in the areas they are put—for instance, coming right back down off the mountains to try to get away from wolves. They may use some areas more heavily while avoiding others. Cattle also crash through fences and are harder to handle. Rimbey says:

The Oregon and Idaho ranchers have mentioned increase in time, labor, and costs associated with managing cattle. There's more travel involved; they have to go more frequently to try to check cattle. There is also more time spent meeting with Fish and Wildlife Service to try to get confirmation on death loss, etc.

The study has confirmed things ranchers suspected regarding behavioral changes in cattle, and it will be interesting to see how the economic impacts calculate out.

For instance, each 1% change in the percentage of calves weaned in Idaho (using conservative prices) is amounting to about a $1,750 change in gross revenue, or nearly $6 per cow… And this is just a starting point of this component of the project.

Wolf Threats to Humans

Casey Anderson at the OX Ranch says that, when wolf numbers are controlled, they have a tendency to stay farther back in the high country. Anderson's wife has found wolf tracks in fresh snow less than 50 feet from the house. Anderson says:

We've had wolves lie on the county road in the snow at night, watching cattle under a light at the end of the barn. You could see where the wolves had come down next to the corrals and were lying in the snow watching the cattle.

151

Data from the collared wolf showed how close these wolves are coming to homes and human activity. "We have several houses here on the ranch. The collared wolf came within 500 yards of one house 307 times that summer," says Anderson. The pack of 12 came within 300 yards of the ranch lodge and spent all day there.

The people who take care of the lodge have three little boys. The wolves were there all day, right above the county road in a little clump of timber, and watched the lodge. We had proof because the collared wolf was there with them in what we call a rendezvous site.

Many ranchers are reluctant to be out doing routine tasks, such as hiking around their pastures fixing fences, without a gun. Says Jay Smith, a rancher near Carmen, Idaho:

This maneuver by our federal government is stealing from me, my children, and our future. It may destroy our livelihood, and our entire lifestyle is also in jeopardy. I have small children, and I like to take them with me to the hills when I check cattle or fences, and now it's a risk.

Ralph McCrea near Leadore, Idaho, tells of one incident a few years ago when his grandchildren were sledding on the hill behind his house.

A couple hours later, I went out to check the cows and found a wolf standing in the herd, just 40 yards from where the kids had been sledding. They had our old dog with them, and if the wolf had arrived a little sooner, it would have been attracted to that dog and the kids. We're being pulled right back into the 1890s when everyone had to be armed to defend against predators.

Wolf advocates say that wolves are shy and stay away from people, but this is proving to be untrue. There have been increasing numbers of incidents the past dozen years in which wolves have threatened, attacked, or killed people. Wolf advocates try to discredit many of these stories, however, as they did an earlier attack of a vacationing family on a pack trip into the Middle Fork of the Salmon River when the wolves were still under full protection.

From Fishing Vacation to Nightmare

Tim and Diana Sundles, who lived on Carmen Creek near Salmon, Idaho, had taken their teenage sons into a remote area where they had been camping for many years. They had ridden the 16 miles from the trailhead with their horses and mules and set up camp. Sundles recalled:

We woke up the next morning before daylight, hearing wolves howling and horses screaming. I ran out of the tent with a flashlight and small revolver, hollering and shooting into the air to run the wolves off. I assumed it was over, that the wolves were afraid and left. We cooked breakfast, sat around camp an hour or so, and then got our fishing gear ready to hike to one of those remote lakes to spend the day, leaving the horses and mules at our camp, with a couple of them turned loose to graze. I picked up my rifle as we started to walk out of camp and commented to my wife that I had an uneasy feeling.

We got about 100 yards where two pack animals were grazing, and a big grey wolf was sneaking up on them. In the back of my mind, I was wondering where the rest of the pack was. I fired a shot over the wolf to scare it, and it turned on me. I didn't know at the time it was the alpha male and radio-collared. When I fired the shot over it, the wolf came straight at me, full speed. I fired two more rounds, trying to hit it, but a wolf running at you through the trees is hard to hit!

The wolf came within about 10 feet of Sundles, then veered off, and circled around the family, possibly because he'd been shot in the foot, as Sundles discovered later, looking at the body.

But he didn't act lame or even slow down. He turned toward my wife on the trail behind me, and I finally had a good broadside shot and killed the wolf. It dropped about 10 feet from my wife. Then I saw the radio collar and ear tags and realized our vacation was over.

Sundles went home and tried to find an attorney familiar with the ESA but had no luck.

The law said you've got to report the shooting of a wolf within 24 hours, but it took us 48 hours to get out of the wilderness so I was already in violation of the law by the time I found out about it.

Sundles was going to keep quiet about shooting the wolf, because of the heavy penalty involved, but then a few weeks later decided to tell people—at a public meeting in Salmon that August when U.S. Senator Mike Crapo was holding a hearing to get local input about delisting the wolves. Sundles said:

If the wolf attack happened to us, it could happen to anyone, and I felt the public needed to know that wolves can be dangerous. One of the big lies the U.S. Fish and Wildlife Service and the environmentalists are telling us is that wolves are not dangerous to humans. If I kept quiet, the next time it might be a little kid, and then I would have felt partly responsible. Here we were, on vacation, suddenly thrust into this situation, and now I'm feeling guilty, like a felon! A lot of folks who are out camping or fishing don't have a gun. If I had not picked up my rifle when we headed out from camp, we would have had a serious problem!

Protected

Wolf packs in Washington are on the rise, migrating in from Idaho, Oregon, and Canada, preying on livestock and game herds. Ranchers in northeastern Washington have been especially hard hit. Len McIrvin, his son Bill, and grandson Justin, who run Hereford cattle on their Diamond M Ranch, have suffered alarming losses, as Len explains:

We had wolf damage starting a few years ago, but not this severe. In 2011, we had 16 head killed; in 2012, it was 40 head. One of our herds on Forest Service land and another herd of 350 on our private land took the brunt of it.

Wolves are efficient killing machines. It's easy for them to take down mature cattle.

In 2011, we had five bulls lost to wolves on our range. In the fall after breeding season is over, they tend to go off by

154

themselves and are easy targets for a pack of wolves. When wolves find one bull by himself, he doesn't have a chance. Though most of the time they rip the animal open, we've had many calves killed that they never took a bite of. They sometimes just chew him up to kill him, and once their plaything is dead or dying, they go on to the next, like a cat playing with a mouse. For a long time, our Fish and Wildlife people refused to believe that wolves were killing those calves. You don't see any blood when a cat kills a mouse, and it's the same with these calves; they are mauled to death.

There may be a few scratch marks on the hide from their teeth and sometimes no marks at all, but often there's no open wound—just a lot of bruising and internal bleeding. The McIrvins found one 500-pound calf dead and couldn't find a mark on it.

We hesitated to call Fish and Wildlife, but we did, and when they opened that calf up to do a necropsy, they saw the muscles under the skin were pulverized, like hamburger. The calf died of internal injuries.

NAR-Wolves

Then there's the problem of relocated wolves, wolves already in the habit of killing livestock, as Len explains:

Wolves are showing up in every little hamlet and community, but it's scientifically impossible for this many breeding pairs of wolves to crop up simultaneously in an area 200 miles by 200 miles. These wolves are suddenly everywhere. The Fish and Game call them NAR-wolves—non-authorized release. One group of wolves in our area has been particularly aggressive on cattle, and I have a feeling they are problem wolves that were gathered up from somewhere else and dumped out here.

Even though wolves have been delisted in the wolf management areas in Idaho, Montana, Wyoming, and eastern Washington, the state of Washington is still protecting them; it is illegal for a Washington rancher to shoot one. Len said:

The eastern third of Washington is part of the Rocky Mountain Management Area where the Feds delisted wolves,

155

but Washington State made them stay endangered, with all the ramifications of federal protection.

And in the western two-thirds of the state, the wolves are still on the federal endangered list, which causes its own set of problems, as explained by Len:

We had help from the State Senate and House of Representatives as we were dealing with our depredation problem, and the county sheriff was there for every kill we found, documenting it. The county commissioners were there, and we got a lot of press from our local newspaper and radio station. The U.S. Fish and Wildlife finally had to recognize the problem and made the decision to eliminate this pack of eight wolves. They killed some, but didn't get them all, and there are plenty of other wolves.

We have very little game left. The only deer and elk we see now are in our back yards. Mule deer that are usually in the high country spent this past summer in our hayfields and pastures by the house. The same thing has been happening in Idaho the past dozen years; elk and deer have been coming down out of the mountains to take refuge on private land, amongst the cattle, and congregating in populated areas.

We try to keep our cattle up high in the good feed on summer range, and the wolves run them right back down to the bottom fences. Even if we make an eight-mile drive one day to take cattle back up, the next day they are down again.

This makes it hard to manage a range. The Forest Service people are not happy if cattle are spending the summer down in the riparian areas, but when wolves keep harassing them, the cattle won't stay in the high country. Len stated:

Ranchers in Idaho, Montana, and Wyoming have been having wolf problems for nearly 20 years, but it's fairly new here—within the last five years. One of our neighbors runs about 80 cattle. The first 17 cows he brought home from the range last fall had only two calves with them. The other

156

15 calves were almost certainly killed by wolves during the summer. His losses are horrific and will probably put him out of business.

The winter before, wolves harassed that rancher's cows all winter; he finally fenced them into a two-acre pasture next to his house and barn. At that point the wolves hadn't killed any yet, but the cows had trampled several newborn calves when they were upset by wolves. "I've estimated our loss last year at about $100,000, but it's hard to know exactly what it is," says Len. Losses go far beyond actual kills, when injuries, weight loss, and lower conception rates in the herd are taken into consideration.

Cattlemen in eastern Washington have tried using non-lethal means to scare off wolves, but these methods haven't worked. Ranchers in Washington are hoping for delisting of wolves as an endangered species before another problem pack has to be removed at taxpayer expense. There are at least seven or eight other packs in this three-county area. If ranchers could be allowed to address wolf depredation as they address other predator issues, major problems could be minimized.

Political Pawns?

Len McIrvin said:

My son Bill went to the state capitol in Olympia and testified on the wolf issue. He said the room was full—more than 500 people, predominantly women—and he could feel their hatred. He said they could not have hated him any more if he had raped and murdered their daughters. They were screaming at him and crying.

They really believe that wolves are harmless and that ranchers are evil in trying to control them. Len recently saw an elementary school assignment, given by a teacher in the Colville School District north of Spokane, Washington.

This was a class assignment paper for third graders to fill out. It was titled "The Three Little Wolves and the Big Bad Pig: Building Vocabulary." The students were to fill in the blanks in these sentences:

"The three little wolves look soft and_____. In order to sneak up on the wolves, the Big Bad Pig came prowling through the trees. The Big Bad Pig grunted because he was big and bad. The wolves were scared and began trembling. Each time one of their houses crumbled, the wolves were determined to build a better one."

We grew up with the Big Bad Wolf and the Three Little Pigs, and this example is a total turn-around. It's a warped mentality that would come up with this for a third grade worksheet in public school.

He feels the wolf introduction into the western U.S. and their protection is NOT about the wolves. It's about depopulating the West.

There are many people—including some high up the government—who want a 90-mile corridor from Yellowstone to the Yukon (Y to Y) for wildlife. They have an agenda, and wolves are part of the plan. Ranchers don't have much say in these issues; all of agriculture is less than 2% of the population in this country. All we do is feed the world, but most people don't care about that.

Before the wolf plan was finalized in Washington State, I was in a meeting and asked the Fish and Wildlife person, "If I go home tonight and see a wolf pulling down a cow, can I shoot the wolf?"

They said, "Absolutely not, unless you want to pay a $50,000 fine, spend 10 years in jail, be a felon the rest of your life, and never be allowed to own a gun." By contrast, if I was sitting at home and someone knocked a window out with a baseball bat and came in after me, can I shoot him? They said, "Absolutely! That's your Constitutional right! You can shoot a human in self-defense, but you can't shoot a wolf that's killing your animals."

They told us that, if they find a dead wolf in our area, they would have a complete investigation. Every firearm the rancher and crew own would be confiscated until ballistic tests had been run. By contrast, right across the state line in Idaho, it's legal to shoot wolves that are harassing or killing cattle, and anyone can buy a wolf tag for the wolf hunting season.

We've had a lot of damage this year, yet we've only seen the wolves a couple times and never had a chance to get a shot. It was usually a brief glimpse in the headlights of a vehicle in the middle of the night or when we didn't have a gun.

The wolf problem will make ranching very tough, especially in areas like Washington and northern Idaho where there's a lot of timber for them to hide in. It's a little easier to hunt them in open areas like eastern Montana. However, it's impossible to keep wolves under control just by hunting them. Idaho's first wolf hunt was a perfect example. Everyone thought it would be great, and the Fish and Game Department sold thousands of tags but it was unable to fill the quotas. Lots of people were eager to shoot a wolf. The hunters bought tags thinking they could shoot a wolf when they were out hunting deer and elk. But they didn't get much opportunity because wolves are elusive. Now the novelty of hunting them has passed, and the wolves are still expanding. In Idaho, the Fish and Game Department can't begin to keep up with the natural population increase.

Cattle and wolves are never going to be compatible. Some people keep saying that wildlife people and cattlemen are going to have to learn to manage for wolves, but that's impossible. It's hard to "manage" when you are always on the losing end.

Ranchers have always had challenges, explained Len:

We've fought cougars, bears, and coyotes killing baby calves, but those problems were nothing compared to wolf depredation. And this has become a political thing. Our state Wildlife Commission says they've had 10,000 e-mails from people who want wolves and only a few from people who don't. They forget that we have a Constitutional right to defend our property. But the Wildlife Commission is swayed by the mob, and there goes our rights.

There were always plenty of wolves in Minnesota, Canada, and Alaska (at least 50,000 to 60,000 in Canada). They were never endangered. Now Alaska is in trouble because wolves have decimated wildlife in many areas. The state was going to start eradicating wolves with planes and helicopters. Then environmental groups and wolf advocates started a boycott on tourism, and Alaska backed down. The wolf is now the spotted owl for the ranching industry.

Conclusion

Even though the wolf was finally delisted, there are still political agendas at hand, and certain elements are fighting for continued protection. Some people believe the wolf issue is a ploy to eventually do away with hunting (since, theoretically, wolves can keep game numbers in check without human management) and then our guns.

Laws that supersede our other laws and rights, such as property rights and the right to protect one's family and animals, are dangerous. There is growing concern among ranchers and many Westerners that the wolf, like the spotted owl, is being used to put unacceptable restrictions on land use, both public and private. The Endangered Species Act (ESA) is a dangerous law. Few laws have been so divisive or driven by such emotional forces. The ESA is being used by preservationists and pseudo-environmentalists to lock up natural resources and halt logging, mining, grazing, and traditional use of private property.

Hero worship puts movie stars, sports, and music superstars in a league of their own, untarnished by their actual failings and foibles. Wolves have come to hold that same god-like status, despite the realities of what living with them is really like.

Historical and current evidence indicates
that one can co-exist with wolves
where such are severely limited in numbers
on an ongoing basis so that there is continually
a buffer of wild prey and livestock between wolves and humans,
with an ongoing removal of all wolves habituating to
people and domestic animals.
The current notion that wolves can be made to co-exist
with people in settled landscapes is not tenable.
Dr. Valerius Geist

Chapter 10

THE WOLF'S FINANCIAL LEGACY

By Rob Arnaud and Ted B. Lyon

Rob Arnaud is a fourth-generation Montanan, who grew up in Manhattan, Montana. His love of the outdoors and hunting led him to study animal science at Montana State University where he graduated with a B.S. in 1980. He has been the contracted wildlife manager and outfitter on several large ranches in the West and has served on a private wildlife board. He is a licensed outfitter in Montana, Wyoming, and California. In total, he provides trophy hunting opportunities on 16 different ranches. He is also the president-elect for the Montana Outfitters and Guides Association and is very active in lobbying for hunter rights and responsible wildlife management. Rob is fortunate to have hunted many of the North American animals.

Editors' Statement of the Issues

Since the Canadian wolf introduction, livestock losses have been substantial. Since only about only one in seven or eight losses leaves enough evidence to prove wolf predation, ranchers are facing crippling financial effects, as well as danger to their families.

Editors' Statement of the Facts

Wolf numbers must be reduced to ensure they represent no human threat. Wolf numbers must be reduced to allow reasonable safety for livestock. Wolf numbers must be reduced to allow wildlife populations to recover in sufficient numbers for public enjoyment. The undeniable fact is that wolf numbers must be reduced so that wolves are not the dominant figure in the environmental equation.

It has been estimated that there may be at least 100,000 wolves in North America today. And with them comes the corresponding and crippling costs that states with burgeoning wolf populations are experiencing.
Rob Arnaud & Ted B. Lyon

What Wolves Cost Ranchers

Coyotes, wolves, bears, and cougars are responsible for livestock losses nationwide totaling more than $60,000,000 annually.[1] In states where wolves are abundant, it's estimated that livestock predation by these animals accounts for at least half of all livestock losses and probably more.

Five states currently are experiencing huge livestock losses to wolves, according to the USDA's 2010 Cattle Loss Report. Considering that only one in seven or eight kills—depending upon the study—can be "positively" identified as a wolf-kill by state fish and wildlife personnel, the percentage of cattle killed in these states by wolves is much higher. Of the reported cattle losses due to predation, according to the 2010 USDA Cattle Loss Report, "confirmed" wolf kills account for the following percentages:

🐾 Minnesota   16.8%
🐾 Wyoming   18.6%
🐾 Idaho   30.0%

🐾 Montana   44.0%
🐾 Wisconsin   58.0%

The economic havoc that Canadian wolves have wrought since their release in 1995 in Yellowstone National Park—that already had native wolves in it—has been devastating. These Canadian wolves were set loose on American ranchers and farmers in the midst of millions of coyotes and a population of black bears numbering more than 100,000 in the lower 48 states to which ranchers, farmers, hunters, and the public had adjusted to decades ago.[2] Considering the rate at which the imported wolves are reproducing, their burgeoning numbers have triggered an economic tragedy for stockmen, businesses associated with hunting, America's world class big game herds, and state wildlife agencies alike.

Study Results

The magnitude of wolf losses facing ranchers caused by a rapidly multiplying wolf population is amplified in a study by Mark Collinge, who found that individual wolves in Idaho are approximately 170 times more likely to kill cattle than individual coyotes or black bears. He also found that individual wolves were about 21 times more likely to kill cattle than a mountain lion.[3]

Between 1987 and 2009, the Defenders of Wildlife, from a private fund, reimbursed ranchers in the Northern Rockies for approximately 4,000 "verified" wolf kills. Officials with the Montana Department of Livestock believe that, for every "verified" wolf kill, most of which must be verified by a DNA lab if any remains are found, another seven head of livestock are killed by wolves that go unreported. Often, scavengers feed on a carcass after wolves leave it. By the time a rancher finds the carcass, the only thing left is the animal's metal ear tag. Summers discovered similar numbers of real wolf kills when he found that 6.3 calves were likely killed for every one calf that was "confirmed."[4] So whatever number the government reimburses ranchers for, true losses are a times-six or times-seven factor.[5]

Apply this times-seven or times-eight factor to the "official" number of cattle and sheep wolf kills (shown below) from 1995 through 2009 in northwest Montana, the Greater Yellowstone Area, and the central Idaho "Wolf Recovery " areas, and you begin to see the enormity of the economic tragedy facing ranchers, America's prized ungulate herds, community businesses that rely annually on hunters and outdoorsmen, and state wildlife agencies that rely on license sales:

🐾 Sheep   2,842
🐾 Cattle   1,278
🐾 Guard Dogs   141

These numbers are occurring wherever wolves roam unchecked and are rising annually. According to the USDA's Wildlife Services, wolves have "officially" killed 4,440 head of cattle and calves throughout Idaho, Montana, and Wyoming.[6] Put a times-seven factor on that number. In Wyoming alone, wolves killed 222 head of livestock in 2009.[7]

Mass Killings

Wolves also engage in mass (surplus) killings of livestock. Wolves slaughtered 120 registered rams one night on a farm near Dillon, Montana.[8] Wanton slaughters like this occur wherever wolves are found. In July 2011 in France, one wolf killed 10 sheep and sent 62 of the panicked animals plunging off a cliff; 30 more from the same flock went missing in the woods.[9]

On August 17, 2013, the Siddoway Sheep Company suffered a substantial loss when the Pine Creek wolf pack attacked a band of sheep on Siddoway's summer allotment six miles south of Victor, Idaho. Idaho Wildlife Services confirmed the kill on August 18. The August 17 wolf attack resulted in the greatest loss of livestock to wolves ever recorded in a single incident in Idaho, surpassing an attack that resulted in 105 sheep killed in a single attack 10 years earlier. A total of 176 sheep—119 lambs and 57 ewes—were killed; many of the animals died from suffocation as some apparently fell in front of the rest, resulting in a large pile-up. Only the hindquarters of one lamb was eaten by the wolves.[10]

Wolf advocates assert that "these are isolated incidents that are blown out of proportion, resulting in needless fears." These are *not* isolated incidents. Mass killings of livestock by wolves are well documented and predictable.

Incurred Costs

Add to the cost of replacing livestock killed by wolves the loss of potential long-term payouts that the slaughtered animals would have produced, the cost of buying more guard dogs, the cost of hiring more men for herd security, and the cost of moving livestock to pastures near ranch dwellings and onto pastures where they must be fed absent their normal grazing range because of wolves, and stockmen—many of whom have ranched for generations—now find themselves saddled with impossible debt and operating costs threatening their livelihood. And none of the costly "deterrents" above stops wolves from killing their livestock.

Still another cost generated by wolves is *Neospora Caninum*, a parasite carried and spread by wolves that causes cattle to abort. Millions have been lost from damages to livestock triggered by this parasite.

Weight Loss

Even the mere presence of wolves stresses cattle and affects their calf production and weight. A study in Oregon found that livestock herds in areas where wolves were found are at least 10% lower in weight during fall shipping. This weight loss is associated with stress caused by wolves on their range and attacks on the herd.[11]

Corresponding with Will Graves, author of *Wolves in Russia: Anxiety Through the Ages*, Lori Butterfield in Joseph, Oregon, wrote:

> *... My brother and I were filming a wolf investigation on a dead cow when we caught on film how a herd of cattle acted after a wolf attack. I venture to say this behavior had never been filmed before.*
>
> *In another video we're working on, Denny, a neighboring rancher, explains how he can't use his stock dogs anymore because the cows get so agitated at the sight of them. My neighbor had to sell a mother cow that had always been fine until wolves attacked and killed her calf 500 feet from their house. The cow went crazy and wasn't safe to be around. Another time Denny was moving cattle from one pasture to the next—like he'd done for 30 years before wolves entered their area—when a grouse flushed near them, and the cows stampeded in the opposite direction.*
>
> *When wolves are around, cattle are continually stressed and on edge. The cattle industry has spent years breeding up gentle beef cattle that are easy to handle. When a top-tier predator like large Canadian wolves is set loose in their midst, they have no defense or chance against them. ... Casey Anderson from the OX Ranch in Idaho, who participates in*

a scientific study of cattle-wolf interactions, has said that lead cows usually will close with, and try to fight, cow dogs. When they do this with wolves, they're the first killed.

Stopping livestock losses to predators like wolves as well as the added expenses of trying to prevent such losses is financially critical to ranchers and farmers.

What Wolves Cost Farmers

Farmers' costs parallel ranchers' costs, the difference being that farmers grow livestock that are not free-roaming. Dairy cattle, pigs, chickens, turkeys, and ducks—all revenue producers for farmers—are easy prey for wolves. For example, Wisconsin's wolf population is growing, and the state's packs are expanding into adjoining states. Verified wolf depredations in Wisconsin from 1976 through 2005 were reported as follows:

🐾 Horses   5   killed (1 injured)
🐾 Sheep   50   killed
🐾 Dogs   99   killed
🐾 Cattle   184   killed (and 7 more injured)
🐾 Poultry   264   killed

Farmers in Wisconsin are more likely to raise dairy cattle in proximity to their barns, and 75 Wisconsin farms lost at least one cow or sheep to wolf depredation from 2001 to 2006. In 2010 alone, according to the Wisconsin Department of Natural Resources, wolf depredations to livestock occurred on 47 farms, surpassing the previous annual high of 32 farms in 2008. Similar patterns are developing in states where wolves are expanding. Following is the total number of livestock killed by wolves in Wisconsin in 2010:

🐾 Cattle   63   (plus 5 injured)
🐾 Sheep   6
🐾 Farm Deer   6

What Wolves Cost Local Businesses

The wolf slaughter of America's historically prized big game herds such as elk, moose, caribou, wild sheep, and deer has put and is putting outfitters out of business at an alarming rate. Hundreds of employees—both full and part time—have been laid off as outfitters close their operations, and hunters that traditionally frequented local motels, restaurants, bars, gas stations, sporting goods, and grocery stores no longer return to communities where this big game has disappeared.

This is a serious loss of revenues for states. The USFWS says the value of hunting for Montana alone is approximately $239 million a year. Studies in Alaska have shown that, when wolves and bear are managed intelligently, prized big game numbers increase and translate directly into more hunters and hunting revenue for the state.[12]

What Wolves Cost State Wildlife Agencies

The natural resources and the agricultural agencies of each state where wolves are found are responsible for making an annual management plan, conducting research, responding to depredation reports, removing individual wolves deemed dangerous, and compensating ranchers, farmers, and pet owners for "official" wolf kills. States with large wolf populations are reporting very serious losses of revenues from hunting license sales due to the disappearance of big game. Summaries of state-by-state losses to wolf depredations follow.

Montana. Montana Fish, Wildlife and Parks biologist Allen Schallenberger states that wolves kill 23 elk *per wolf* during the six winter months. Assuming that wolves will kill at about the same rate throughout the year, wolves kill an average of 46 elk *per wolf* per year. If there are 600 wolves in Montana, then 27,000 elk are being killed by wolves each year in Montana.[13]

Based on the Schallenberger study, Montana Senator Joe Balyeat, who is also a CPA, came up with a very simple method of calculating the total cost of lost wildlife due to wolves in Montana. In a 2010 presentation to the Montana Senate, Senator Balyeat calculated the cost of lost wildlife devoured by wolves in Montana to be $124 million a year. Balyeat placed the value of each elk at $4,500, which is based upon values established in Montana statutes for bull and cow elk restitution for illegal takes—$8,000 for bulls and $1,000 for cows for an average of $4,500 per elk. The 600 Montana wolf count is considered very low—there are perhaps as many as 1,500 wolves in Montana alone—but using the 600 wolf count x 46 elk per year x $4,500, the total value of elk killed by wolves each year in Montana is approximately $124.2 million.

Montana Fish, Wildlife and Parks estimates that it costs an additional $1 million a year just to manage wolves in the state.[14] This federally funded budget has had to be increased by 8% since 2005, while the Montana Fish, Wildlife and Parks' budget for all big game monitoring has declined by 15% since 2006. *Currently, the wolf program budget has become two-thirds the size of the entire state's big game program.*[15]

Balyeat points out that his wolf cost model assumes wolves eat only elk. If wolves consume bighorn sheep (with a $30,000 restitution fine for an illegal take) or moose (with a $6,000 fine), the value of Montana's lost wildlife would greatly increase. If wolves consumed only antlered deer or buffalo (with a $500 fine), it would decrease. Elk were used as the norm because the number killed by wolves is best documented.[16]

Idaho. Idaho's Wolf Conservation and Management Program is expected to cost about $1 million annually.[17] Taking the Idaho-Montana-Wyoming area low estimate of 1,500 wolves, wildlife losses caused by wolves per year would be 1,500 wolves x 46 elk per year x $4,500, or $310.5 million annually. If you took the higher wolf estimate of, say, 2,500 wolves for this tri-state area, the cost of wildlife losses would be $517.5 million—or right at a half billion dollars annually. These losses, based on restitution fines for illegal takes, do not include wolves outside the tri-state area. Nor do they include lost revenue traditionally generated by hunters for outfitters, motels, restaurants, bars, gas stations, and sporting goods and grocery stores or lost state revenue from a drastic decline in hunting license sales in recent years when hunters ceased frequenting areas void of big game.

Idaho Senator Gary Schroder requested Idaho Department of Fish and Game to conduct a study of what wolves have cost Idaho in hunting revenue. The report put an economic value of each elk taken legally by hunters at about $8,000, including direct and indirect benefits generated by hunter revenue. The 1,903 animals taken legally by hunters represented an economic loss of approximately $15 million annually when taken by wolves.[18]

Another 2009 economic analysis of wolf management indicates that Idaho could be losing up to $24 million annually in hunting revenue due to wolf depredations of wildlife herds.[19]

Wyoming. In 2004, Wyoming Game and Fish Department spent just under $119,000 to manage the gray wolf in Wyoming, *even though the state did not have "jurisdiction" to manage wolves.* Game and Fish directors originally estimated that wolf management costs for Wyoming would approach $1,000 per wolf annually after delisting.[20] In 2007, Wyoming officials revised their estimate, stating it would cost more than $2.4 million annually to manage the state's wolf and grizzly programs during the year after delisting and $2 million for each year thereafter.[21]

Oregon. Oregon has a growing wolf population in the northeastern part of the state. The state wolf management plan projects that Oregon ultimately will have 14 pairs of wolves and a total population of 207 wolves. The 2009 Oregon Wolf Conservation Management Plan estimated that these 207 wolves would translate into 4,844 deer and 1,615 elk killed annually. The total loss in hunting benefits (license fees and all other expenditures by hunters) to the state from 207 wolves was estimated at more than $2.3 million per year.[22]

Wisconsin. The Wisconsin Department of Natural Resources has paid more than $1 million in reimbursements to residents who have had livestock and pets killed by wolves.[23]

Conclusion

Wolf populations are expanding or being "reintroduced" into other states such as Washington, Utah, Colorado, Arizona, New Mexico (with spillover into western Texas), Michigan, New England, and Minnesota, where an estimated 3,000 wolves now reside. It has been estimated that there may be at least 100,000 wolves in North America today. And with them come the corresponding and crippling costs that states with burgeoning wolf populations are experiencing.

🐾 🐾 🐾 🐾 🐾 🐾 🐾 🐾

[1] http://news.yahoo.com/ravenous-wolves-colonise-france-terrorise-shepherds-064719014.html
[2] http://digitalcommons.unl.edu/cgi/viewcontent.cgi?article=1015&context=wolfrecovery
[3] Lehmkuhler, J., G. Palmquist, D. Rud, B. Willgrey, and A. Wyderem, "Effects of Wolves on Farms in Wisconsin Beyond Verified Depredation," Wisconsin Wolf Science Committee, WI Department of Natural Resources, Madison, WI.
[4] http://www.mnforsustain.org/wolf_mech_wolf_recovery_costs.htm
[5] See chapter by Laura Schneberger.
[6] http://www.dfw.state.or.us/Wolves/management_plan.asp, page 101.
[7] http://dnr.wi.gov/org/land/er/mammals/wolf/pdfs/wolf_damage_payments_2010.pdf
[8] http://westinstenv.org/wildpeop/2008/11/17/montana-fwpd-wolf-management-fiasco/
[9] Personal communication from Senator Balyeat, 2/13/2011.
[10] http://www.tetonvalleynews.net/news/wolves-cause-death-of-sheep-near-fogg-hill-forest-service/article_f79754fc-08eb-11e3-82d9-001a4bcf887a.html
[11] http://www.capitalpress.com/content/ml-wolf-plan-sidebar-2-100810
[12] http://www.nap.edu/openbook.php?record_id=5791&page=1

13 http://www.journalnet.com/news/local/article_639aacda-1232-11df-87ef-001cc4c03286.html

14 http://westinstenv.org/wp-content/ID%20WS%20FY%202009%20Wolf%20Report.pdf

15 http://westinstenv.org/wildpeop/2009/02/09/the-high-costs-of-wolves/

16 http://www.newwest.net/topic/article/wyoming_lawyer_environmental_groups_using_taxpayermoney_for_legal_fees/C37/L37/

17 http://fwp.mt.gov/wildthings/wolf/wolfQandA.html

18 http://westinstenv.org/wildpeop/2009/02/09/the-high-costs-of-wolves/

19 Northern Rocky Mountain Wolf Recovery Program 2011 Inter-Agency Annual Report.

20 http://archive.sba.gov/advo/laws/sum_eaja.html;
http://www.boone-crockett.org/images/editor/ND_EAJA.pdf

21 http://westinstenv.org/wildpeop/2009/02/09/the-high-costs-of-wolves/

22 http://www.nrahunterrights.org/blog/Default.aspx?id=482

Scott E. Read / Shutterstock.com

Chapter 11

MEXICAN WOLF IMPACTS ON RURAL CITIZENS

By Laura Schneberger

Laura Schneberger ranches with her husband and children in southwestern New Mexico on a historic cattle ranch located in the Black Range of the Gila Forest. She spends as much time as possible educating the public and writing articles on the realities of wolf reintroductions and their impacts on families and small communities that depend on livestock production for an economic base. Living in the middle of wolf and mountain lion territory has made her a much sought-after expert of wildlife activity in the Southwest.

Editors' Statement of the Issues

A large part of the Mexican wolf recovery project hinges on livestock instead of wildlife as the preferred prey. Low wildlife populations and the increased willingness to habituate has changed wolf behaviors to pose frequent and systematic risks to humans.

Editors' Statement of the Facts

There are no possible justifications that the Mexican wolf recovery project should continue on its current course. Well-documented wolf/ human encounters establish the dangers inherent to increasing or maintaining the Mexican wolf population, and must be the basis for regarding it for what it is—a large and vicious killer whose presence is not to be allowed in vicinities with humans and livestock.

It is disturbing to see that there is a consistent trend from government agency personnel working within the program to promote the extremist notion that wolves on the ground are genetically special and cannot be removed or controlled. This is contrary to all scientific and policy documents available.

Laura Schneberger

Mexican Wolf Recovery Project

The Mexican wolf recovery project did not begin with the passage of the Endangered Species Act in 1973. Nor did it happen in 1976 when the gray wolf was listed as an endangered species. It did not take hold until 1979 when the U.S. Fish and Wildlife Service (USFWS) approved a recovery team to assist the agency in mapping out a potential recovery strategy for the Mexican wolf, the smallest and rarest subspecies of gray wolf in North America. The USFWS approved the subsequent Mexican Gray Wolf Recovery Plan in 1982. This plan called for captive breeding and establishment through reintroduction of two viable wild populations of the Mexican wolf in the Blue Range Wolf Recovery Area—a 6,800-square-mile area that encompasses the Apache National Forest in Arizona, the Gila National Forest in New Mexico, and 2,400 square miles managed by the White Mountain Apache Indian tribe.

Unlike other predator recovery plans, this one failed to contain numerical criteria for recovery and delisting of the Mexican wolf, and when the actual reintroduction project was on the ground, it was put on the back burner. The majority of biologists believed that success would not be feasible with the genetic limitations that existed due to the small number of Mexican wolves, none of which were in the wild. A determination was made that this wolf would never be recovered enough for delisting.

Lack of action on the Mexican Wolf Recovery Plan by USFWS provoked litigation by an environmental organization, then called the Southwest Center for Biological Diversity, to force immediate implementation of the recovery plan. This lawsuit resulted in a settlement between the plaintiffs and the USFWS, with undisclosed conditions and parameters. Although the USFWS agreed to the settlement in federal court, neither the public nor the state agencies has ever seen the terms of that settlement. To this day, USFWS claims there was a court order to proceed with the program, though clearly there was not. Somewhere under those hidden terms, "no legal win, no loss," were the words, "agreed to attempt reintroduction of the Mexican wolf, *Canis lupus baileyi.*"

The current Mexican wolf program, based on the 1982 agreement, is woefully out of date. It is not scientifically current and lacks even the basic backbone of most recovery plans—delisting criteria that would eventually remove the Mexican wolf from Endangered Species Act protection. By 1996, a proposed experimental rule was written, and by 1998, the Final Environmental Impact Statement (FEIS) was published directing the program's beginning. In 1998, designation of a "Nonessential Experimental Population Rule"—with the required section 10(j) special rule on managing the reintroduced population of large predators allowing lethal control and authorized take—was written. On March 29, 1998, the first 11 wolves were released into the Blue Range Wolf Recovery Area.

Wolf Lineage

At the time of the 1982 agreement, the captive breeding program consisted of three different lineages: McBride, Aragon, and Ghost Ranch. The only certified animals were the five Mexican wolves captured in Mexico in the late 1970s and early 1980s by Roy T. McBride. McBride, a trapper who often contracted for the federal government, spent years in Mexico hunting out the last of the wild packs of Mexican wolves. He came back with a single female and four males, all from the same pack. McBride was in a unique position for this project with both expertise in trapping wolves and a Master's degree in Biology.

During the initial planning for captive breeding of the lineages, it was determined that only the lineage captured by Roy McBride in Mexico was certifiable as "pure" Mexican wolf.[1] In 1997, controversy arose when a captive pack at Carlsbad Caverns National Park designated for release was found by McBride to be largely composed of wolf-dog hybrids.

Zoo employees initially argued that the animals' odd appearance was due to several generations of captivity and diet, but diet and captivity alone could not account for blue eyes (that wolves do *not* have) or curly tails. Scientists involved in the captive breeding program and the recovery team decided to euthanize the line. Unfortunately, there remained a few animals from the Ghost Ranch lineage in private collections around the Southwest.

With only one purebred female, the USFWS began seeing inbreeding repression in the McBride lineage. Some of her male offspring showed signs of chryptorchidism (un-descended testicles) and were not fit for breeding, although crossings with her sons and brother did supply more females for further breeding. The program seemed dead in the water, but in the mid-1990s, David Parsons, the Mexican Wolf Recovery Coordinator, made the startling decision to include both the Ghost Ranch private collection and the Aragon lineages in the captive breeding program in an attempt to improve the genetic diversity of the animals. McBride, who was livid at the idea of including what were considered by all experts involved to be wolf-dog hybrids, challenged Parsons on this decision. On June 2, 1997, just six months before the first captive wolf release, McBride sent a letter outlining his concerns about the decision.

Dear Mr. Parsons: In reading the recent status report of April 1997, I was shocked to see that the wolves from the Ghost Ranch lineage were being included in the captive breeding program. In the early day of the Mexican Wolf Recovery, the origin and genetics of the Ghost Ranch animals were discussed and investigated ad nauseam. *In fact, the conclusion by all members of the early recovery team was that the animals were wolf-dog hybrids. This was the primary factor behind the decision to seek and capture the remaining wild population, because it was the only pure genetic stock available.*[2]

McBride's letter goes on to discuss the genetic and legal ramifications of the inclusion at length, citing the flawed science behind the decision as "right out of the twilight zone." He questioned whether the ESA would even protect the animals that would be created by USFWS if the Ghost Ranch lineage were included in the breeding program. He was adamant that the Ghost Ranch and Aragon lineages were cosmetically, genetically, and completely different from the original McBride wolves captured in Mexico.

Twilight zone, indeed! When geneticists at the University of California-Davis confirmed that the so-called "Chupacabra dog" coyotes found in Texas were related to the Mexican wolf, everyone laughed. But neither McBride nor Parsons could have known of their existence when the original argument erupted over Ghost Ranch. In 2009, DNA tests on a startlingly ugly, blue-eyed coyote-cross-dog—jokingly known as the Chupacabra in south Texas—proved conclusively that this anomaly was genetically related to DNA on file in the Mexican wolf program.[3] Several specimens of this hairless blue dog with blue eyes are available, and the speculation is that, although there is yet no proof, they are likely related to the dog that is one of the unknown founders of the Ghost Ranch lineage. The odd-looking coyote hybrids show marked physical similarities to the Mexican hairless dog or "Xolo," an ancient breed known to have occupied South America for more than 3,000 years. Ghost Ranch "wolves" also had some specimens with blue eyes. However, there has been no investigation into the relationship between either the Texas blue hairless (Chupacabra) or the Xolo (also called Chupacabra) or the different lineages of what are now called *Canis lupus baileyi* (Mexican wolf) as of yet.

McBride's concerns were ignored, and in 1997, he ceased working with USFWS on wolf-related issues. Unfortunately, none of the original specimens of either Ghost Ranch or Aragon exists for modern DNA testing, and the original records of the recovery team that McBride was a part of have disappeared. Instead, USFWS set up a committee to review the Ghost Ranch and Aragon lineages and to rubber stamp their approval, although even that team also appeared to have serious misgivings. McBride writes:

As of December 7, 1995, Mr. Parsons had yet to view any living descendent of Ghost Ranch lineage firsthand. Nevertheless, and without examining any of these animals firsthand, Mr. Parsons pronounced them fit, to be purely

"Mexican wolf," and to be appropriate for inclusion with the certified lineage in July of 1995. Mr. Parsons' determination was based entirely on the results of Hedrick's and Wayne's yet unpublished 1995 studies, which remain not generally available for public review as of January 1996.[4]

One of the first lawsuits against the implementation of the Mexican wolf program was based on the inclusion of hybrid wolf-dogs into the gene pool. This argument was swiftly lost in federal court, and the historical samples that may have proved Ghost Ranch and Aragon founders were dogs or wolf-dog crosses were never found again.

The DNA baseline for the Mexican wolf is now from the three combined certifiable lineages. It must be explained that the only way to determine wolf DNA as different from dog DNA is to have family markers. Mexican wolves are purebred because they are all related to wolves in the captive breeding pool, *not* because they have distinctly different genes than dogs. In other words, it did not matter that they were hybrids, as long as the USFWS claimed they were not and historic records were missing; then the program could go forward as if they were pure wolves.

The unpublished study that Parsons used to include non-wolf lineages was eventually published, but the committee still had misgivings about the methodology. The study never even used coyotes from Mexico or the Southwest for their comparisons with the questionable lineages, but it does admit:

Wayne et al. (1995, p. 6) cannot eliminate the possibility that the Ghost Ranch lineage originated from other North American grey wolves or a dog whose offspring had backcrossed to wild wolves for several generations. And Wayne et al. (1995, p. 6) cannot eliminate the possibility that the Aragon lineage originated from other North American grey wolves or a dog whose offspring had backcrossed to wild wolves for several generations.[5]

At the time and probably still today, there is no other person alive who knows more about wild Mexican wolves than Roy McBride, yet the Service ignored his plea not to combine these wolves with the McBride lineage because the genetics had been fouled and the likelihood of problems such as a propensity for livestock predation would likely be substantial.

These facts may very well explain why USFWS has experienced higher than expected livestock predation, human habituation issues, and the presence of hybrid pups on at least three different occasions in the Mexican wolf reintroduction area.

The entire issue of the existence of the Mexican wolf needs to be reviewed by independent scientists and geneticists. Many of the un-wolf-like behaviors that have been exhibited by this population is, in part, due to the fact that this court-approved, rubber-stamped population is not pure Mexican wolf and that their dog genes have likely impaired their ability to function, as did historic Mexican wolf populations. This leads to the question as to whether these wolves should even be considered an endangered species.

Modern Mexican Wolf vs Modern Rancher

The Mexican wolf reintroduction program launch was also the beginning of almost two decades' worth of struggle for equality and justice for ranching operations and small rural communities in the recovery area. The goal of 100 wolves originally listed as the initial goal for the Blue Range Wolf Recovery area is clearly an over-estimate. In over a decade, there are 58 wolves in the wild in spring 2012 according to USFWS.[6]

Regardless of the number of Mexican wolves, a substantial number of those now on the ground have left the wilderness recovery area—preferring to eat livestock, pets, and garbage and to frequent homes and campgrounds. This may seem to lead anyone not connected to the relocation program hierarchy to believe that wild prey populations are not sustaining the food needs of the current population, particularly in the wilderness areas.

Logically, this leads to the belief that, at any one time prior to the turn of the century, there were never 100 Mexican grey wolves in southwest New Mexico and Arizona combined. Historical accounts by early explorers such as Emory and Carson and by various Spanish expeditions rarely mention encounters or even sightings of wolves. Further, the same early expeditions by Europeans found the area so lacking in game species that the threat of starvation loomed almost constantly. Kit Carson, who led several expeditions through the area, survived by eating horses he traded out of the local Indian populations. The lack of game and the few documented sightings of wolves by these folks clearly show that the Mexican wolf was, at best, rare in the area and probably rare throughout its entire range.

Where did the USFWS get its population goal estimates? By the turn of the 20th century, Mexican wolves were relatively common, likely because a new prey animal had been introduced to the area. Livestock, in the form of both sheep and cattle, had been steadily increasing in country set aside for ranching that is now considered the Blue Range Wolf Recovery Area. The ranching businesses contributed meat needed by the mining towns springing up all over the region, but that same meat led to an increase in breeding capability and larger litters from the few wolves that were historically documented on that area. The same situation occurred in California at the turn of the century when livestock became prevalent in the territory, and the grizzly population exploded with the expanded food supply.

With increased nourishment, the Mexican wolf populations did increase significantly as did increased wolf / livestock conflict. Early governments authorized the PARC Service (early USFWS) to begin the removal of the Mexican wolf. Although several dozen wolves were indeed killed from 1900 to the 1970s, this number was the culmination of 70 years and should not have been used as a population goal, as it in no way represented a natural ecological number to estimate historic population levels.

Note: The gray wolf recovery goal set for Montana, Wyoming, and Idaho was 100 wolves and 10 breeding pairs in each state. Wolves prey on large ungulates (hoofed mammals) and snowshoe hares. When they are available, elk are preferred food for wolves. According to the Rocky Mountain Elk Foundation, New Mexico has a total elk population of 80,000 (over half of which are found several hundred miles of desert north of the recovery area), and Arizona had 17,500 elk in 2009. Wild elk populations in western New Mexico and Arizona are not large enough to sustain 100 wolves, if the wolves would only prey on elk.[7]

Ranch and Rural Home Impacts

The expanding wolf distribution has caused an increase in wolf–human encounters and generated concerns among wolf managers and conservationists... This may sound like a tabloid headline, but the attacks were well documented by wolf authorities. Several factors may have led to the attacks including a lack of available wild prey, domestic livestock that were well protected, and many small children playing in the vicinity of the wolves. The common factor among nearly all reported wolf attacks was that wolves had become increasingly bold around humans (perhaps because of food scarcity, or possibly as a new strategy to exploit resources brought by humans into wilderness areas). North American wolves involved in recent attacks were repeatedly seen stealing articles of clothing and gear, exploring campsites, and sometimes obtaining food items—behaviors nearly identical to those reported by early frontiersmen.[8]

The current DNA mix of what are now known as "Mexican wolves" supports all of the above—so much so that many adults and children residing in the Blue Range Wolf Recovery Area (BRWRA) have been seriously impacted by wolf encounters.

Often the general public, particularly those that do not live in the Southwest, are unaware that, while the BRWRA is a large area, there are hundreds of private land in-holdings in it where people live, work, and recreate in what the USFWS shows as a vacant, empty, green spot on a map; 66% of the BRWRA is open to raising cattle, mining, and forestry. All of these private in-holdings are over 100 years old, and a large percentage of them are working family livestock operations, hardly good habitat to be releasing and building up a large population of predators.

Following are some stories of what it has been like to live with Mexican wolves to illustrate the realities that non-residents seldom ever experience.

🐾 In 2007, Mary Miller and her husband Mark were forced to witness their eight-year-old daughter Stacy running from a wolf attack on a family dog, an attack that occurred immediately adjacent to the child and likely was an attempt by the dog to protect the child. The dog recovered with extensive veterinary treatment. Two months later, the same wolf killed Stacy's black horse "Six" in his corral while the family was away from home on a grocery trip.

🐾 In 2005, Carlie Gatlin suffered a concussion in an auto wreck and was forced to walk home from a wrecked vehicle with two small children. Her son was bleeding from a head injury, and her daughter was small enough that she had to be carried. Her husband, after having called her destination and was told she was not there, drove over and picked up Carlie and the kids. When he took Carlie to the hospital for treatment the next morning, the tracks of the Luna pack overlapped hers and her son's in the snow beside the road. The Gatlin children have since suffered through several incidents of maiming and killing of their family dogs by wolves.

🐾 Ivy Schneberger, 13, was riding her mare a half mile from home when the Sycamore pack of two animals attempted to corner her. She fired off a round from her single shot .22, and they finally meandered away, choosing instead to kill the family's calves for several weeks before the first legally killed Mexican wolf, AF 1155, was shot on that ranch. This is one incident where the wolf was considered a threat to human safety. One story circulated that, a few days prior to the Schneberger incident, wolves had threatened one of the program volunteers near their wilderness release site some 45 miles to the southwest. But when the annual report came out, both incidents were downplayed, and the wolf was categorized as having been shot for livestock kills, not human safety and removal problems.

🐾 AF 1155 was the same wolf that the Schneberger family had dealt with three years previously during the winter of 2000 when her mate left her to starve at their home and agency personnel chose to allow her to stay there from December until April. While living on the Schneberger ranch, the wolf subsisted on dog feces from their kennels and occasionally scavenged off coyote kills in the area. After she had been at the ranch headquarters off and on for three months, the family's milk cow died on a tributary leading into the home, which provided an excuse for the agency to justify to the public about the behavior of the obviously habituated wolf that fed on the dead cow.

🐾 J.C. Nelson was 14 years old when the Luna pack, five wolves, crowded him against a tree and circled him for 15 minutes. J.C. made no move to run, instead choosing to find as much cover as possible in an area damaged by wildfire. He was only able to back against a charred Ponderosa pine. The pack apparently smelled his rifle and, having better things to do, eventually left. This incident was reported to USFWS by Catron County law enforcement. USFWS investigated by sending volunteers to try and entice the pack to copy the behavior. This did not occur, and the USFWS chose to deem the incident unimportant. J.C. said he was afraid to shoot the wolves because he thought his father would lose his grazing allotment if he did.

Proximity

Proximity is a problem when dealing with Mexican wolf presence. With the onset of the 2012 breeding season and the rise in the wolves' population numbers, sightings, close encounters, and home encounters once again created a difficult situation for managers of the Mexican wolf program, and they did not get much slack from local governments and citizens.

In December 2011, the program issued its first lethal control order after a female wolf with a long track record of livestock depredations and human habitation was found circling a private home at regular intervals where small children were exposed to her close presence. The same wolf had birthed a litter of hybrid Labrador pups the prior spring, and USFWS are still on the lookout for the one Mexican wolf-dog hybrid that got away. They have not found it. Presumably, it will add to the genetic mix that is the "rare Mexican wolf." The remarkable thing about this control action is the fact that—despite dozens of human safety encounters since the beginning of the program, many of which involved Mexican wolves' attraction to children—this was the first time the agency admitted that lethal control was warranted for human safety reasons.

The encounters with wolves did not end after AF 1105 was removed from the picture, nor did the controversy over the necessity to remove her. Despite claims from radical environmental organizations that the wolf was merely lonely and only needed to find a male and that there just were not enough male wolves in the wild for her to mate with, the next three encounters at homes and highways in the same area were with male wolves looking for a mate.

Numerous photos show that these animals were clearly very close to people. Also, they were in the area where pairing with AF 1105 was possible and feasible. Instead, the wolves appear more interested in easy prey or a handout at a home than pairing with a female that was making itself readily available. It makes a reasonable person wonder what these big males are breeding with since they showed little interest in AF 1105. Possibly coyotes? It is possible as the canine DNA is nearly identical for all species with the exception of family marker identification.

It is disturbing to see that there is a consistent trend from government agency personnel working within the program to promote the extremist notion that wolves on the ground are genetically special and cannot be removed or controlled. This is contrary to all scientific and policy documents available. This biased advocacy is more evidence that not only are there close ties between the program and the most extreme environmental advocates for wolves in the Southwest, but also USFWS people appear to be coordinating media and press release strategies with those same organizations.

The behavior of the wolf population that the federal agencies are in charge of (as well as the radical wolf advocates) is such that local governments in New Mexico and Arizona counties that contain wolves are examining their options to protect human safety in events such as those that occurred at Beaverhead and along Highway 59 and other nearby rural homes and communities. Human health and safety are two things that the agencies themselves are supposed to uphold over any policy that they have in place including promoting increases in the wolf population. With the large number of incidents involving children in the BRWRA, the counties feel they must be ready to step in and do the job the federal government is somewhat lackadaisical about doing. We do not know what will happen if Catron County, New Mexico, kills a wolf in a constituent's yard, but the majority of the small human population in the county are supportive, desperately so, of the idea.

Middle Fork Wolf Pack

Megan Richardson, who lives with the habituated livestock-killing Middle Fork pack coming to her home at regular intervals, has to wonder if the noises her small baby son makes draws the wolves even closer. The wolves can be seen coming into her driveway in photos from game cameras she has installed. She puts it bluntly: "Is it going to take someone getting seriously hurt or killed before something is done?" Megan deals with the situation by keeping game cameras out in the yard to document the animals' frequent presence on her land.

The Middle Fork wolf pack has been implicated in dozens of unconfirmed livestock deaths, and ranchers are frustrated with the uncontrolled killing. The Middle Fork pack was the first and perhaps only wild-born Mexican wolves to pair and raise successful litters on their own. For the first four years of their existence, there is no record of them killing cattle, and they stayed in the wilderness.

In 2007, FWS discovered the female had a broken leg that had healed poorly and was cosmetically unattractive, although not a deterrent to her hunting ability. They trapped her, took her into captivity, and amputated the leg. She was in captivity for over a month. Her pack, perhaps sensing that she was alive, made wider and wider circles out of their normal territory looking for her. By the time the female

Wolf-proof school bus shelter.

Young boy in Catron County waits for the school bus in a wolf-proof shelter.

was placed back into the wild, the wolves had discovered a livestock operation miles to the north. Later that year, they began killing calves. Sometime later, the agency also chose to amputate the leg of the alpha male that had been damaged in a trap likely belonging to the same personnel that had trapped the female. The Middle Fork pack was never good again, and USFWS once again contributed habituation behavior of wild-born Mexican wolves that had previously displayed normal behavior patterns.

In spite of a massive hazing effort that took place on the Adobe and Slash Ranches in 2009 in southwest New Mexico, the Middle Fork Mexican wolf pack was allowed to kill and maim livestock as if USFWS owned the cattle themselves. Early that spring, the pack was suspected to have killed several yearlings, but none were found early enough to get a confirmation on the kills. Ranch manager Gene Whetten grew more frustrated by the day.

> *We work our tails off managing this place for elk and livestock, and we do a good job here. However, when you read the papers, you see the implication that something is managed badly on the ranch and that is why there are wolves here killing our stock. We have even read where someone makes the claim that there were 16 livestock carcasses that we supposedly left lying around the den. I guess they meant deliberately. We did not even know where the den was until the first confirmed kill happened. If there were yearling kills at the den, it was wolves that dragged them there after they ran the elk out of the country.*

Whetten is referring to the tendency of the USFWS to release information to the environmental organizations on wolf incidents, allowing them to filter news releases regarding the program, particularly livestock conflicts. The agencies hang back on their own press releases. When caught between the environmental extremists' false accusations against the rancher and the facts, program spokespeople simply refuse to correct deliberate misinformation they know is not true. This puts them in a good position to take as much from the rancher as they please and do nothing they are required under the rules and under the law to do.

Adobe Ranch Case Study

From 2000 to 2003, the Adobe Ranch knew of only two wolves on the ranch. In 2004, the number of wolves increased to nine; in 2006, the total dropped to six. By the fall of 2007, 14 wolves (three packs) were known to be on the ranch.[9] Wolves were also in close proximity to the ranch headquarters and branding pasture beginning in February.

This level of wolf activity led to eight confirmed cases of livestock depredation and one probable case. Total depredations for 2007 included 13 confirmed animals, 1 probable animal, and 4 possible animals on the Adobe Ranch. The Adobe Ranch alone accounted for 46% of the total confirmed depredations reported to the Bailey Wildlife Foundation Wolf Compensation Trust in New Mexico for 2007. In addition, 50% of the possible depredations and 100% of the probable depredations for 2007 occurred on this ranch. Although depredations were confirmed, no compensations were paid by the BWFWC in 2007 and 2008.[10]

The effect of wolves on livestock, however, goes much farther than kills. Stress placed on cattle and sheep by simply the presence of wolves results in weight loss, which translates into income lost. Weaning weights, shipping weights, and site-specific precipitation were available for the Adobe Ranch from 2002 to 2007. Growing season precipitation was correlated to steer performance, as forage production is closely related to growing season precipitation. Cumulative precipitation from April through October was considered growing season precipitation. There was little variation in steer Average Daily Gain (ADG) from 2002 through 2006, with an average of 0.08 pounds per day, which is considered normal performance in the region when fall weaned calves are retained.[11] However, ADG in 2007 was much lower than in previous years when calves were managed similarly between weaning and shipping, falling well below the lower limit of a 99% confidence interval of -0.75 pounds per day. Using actual market values from the Clovis Livestock Auction in Clovis, New Mexico, the cost of the estimated impact of weight loss in 2007 was -$108.83 per steer weaned.[12]

Overall, impacts of wolves on ranches is not only significant, but is also destroying ranch viability. (See Chapter 16: Collateral Damage by Jess Carey.) Actual data on current losses is virtually ignored by USFWS.

Prey Preference

A 2006 study using GPS collars to analyze the prey preference of Mexican wolves showed the agencies' bias and flawed scientific techniques. In a GPS collar study to determine the preferred prey of Mexican wolves, the Nantac pack male was implicated in 13 livestock kills in 2006 over several weeks. The data gathered during the livestock kills was not used in the study. Instead, the study reflected that the majority of Mexican wolves killed elk. This occurred simply because the Nantac male with a GPS collar was shot for livestock depredation because the agency could not stop him any other way. The pack's kills were not included in the study simply because the agency did not want it to reflect a livestock surplus killing spree, which they consider abnormal wolf behavior. In one 24-hour period, the two Nantac wolves killed two cows and their calves and ate nothing.

More importantly, the USFWS needed the study to reflect elk as the major food source in order to ask for continued funding for the program and to continue federal planning efforts in the scoping process for an eventual rule change. Proving that wolves surplus-killed privately-owned livestock was simply not on their agenda. Had the Nantac pack's livestock kills been left in the study, USFWS would have had more accurate science.

Carcass Removal

The Mexican wolf management program, from the beginning, has been complicated by environmentalists. The Center for Biological Diversity (CBD), which originally sued to push the recovery forward after biologists had voted against it, demanded that ranchers remove dead livestock or have no wolf damage control occur on their allotments or land. Ranchers balked at this demand in part due to costs, logistics, and the fact that most allotments have areas that are completely inaccessible for such maneuverings. The ground is always frozen in the winter, leaving them with no way to bury carcasses. This demand is simply unattainable.

Neither the experts nor the science agrees with the CBD. For example, the University of Minnesota conducted a study in early 1999 to determine if any livestock management practices could prevent wolf depredation. The study could find no management practices certain to prevent wolf depredation. The only method proven to prevent wolf depredation was removing the depredating wolves from the farm.[13]

Wildlife biologist, Ed Bangs, who until 2012 was the Wolf Recovery Coordinator for USFWS and the principle author of the

1993 Environmental Impact Statement for the wolf recovery program in the Northern Rockies, wrote in a 2006 e-mail that he did not believe that removal of carcasses would make a difference in wolf behavior:

> *I thought the idea that wolves eat a dead cow, think beef tastes great, and then start attacking cattle is mythology—as eating carrion and killing prey are two totally different wolf behaviors. Wolves often scavenge all they can. However, I do think that having bone piles next to calving pastures can increase potential for conflict by attracting wolf activity in the vicinity of livestock [research shows cattle near wolf dens are more likely to be the ones attacked simply because the level of wolf activity and interacting with livestock are highest there] and that anything that helps wolves become more familiar with livestock can increase the chance they might test them as prey. But normal range practice out here makes it nearly impossible to find and bury [or blow up for human safety concerns as they do for grizzly bear issues and livestock carcasses along trails] every carcass, so if livestock carcass disposal is within "normal" and traditional livestock husbandry practices, we don't consider them an attractant that we would withhold wolf removal. We do advise ranchers not to have a bone yard next to livestock, and we have removed carcasses in pastures to prevent wolves from coming back into concentrated livestock to feed on their kills. But feeding on livestock carcasses is a very different thing than attacking livestock—one doesn't necessarily lead to the other.[14]*

In spite of the lack of scientific evidence or publications that confirm livestock carcass removal will benefit the program, it is still being pursued by Region 2 USFWS as an option. In addition, there is the notion that the mere presence of livestock on ranches more than a century old is impacting the success of the wolf recovery program.

Recent changes in the program seem to have all been aimed at the cessation of mitigation for problem animals. The only exception in the past five years has been the removal of AF 1105 for human safety reasons. Defenders of Wildlife has stopped reimbursing the immediate cost of ranchers' confirmed livestock kills as they also did in Wyoming when bears became established and the organization felt ranching should become a thing of the past in those areas. New

Mexico is far too poor to appropriate state funds to reimburse ranchers for the take of their livestock by wolves. It is hard not to conclude that the promises outlined in the 1996 EIS and final rule have simply been discarded when it became apparent that wolves or wolf hybrids will wipe out ranchers and ranches if left unmanaged long enough. Let us look at some specifics.

Conflict of Interest, Collusion, and Corruption

Complaint: Wolf team "full cooperators" do not include local organizations or local people and do not efficiently mitigate problems arising from the program. (Note: Technically, "full cooperator" means any government agency, federal or state, that is involved in planning and/or implementing a program such as the Mexican Wolf Recovery Program. Unofficially, it also means any environmental group or possibly even an agricultural group with a bank account and/or team of lawyers that sits on a recovery and compensation team. Individuals—even those who have a vested interest in managing wolves and/ or in being compensated for take by wolves—are not considered "cooperators" and must participate through their local, county, or state representative.) Instead, non-government organizations (NGOs) such as the Turner Endangered Species Fund (TESF) and Defenders of Wildlife (DOW) enjoy full cooperator status. This status is unofficial, but the results of their activities in the program speak for themselves. It has also become apparent that other environmental groups funded by TESF have internal knowledge of the day-to-day operations and internal planning of the wolf reintroduction and recovery. The draft of the program's five-year review was released to environmental groups before being made available to the public or local government.

Dave Parsons—former program coordinator, author of the original fatally flawed EIS, and the person behind the introduction of the Ghost Ranch hybrid genetics—is now the coordinator of the Southern Rockies Wolf Restoration Project, another TESF grantee. Involvement and support of the ideology of these organizations are, in part, the reasons why the original EIS projections have never been met. It was apparent from the beginning of the program that Parsons had exerted entirely too much influence and was promoting a plan that would eventually be detrimental to the livestock industry. It was also apparent that his team was determined to ignore socio-economic comments, barely referring to those issues or arbitrarily discarding the notion that the program would harm people. In fact, USFWS claimed in the EIS that there was not going to be significant harm done to the region, although they had the data from historical Mexican wolf damage records, such as the following:

The estimate of economic damage in New Mexico caused by 40 to 50 wolves in 1918 was $60,000—equivalent to about $960,000 in 2007 dollars.[15] From 1915 to 1920, wolf-induced economic losses were estimated at half a million dollars—comparable to $9.4 million in 2007 dollars.[16] In a 1921 U.S. Department of Agriculture news release, the Bureau of Biological Survey estimated annual economic losses in livestock of $20 to $30 million ($205 to $308 million in 2007 dollars) to all predators throughout the West. According to Brown (1992), average destruction by predatory animals during this same period was estimated to be $1,000 worth of livestock annually ($10,000 in 2007 dollars) for each wolf and mountain lion, $500 ($5,000 in 2007 dollars) for each stock-killing bear, and $50 ($500 in 2007 dollars) for each coyote and bobcat. He also illustrated cases where substantial damage was caused by just a few predators. For example, one wolf in Colorado killed nearly $3,000 worth of cattle ($30,000 in 2007 dollars) in one year, two wolves in Texas killed 72 sheep in two weeks, one wolf in New Mexico killed 25 head of cattle in two months, and another wolf killed 150 cattle valued at $5,000 ($51,000 in 2007 dollars) during a six-month period.[17]

The 2010 Mexican Wolf Recovery Team had room for only three to four members that represent economic, livestock, or hunting interests. Those members are the only members of the recovery team that are true volunteers and not reimbursed or paid for attendance. USFWS, TESF, DOW, and other NGOs as well as the state wildlife departments pay other team members. The 2010 Recovery Team was not much different from the 2003 team. There are so many divisions to the team that only the top tier will have significant input into the program. The following is from the official recovery team page in the USFWS website for the program:

The Southwest Region has initiated the revision of the 1982 Mexican Wolf Recovery Plan. In December 2010, we charged a new recovery team with the development of a revised recovery plan for the Mexican wolf. The team includes a

Tribal Liaisons Subgroup, Stakeholder Liaisons Subgroup, Agency Liaisons Subgroup, and a Science and Planning Subgroup. When completed and approved by the Service, the plan will include objective and measurable recovery criteria for delisting the Mexican wolf from the List of Threatened and Endangered Wildlife and Plants, management actions that will achieve the criteria, and time and cost estimates for these actions.[18]

Is this collusion, corruption, or simply cooperation? Currently, there are no actual stakeholders involved in the recovery planning, only the Stakeholder Liaisons Subgroup, which is a transparent method of keeping impacted parties from putting information into the recovery plan that will protect and assist actual stakeholders.

It appears that far less influence by non-affected wolf advocacy parties is necessary to further the goals of the ESA in reference to that specific and emphatic statement in the law that the ESA *"will not be used to engineer social change."* This is recognized in various federal documents; for example the 2011 USFWS publication "ESA Basics: More than 30 years of Conserving Endangered Species" states:

Two-thirds of federally listed species have at least some habitat on private land, and some species have most of their remaining habitat on private land. The FWS has developed an array of tools and incentives to protect the interests of private landowners while encouraging management activities that benefit listed and other at-risk species.[19]

Wolf activists, however, seem to believe that it is necessary to specifically foster changes in the economic and social structure of the region and to keep required and necessary wolf control and management at a dangerous minimum. The NGO involvement promotes keeping problem wolves on the ground and further harms the local communities that are relatively defenseless. This is a clear violation of the ESA. USFWS should be developing protocol to enhance affected party participation in decision-making and management and to limit non-affected NGO interests whose goals are contrary to the ESA's requirements for avoiding social change.

190

Collusion with NGOs

The technical team of Paul Paquet and Mike Phillips, both advisors to the NGO Southern Rockies Wolf Restoration Project, made three-year review recommendations. There was no data collected the first three years for the tech team to go on. The scientists admitted as much and stated as much publicly. They made socio/political recommendations instead:

🐾 Force ranchers to remove livestock carcasses, holding U.S. Forest Service permit removal over their heads, and

🐾 Institute boundary removal, allowing wolves to spread beyond the original Blue Range Wolf Recovery Area.

These recommendations were admittedly socio/political with no scientific basis. Additionally, manipulation of problem wolves and control of depredating wolves were claimed in their report to be counter-productive to reintroduction efforts.

Due to the non-scientific nature of the technical recommendations, then wolf reintroduction coordinator Brian Kelly put together a rather large team of stakeholders to make recommendations for changes to the program in a way that would promote the program, yet eliminate or mitigate the problems and make things work for the majority of the stakeholders. It was a good idea, but the NGO participants were not happy about it. The individuals invited to this large working group comprised many different facets of the local communities—economic, human dimension, environmental, and scientific—as well as many wolf experts and biologists. The document developed from that group represents the last time a concrete attempt was made to cooperate with local communities and governments. However, it never went into effect.

The conclusion to be drawn here is simply that the recommendations were not politically expedient to meet the unstated goals of the agencies and, more importantly, the cooperating NGOs, because they cooperated with both sides of the wolf issue. Because of this, the recommendations were shunted aside when lead agency status changed in the spring of 2003. Brian Kelly, former USFWS Mexican wolf team leader, left the program, and with him went the public input into the program.

This removal of public input allowed the agencies and NGOs to ignore their obligations to stakeholders as set forth in the final Environmental Impact Statement and Final Rule. There was no one involved at upper levels to hold their feet to the fire and be fair to the local stakeholders. What happened next was entirely predictable. Suddenly, the three-year review recommendations by Paquet and Phillips were massaged into the basis for the new five-year review by new wolf managers. The ranching and community sustainability swiftly took a back seat to wolf recovery.

For the past several years, there has been very little effort to follow the final rule as it relates to complying with wolf management, habituation control, or livestock protection. Instead, funding is focused on planning total recovery in the whole recovery area and more releases and on promotional programs marketing wolf reintroduction to the uninformed and unaffected public.

Allowing the NGOs to serve as experts in the wolf reintroduction and recovery has given them unique power over the landowners in the areas affected by the Endangered Species Act. The NGO scientists have the full force of the federal government behind their plans. When rural people are forced to sue the USFWS, judges always defer to the expertise of the USFWS; however, in reality they are deferring to the expertise of the NGOs, who are heavily involved in structuring the day-to-day management and planning of the program. There is no input or re-dress for affected people, either at the agency level or in the courtroom.

Management and Personnel Bias Against Ranching

H. Dale Hall, USFWS regional director up to 2006, admitted that, when he was first placed in position, members of his wolf staff brought him an anti-grazing, NGO-published book and told him that grazing should be removed to make way for the wolf recovery. There has been no exposure of this action; nor were there any personnel changes. These persons were not removed from the team and placed elsewhere in the agency. Clearly, there are still many wolf program employees that are actively promoting an anti-grazing agenda.

Reintroduction of Mexican wolves into their historic habitat or historic DPS area would have required reintroduction into northern Mexico, extreme southern Arizona and New Mexico, and southwest Texas. Mexican wolves have been released into Arizona and New Mexico. Five wolves were released in Mexico in the state of Chihuahua; four of them were either killed by poison or died within the first six months. Suspicious death and disappearance are also factors in the U.S. side of the program. Cornered people simply do not respond well to threats.

Politically-Motivated Decisions

Mexican wolves have been shown through DNA analysis to include DNA from three different wolf species from North America as well as coyote and domestic dog genes, again showing that Ghost Ranch and Aragon were completely inappropriate additions. Hybrid wolves are known to be far more aggressive and to kill more livestock and domestic animals than non-hybrid wolves, yet the Mexican wolves we now know were created in a kennel and deemed a "pure" subspecies in a courtroom.

USFWS numbers indicate there are more than 300 Mexican wolves with the same DNA in the captive breeding program available for release. Although they are naïve animals without a history of livestock kills, any discussion of removing depredating packs has hit a brick wall within the program, mainly due to the Center for Biological Diversity's insistence, through lawsuit threats, that each individual wolf is important to the gene pool. This is not a scientifically valid position as the rule states clearly that only genetically redundant wolves are to be used in the release program. Yet that false premise has recently been taken up by the USFWS as if that claim had legitimacy, when in truth it is merely another politically-motivated decision to kowtow to extremists in order to keep to their own agenda.

Wolf Habituation

The seriousness of the habituation of Mexican wolves can be seen in the documented incidents involving children, many backed up with photographs.

- 🐾 Child encounters wolves horseback.
- 🐾 Child finds pet dog slaughtered by wolves on private land.
- 🐾 Child is surrounded by wolf pack on hunting trip.
- 🐾 Children are followed by wolves from bus stop.
- 🐾 Dog defends child in yard from wolf.
- 🐾 Child's horse is slaughtered by Aspen pack in corral.
- 🐾 Child watches a wolf kill a kitten in one yard.

- 🐾 Children are kept captive in rural communities due to wolf presence.
- 🐾 Children are subjected to wolves coming in their yard while mother unloads groceries from her truck.
- 🐾 Children call in a wolf by crying when stopped on the side of a rural highway due to motion sickness.
- 🐾 Wolves lay in yards and defecate in places where children play, possibly as a method of marking territory where there are dogs present. (This introduces possible exposure to hydatid disease found in wolf feces.)

USFWS wolf managers still insist that nothing is amiss in this program. Some of them have even called these accounts hyperbole and blame the parents for instilling fear into their children. Fear is not always a bad thing. J.C. Nelson responded to that fear instilled in him and reacted calmly; had he reacted to his natural fear and run from the Luna pack, he likely would have been killed.

Little to nothing is done to mitigate these types of behaviors. Does the name calling of a family matter if their children are deliberately being placed in harm's way by a mismanaged, government-sponsored program if the community was adamantly against it to begin with? Not really, not on a local level since the majority of rural residents impacted understand the circumstances that led to the incident. Marginalizing the family is a common method of downplaying the incident that actually occurred. The Mexican wolf reintroduction area residents share an attitude that is not all that unusual in other cultures where wolves are actually present. Proximity to wolves affects your mindset.

As the chapters in this book on wolf attacks and wolves in Russia clearly show, in Norway, Finland, Sweden, France, Spain, Russia, Karelia, and throughout Asia, historically documented wolf attacks on people were common enough that on occasion they became epidemic. In fact, in a 2003 study in Scandinavia, 85% of the deaths that were verified were children with no adult present.[20]

These incidents lead researchers to ask another question: Why did wolves become child killers? Researchers came to believe that wolves understood that children were vulnerable prey and made efforts to avoid adults to hunt them. In fact, most of those killed were doing some very familiar things such as herding livestock near a house or farmyard, in a forested area, or in outlying barns or fields.

This sounds a lot like the Mexican Wolf Blue Range Reintroduction Area community and culture right now. None of the talking points about wolf behavior or the demonization of their parenting abilities really fazes rural people actually living among and dealing with Mexican wolves, especially those who are suffering under a captive breeding program that is creating habituation. Rural people instinctively know there is danger and distrust those who downplay their real experiences. Rural mothers arrive at the same conclusion: if their children are followed home from the bus stop by a wolf, or if they encounter one while riding horseback, or if a house dog is killed by a wolf, there is an immediate and very real danger. They keep their kids confined so we are actually seeing rural residents changing their behavior due to this program.

Perhaps the question shown by the Scandinavian study should not be why are people afraid of wolves, but rather, in the face of such overwhelming evidence, why do so many people appear to be unafraid of wolves.

Conclusion

The current situation in the United States as far as recovery of wolf populations in historic habitat is concerned is very bleak for rural Americans. States such as Montana, Idaho, Wyoming, Oregon, and Washington are being overrun with an unmanaged and possibly overprotected population of gray wolves. Wolves are spreading into Utah, Illinois, Indiana, and Colorado. While the Southwest program has been floundering along for many years without significant successes such as those seen in the Northern Rocky Mountains, the very fact that the Mexican wolves are not listed as a subspecies indicates that Congress never intended to allow them significant status for full recovery, instead preferring to allow them a footnote in history as a small population of gray wolves with distinct features adapted to the arid regions of the Southwest and Mexico.

The agenda of the environmental movement and federal bureaucrats, however, is significantly different. The delisting of the gray wolf by Congress was the beginning of the end of federal funding for wolves, and these entities are losing their cash cow. It makes the Mexican wolf that much more important as a tool for federal programs, as well as for manipulating landowners and federal lands ranchers, and a replacement poster animal for the previously lucrative gray wolf of the Northern Rockies and the Midwest.

Editors' Note: As this book goes to print in December 2013, several important developments have occurred that need to be noted:

🐾 On June 13, 2013, the U.S. Fish and Wildlife Service (USFWS) proposed to remove Endangered Species Act protection for most gray wolves across the United States. The USFWS is also proposing to continue federal protection of and to expand recovery efforts for the Mexican gray wolf in the Southwest by designating the Mexican wolf as an endangered species and modifying existing regulations governing them. Under two agreements reached August 26, 2013, between USFWS and the non-profit, pro-wolf group, the Center for Biological Diversity, USFWS has agreed to finalize a rule to allow the direct release of captive Mexican gray wolves into New Mexico and to allow Mexican wolves to establish territories in an expanded area of New Mexico and Arizona.[21]

🐾 Directly related to the above news, on September 9, 2013, John Harris, Chairman of the Arizona Game and Fish Commission, wrote a letter to the U.S. Department of Interior Secretary and the USFWS Director, which expresses everything that is wrong with the entire wolf project. Following is most of the letter (see Appendices for entire letter as well as its resolution):

...(We) express our disappointment at the refusal to hold public scoping hearings in Arizona regarding the proposed expansion of Mexican wolf conservation, an expansion that will almost certainly deliver a preponderance of the program's social, financial, and biological costs to the citizens of Arizona. As your partners in this effort since the program's birth, the passion of the resolution's language reflects our regret at what seems like an opportunity being missed by the Department of Interior and the USFWS to show their respect for those who share the working lands of Arizona with this species and who live with the consequences of this program, potentially in perpetuity. The best meetings in Sacramento, Albuquerque, or even the nation's capital will not suffice as substitutes for meeting on the affected land with the affected parties.

Each species, through individual selection,
evolves its own distinct behavior-habitat strategies to persist,
but that does not guarantee survival.
Each species walks its own road down through time.
There is no magical balance of nature.

Dr. Arthur Bergerud

🐾 🐾 🐾 🐾 🐾 🐾 🐾 🐾

[1] Hedrick, Philip W., Philip S. Miller, Eli Geffen, and Robert Wayne, "Genetic Evaluation of the Three Captive Mexican Wolf Lineages," *Zoo Biology 16,* 1997: 47-69.

[2] http://graywolfnews.com/pdf/the-courts-were-wrong-these-wolves-are-hybrids.pdf

[3] http://www.cuerochupacabra.com/id25.html

[4] http://graywolfnews.com/pdf/the-courts-were-wrong-these-wolves-are-hybrids.pdf

[5] http://graywolfnews.com/pdf/the-courts-were-wrong-these-wolves-are-hybrids.pdf

[6] http://www.fws.gov/southwest/es/mexicanwolf/BRWRP_notes.cfm

[7] http://www.rmef.org/NewsandMedia/NewsReleases/2009/ElkPopulations.htm

[8] Boyd, Diane K., "Wolf Habituation as a Conservation Conundrum" Case Study; http://www.sinauer.com/groom/article.php?id=24

[9] Adobe Ranch management, personal communication, 2008.

[10] http://www.defenders.org/resources/publications/programs_and_policy/wildlife_conservation/ solutions/full_list_of_payments_in_the_northern_rockies_and_southwest.pdf

[11] Mathis, C., personal communication, 2008.

[12] http://aces.nmsu.edu/pubs/_ritf/RITF80.pdf

[13] Mech, L. David, Elizabeth K. Harper, Thomas J. Meier, and William J. Paul, "Wolf Depredations on Cattle," *Wildlife Society Bulletin* 28:3, Autumn 2000, pages 623-629.

[14] Bangs, Ed, email excerpt dated September 1, 2006, from correspondence with other wolf managers in Mexican Wolf Program referring to call from reporter concerned about livestock carcasses attracting wolves.

[15] Brown, D.E., *The Wolf in the Southwest: The Making of an Endangered Species*, Tucson: University of Arizona Press, 1992.

[16] *Ibid.*

[17] http://aces.nmsu.edu/pubs/_ritf/RITF80.pdf

[18] http://www.fws.gov/southwest/es/mexicanwolf/

[19] http://www.lcie.org/docs/regions/baltic/linnell%20azl%20wolf%20attacks

[20] Linnell, J.D.C., Erling J. Solverg, et al., "Is the Fear of Wolves Justified? A Fenniscandian Perspective," *Acta Zoological Lituanica,* Vol. 13, 2003

[21] http://ens-newswire.com/2013/08/26/mexican-gray-wolves-gain-protection-in-arizona-new-mexico/

FloridaStock / Shutterstock.com

Chapter 12

COLLATERAL DAMAGE IDENTIFICATION

By Jess Carey

Jess Carey has lived in Reserve, New Mexico, for 35 years. Being trained in woods skills by his father over 50 years ago has been an asset in his job as Catron County Wildlife Investigator, which he has done since April 2006. After graduating from high school, Carey did a three-year tour of duty in the U.S. Marine Corps. Carey served as the elected Sheriff, Undersheriff, and Investigator for the 7th Judicial District Attorney's Office in Catron County.

Editors' Statement of the Issues

Several critical factors have caused the current Mexican wolf to be particularly prone to habituation and livestock predations: wildlife populations are limited, population base is crossbred with dogs, and wolves are raised in captivity and then released.

Editors' Statement of the Facts

Released habituated wolves cause unacceptable rates of depredation to livestock and they grow increasingly more aggressive towards humans living in their adopted areas. The losses to livestock are catastrophic to the ranchers' livelihood and the dangers posed to humans are unacceptable. The staggering financial losses suffered by predation must stop. The wolf should be eliminated from those areas.

Wolves kill by consumption; they eat their victims alive.

Jess Carey

Wolf Preferences

Created in 1924, the 558,065-acre Gila Wilderness in western New Mexico, which was part of the Gila National Forest, was the first designated wilderness area in the world. The Gila Wilderness is also located in Catron County, the largest county in New Mexico. There is little human activity and little livestock grazing in the Gila Wilderness compared to the human use, activity, and livestock grazing in the surrounding Gila National Forest, which is fragmented by homesteaded family ranches, subdivisions, and isolated homes.

Since 1998, the U.S. Fish & Wildlife Service (USFWS) has released numerous Mexican wolf (*Canis lupus baileyi*) packs into the interior of the Gila Wilderness as this is prime wolf habitat with wildlife for food and timber for cover. Not one of the released wolf packs, however, has stayed there. The released wolf packs leave the Gila Wilderness within a short time and travel to human activity— ranches, homes, and communities—where they interact with people and their livelihoods in a number of mostly negative ways.

Wolf Depredation

I was hired to investigate wolf depredation for Catron County, New Mexico, in April 2006, as there were major complaints from resource owners that the results of depredation investigations by federal agencies were lacking. Without such confirmation, no reimbursement for depredation losses can be paid. Additionally, no investigations of wolves were being conducted around children, in yards, and at homes, except by the wolf recovery team, and they omitted reports and did no documentation.

In the past, it often took days before an agency representative came on site to assess the kill to see if it could be unequivocally confirmed a wolf depredation. However, after several days, it is often difficult or impossible to tell for certain whether wolves or other predators are responsible for the cause of death due to lost evidence by scavenging canines, birds, insects, and advanced decomposition. This is why a study by the USFWS concludes that only one in seven wolf depredation livestock incidents by wolves can be proven.

After I started as the County Wolf Investigator, the depredation confirmations of wolf-caused deaths of livestock and pets doubled. This was, in part, because I could respond to the reports immediately, which caused other agencies to respond in like time.

To put it bluntly, despite what you might hear, these Mexican wolves are destroying family ranchers' ability to survive, resulting in their selling off their ranches for a fraction of what they once were worth, if they can sell them at all.

The conflict between wolves and humans started during the colonial days when wolves killed settlers' livestock and dogs. Nothing has changed since that time, and wolves will continue to kill livestock and pets no matter what non-lethal scheme is used by the USFWS wolf recovery team. This is a scientific fact.

When I looked for a title for this chapter, I had to look at the folks most impacted by Mexican Wolf Recovery—the many rural family ranchers, who have lost their peace of mind, lost their dreams, lost their pursuit of happiness, lost their livestock, and lost their ranches. "Collateral Damage Identification" seemed appropriate. The damage in question is due to wolf-caused livestock kills, attacks, and harassment, with little or no compensation.

Some Background

The Mexican wolf (*Canis lupus baileyi*), a subspecies of the gray wolf (the smallest, rarest, and most genetically-distinct wolf subspecies), once ranged the Sonoran and Chihuahuan Deserts from central Mexico to western Texas, southern New Mexico, and central Arizona. They may have once ranged as far north as Colorado. Most wolves are bigger than a German Shepherd and weigh 70 to 90 pounds, but some are smaller and weigh 45 to 50 pounds. The head of the wolf is blockier than a coyote, and they have a broader nose than a coyote; also, the ears are more rounded. The front feet are larger than the rear feet. Colors range from a grizzled gray, reddish-brown, whitish mixture to reddish-brown. Their behavior is highly unpredictable.

The Mexican wolf was driven to extinction in the wild by the 1950s as wild herds of deer and elk declined with increased settling in their range and wolves began preying upon livestock and pets. In 1976, the Mexican wolf was listed on the Endangered Species Act.

In March 1998, the USFWS began releasing the Mexican wolf in the Blue Range area of Arizona. Eleven wolves were initially released. The restoration program is based on raising wolves in captivity in 49 different breeding facilities in the U.S. and Mexico and then releasing them into the wild in hopes that they will establish wild breeding. There are currently about 300 wolves in the 49 breeding facilities. Prior to being released, the wolves are sent to release sites such as Sevilleta National Wildlife Refuge in Socorro, New Mexico, where contact with people is minimized in an attempt to discourage habituation. The goal of the program was to establish a wild population of 100 Mexican wolves in the Apache and Gila National Forests of Arizona and New Mexico by 2008. As of January 2012, the "wild" population is 58 with six breeding pairs.[1]

Since the outset, this program has been fraught with problems as the wolves arrive habituated to human feeding when in captivity. When they are released into the wild, the agencies continue to feed them with horse meat and road kills dropped beside roads—"supplemental feeding"— because these wolves do not have the hunting skills of wild wolves. The result has been wolves standing beside roads, expecting handouts like bears used to do in Yellowstone, and wolves preying on livestock and farm animals and approaching people. Some habituated wolves will stand and look at you even after you fire a firearm into the air.

In addition, Mexican wolves have taken to breeding with house dogs and stray dogs, resulting in three confirmed litters of wolf-dog hybrids that are even more prone to prey on livestock and approach people. Wolf-dog hybrids are not protected by the Endangered Species Act; however, telling the difference between a wolf and wolf-dog on sight is hard, if not impossible, depending on the breed of dog. Wolves can breed with any other *Canis* species, and wolf-coyote hybrids are also hard to distinguish. Only DNA analysis can give conclusive proof.

The Endangered Species Act only protects purebred species. It is nearly impossible to tell if wolves, coy-wolves, wolf-dog hybrids, or even some animals with wolf, coyote, and dog DNA are responsible for a livestock depredation or a dead pet dog, if tracks are all you have to go on.

One case of a suspected wolf-dog hybrid occurred north of Luna, New Mexico, on December 27, 2006. I was notified that feral dogs had just killed an elk calf. The elk calf had been killed by bites in the throat, just under the jaw. The calf had been caught against the barbwire fence and killed. A small hole was opened in the throat area where the dogs started feeding on the calf. Just after daylight, three dogs were feeding on the elk calf carcass. When the dogs detected the presence of humans, they ran. Two of the dogs were killed, and the third got away.

On December 29, 2006, Wildlife Services' J. Brad Miller came to my residence to inspect Canine #1. He stated that Canine #1 had unusual characteristics and should be DNA tested as he felt it could be a wolf-dog hybrid. I gave Mr. Miller DNA material from samples taken on December 27 of Canine #1C and Canine #2C as well as a CD with the photographs that I had taken at the scene. I understood that Mr. Miller would turn over these items to Dan Stark at the Alpine office and that Mr. Stark would send out these samples for DNA tests.

On December 31, 2006, Dan Stark came by my home, and I gave him a blood sample of Canine #1 to send in for DNA analysis. Also Mr. Stark reaffirmed that the results of the DNA report would be given to Catron County. I requested the DNA test results from USFWS Wolf Field Coordinator John Oakleaf several times in 2007, 2008, and 2009. Mr. Oakleaf stated that the lady at the DNA lab could not find the results.

Finally, in September 2009, Mr. Oakleaf stated that Dan Stark had thrown the DNA samples away and that they never had been sent to the DNA lab. I told Mr. Oakleaf that this was *not* satisfactory behavior for a USFWS biologist and emphasized the importance of knowing if Mexican wolves are breeding with domestic dogs producing hybrids.

As a result of experiences and conditions such as those I've just described, on June 20, 2011, the New Mexico Game and Fish Commission voted unanimously to discontinue collaborating with the Mexican Wolf Recovery Program.[2] They also approved trapping in the area in New Mexico where wolves now reside. So now it's solely up to the USFWS in New Mexico to manage wolves.

On January 13, 2012, during the Arizona Game and Fish Department's nongame activities briefing to the Arizona Game and Fish Commission, the Commission voted unanimously to amend its policy on the release of Mexican wolves in eastern Arizona. If the USFWS wants to release more wolves in Arizona, Game and Fish's director will now have the authority to approve a wolf release in cases where an animal is lost from the population due to an unlawful act. When a wolf is lost to any other cause of mortality, the commission must approve a release.[3]

Nonetheless, the wolves are in and around the Gila, so it is necessary to investigate all wolf-human interactions, especially suspected depredations, as there currently is no way to control Mexican wolves unless it can be proven they are killing livestock and/or menacing or attacking people.

Confirming Wolf Depredation

Livestock killed by predators usually can be distinguished from those dying from other causes by the presence of external hemorrhaging; subcutaneous hemorrhaging and tooth punctures; damage to the skin, other soft tissues, and skull; blood on the soil and vegetation; and carnivore tracks, scats, or territorial marks near dead animals. Wolves primarily attack cattle on the hindquarters, targeting tail, vulva, lower thighs, hocks, and hamstrings. Occasionally they will attack on the neck, face, and jaw, behind the front legs, in front of the rear legs, and on the belly.

Wolves will run cows, calves, and yearlings—stressing the animal until it cannot stand; normally, there will be capture bite and rake marks on the skin with corresponding hemorrhage. Newborn livestock killed by predators and partially consumed can be distinguished from stillborn livestock by characteristics not found in stillborn animals, which include a blood clot present at the closed end of the navel, pink lungs that float in water, fat around the heart and kidneys, milk in the stomach and intestines, milk fat and lymph in the lymphatic vessels that drain the intestinal tract, a worn soft membrane on the bottom of the hooves, and possibly soil on the bottom of the hooves.

Normally, when wolves kill new calves, there is little left of the carcass; possibly a few small bones or a piece of the skull remain, but usually just a bloody place on the ground is all that remains because the calf is totally consumed, including the hooves. Even if wolves are confirmed in the immediate area, the cause of death of the livestock is unknown unless the kill can be positively documented based on strict depredation confirmation standards. What makes confirmation of a wolf kill very difficult includes the following points:

- 🐾 Missing livestock with no remains, resulting from wolves eating the whole carcass of calves including skull, hooves, bones, and hair;
- 🐾 Coyotes or other scavengers consuming remainder of calf carcasses;
- 🐾 Calves, yearlings, and cows not being found in rough, remote terrain;
- 🐾 Advanced decomposition, rapid and severe in summer weather;
- 🐾 Insect infestation;
- 🐾 Scavenging birds;
- 🐾 Other scavenging carnivores;
- 🐾 Weather conditions;
- 🐾 Rocky, hard ground conditions that limit impressions; and/or
- 🐾 Untimely carcass detection.

If a larger calf is killed, there are remains left much of the time, but there may not be capture bite marks. The reason is that often, when attacked, the calf is bedded; the wolf pins the calf down, and the feeding begins. The wolf does not have to bite the calf to capture it.

Wolves kill by consumption; they eat their victims alive. Those that die do so from stress and tissue and blood loss. In 266 wolf depredation investigations, I have never documented a lethal bite site on cattle carcasses. Some cattle are "stressed down" (when wolves chase grown cattle until they are exhausted); the wolves eat 20 pounds from the victim (the injured cow, calf, or yearling) and then leave, often while their victim is still alive. Most cattle die at the feeding site. Some survive after the wolves have eaten their fill. Such animals are not dead and walk around with their rear-ends eaten out; victims with massive tissue loss have to be put down by the resource owner.

Research on Five Ranches

As part of my job as County Wolf Investigator, I conducted a study of the impact of wolves on five ranches located within the Blue Range Wolf Recovery Area in Catron County. The study includes data gathered from 2005 to 2010—both on-site investigations and record research. These ranches all were identified as having wolves denning in or near calf-yearling cattle core areas. When I came on board, the relationship between high calf loss and proximity to denning Mexican wolves was not well understood.

On one ranch, wolves denned in 2005, 2006, and 2007. On the other ranches, denning was in 2008 and 2009. The study compared confirmed wolf-livestock depredations to actual losses, and it identified other monetary losses to the resource owner. The study—which was finalized on March 1, 2011—compares the following factors on the five ranches:

- 🐾 Historic pre-wolf normal calf losses;
- 🐾 Confirmed and probable wolf-livestock depredations;
- 🐾 Actual livestock losses;
- 🐾 Compensation paid by Defenders of Wildlife.

One of the first discoveries was that coyotes swarm to areas where Mexican wolves are continually killing livestock, contributing to the removal and destruction of evidence of the predations by wolves, as well as increasing depredation. This may not be consistent with wolf-coyote interactions in other parts of North America, but it is true for Mexican wolves. Perhaps this relatively peaceful co-existence is associated with wolf habituation.

The Magic of a Clean Slate

Prior to this study, the focus of predation studies was on proving, beyond the shadow of a doubt, that solely wolves had initiated and killed livestock because of the funds for reimbursing ranchers whose cattle are killed by wolves; if it can be proven that wolves are killing livestock, the wolves can be removed. In the past, the "Three Strike Rule" governed removing depredating wolves. But when the wolves started to kill large numbers of livestock, the rules were changed to protect the wolves, and there is no Three Strike Rule anymore. Additionally, one year after a committed confirmed depredation, that depredation is removed from the offending wolf's record. In other words, each year the depredating wolves start with a clean slate. The Middle Fork Pack has a history of 16 confirmed livestock depredation and is still on the landscape.

We now know that irrefutable proof is often difficult or impossible to determine, especially if the kill site is not immediately investigated as many scavengers quickly convene on wolf kills, and some wounded animals wander away and are never found. This is consistent with a 2003 USFWS study by John Oakleaf, Field Coordinator of the Mexican Wolf Recovery Program, which concluded that only one in seven wolf kills of livestock can be conclusively proven to be due to wolves.[4]

Pathological Fatigue

We also now know that wolves kill cattle by stress. Wolves will run, attacking and biting, grown cattle for a long period until they are exhausted, resulting in "pathological fatigue." Pathological fatigue interferes with the activity of every gland in the cow's system; its principle effect is to destroy the capacity of muscles and nerves to perform the work natural to them. A chemical change takes place in the muscles, releasing toxic substances including lactic acid, creatine, and carbon dioxide. These toxic substances are acids that cause a state of fatigue in the cow's muscles and system. During rest following fatigue, these acids are neutralized by alkaline of the blood and internal secretions, which restores freshness, strength, and tone to the muscle. However, based on my extensive experience, I feel that, once a cow's system has been saturated to a certain point "beyond recovery" from these toxic substances, there is no ability for the cow's system to neutralize; the cow's system shuts down, and it dies. I have seen healthy cows in prime condition that appear to have just fallen over dead. They lay on their sides with no indication of leg movement, no sign that the hooves disturbed the ground, and no ground litter at all. Also noted were no noxious plants in the area.

An example of wolf-caused pathological fatigue is Case #AP-030. This cow was black and heavy, approximately 1,200 pounds. I documented the torn-up ground around the dead cow. It looked like a small circle racetrack, and its hooves had gouged the ground, uprooting the vegetation. The cow was bitten at the root of the tail and on down the tail about 12 inches. Upon necropsy, there were bite sites with corresponding hemorrhage; the canine spread (bite size) was consistent with a wolf. The bitten tail did not kill the cow, and there were no lethal bite sites. This was a stress death caused by continued harassment chasing and was confirmed as such. Also the wolf was located on a nearby mountain and was documented by aerial telemetry. If there were no bite sites, hemorrhage, or a canine spread, this cow's death would have been "unknown." I feel there have been numerous depredation cases that were stress deaths, but they were classified "unknown" based on the best available evidence at the scene.

In 12 confirmed wolf-killed yearling calves on one ranch, five did not die at the attack and feeding site. They traveled for some distance after being fed upon by wolves. Four yearlings were found alive and walking around with massive tissue loss. One yearling was found dead, and the scene lacked evidence of an attack and feeding site. Dried blood found on the hind legs indicated the yearling was bleeding while standing upright and walking.

It's not just that wolves are killing livestock. Wolf-caused chronic stress in cattle produces non-compensated economic losses. Wolf-caused chronic stress disrupts the cows' breeding cycle, causing them to abort calves, birth weak calves, undergo weight loss, and become more susceptible to disease. Additionally, disfigured livestock bring less money at the sale, and there is the cost of vet care for wolf-injured animals. All of these things happen at the cost of the resource owner and without compensation. Catron County has documented on one ranch that 36% of the depredated yearlings that were confirmed as having been attacked and fed upon by the Middle Fork Pack were still alive after the initial attack. All these animals had to be put down.

Investigating Procedures

Once notification is given by the resource owner or others that may have found a livestock carcass suspected of predator depredation in Catron County, USDA Wildlife Services and I respond to the scene to perform an investigation to determine the cause of death of the animal.

Arriving at the scene is like being at a crime scene. Canine tracks can be destroyed by people walking within the scene. Other livestock and scavenging birds can also destroy tracks, etc. You yourself can destroy tracks if you do not take the precaution to look where you step. The best procedure when entering the attack scene is to protect the evidence, such as canine tracks, as you find them by covering these tracks to prevent other livestock, animals, and people from trampling them. You cover the carcass with a tarp rocked around the edges to prevent scavenging canines and birds from feeding on it, and you cover blood trails or droplets of blood leading to the carcass if rain is imminent.

Timely carcass detection and notification are key to depredation investigations to determine the cause of death. Lost or destroyed evidence can result in a non-confirmation. Calf carcasses left uncovered in the field will disappear during the night. If you do not have a tarp, hang the calf high up in a tree; if there's no tree, mark the area, bring the calf in, and store it so dogs cannot get to it.

All scene evidence is photographed. The carcass is photographed—the head, back, rear, and belly. The injuries are photographed—the attack sites on the carcass, bite sites, feeding sites, and impact injuries. Scavenging canines and birds are also noted.

Measurements are taken to document predator tracks and scats. A diagram is drawn to reflect attack and feeding site, drag marks, carcass site, blood trails, and predator/victim track location and direction of travel. Barbwire fence are checked; bottom and second strands are checked for hair caught in the barbs when predators pass under or through them. A predator's identification can be made with this transfer evidence (hair).

Dirt roads are checked for predator tracks, scats, and any sign of predators as you near the area of the carcass. If tracks are located on the roadway, they are marked and protected so no one drives over them. Other cattle nearby are observed for unusual behavior—calling and alert, defensive, and frightened—as well as for injury bite sites and impact wounds from running into barriers or barbwire fences.

The area is also checked for wolf collar signals using a ground telemetry receiver. If a signal is picked up, the corresponding wolf number is noted. The scene around the carcass is searched to identify the attack site, feeding site, drag marks, tracks, scats, blood trails, trampled and/or uprooted vegetation, torn up ground, broken fences, etc. The scene could be less than 50 yards to several hundred yards in size.

Once everything is documented, the investigation focuses on the carcass, and a necropsy is performed. The percentage of carcass remains is noted, as well as disarticulation of limbs and bones. Some carcass remains are just dried skin and bones; these have to be soaked in water three to five days to soften the skin; compression bite sites on the skin still remain. A compression bite site can only be confirmed if the victim was bitten while alive.

First the hair is clipped from the skin of the carcass to detect bite sites and rake marks. Without clipping the hair, you cannot see the bite and rake marks. Photographed measurements of all canine spreads (bite size) are documented. The skin is removed to document a bite site's corresponding hemorrhage and deep hemorrhage in the muscle tissue and injuries. Most times there are no internal organs left inside the carcass for assessment. The skin is held up to the sun and photographed to document bite sites and rake marks with hemorrhage in the skin.

Following are results from my study of five ranches located within the Blue Range Wolf Recovery Area:

🐾 Ranch A is a cow/calf operation. Records of average annual pre-wolf introduction losses were 16%. The herd consisted of 300 head. Herd makeup: 20 bulls, 25 replacement heifers (not expected to calve), 0 steers, and 255 production cows. 255 production cow numbers x 16% average pre-wolf annual calf losses = a 41 head loss. 255

– 41 = 214 fall calf crop number, representing an 84% calf crop. Losses pre-wolf-introduction were attributed to calving, open range cows, coyote predation, and winter weather. In 2008, the San Mateo Pack denned in calf core areas on Ranch A.

Pre-wolf annual losses $24,600
2008 losses $71,400
2009 losses $60,000
Defenders of Wildlife compensation paid $600
Rancher sold off remaining herd and went out of business.

 Ranch B is a cow/calf operation that adjoins Ranch A. Records of average annual pre-wolf-introduction calf losses were 2.5% for 3 years running with an average annual loss of 4 to 6 head of calves per annum. The herd consisted of 256 head. Herd makeup: 18 bulls, 30 replacement heifers (not expected to calve), 5 steers, and 203 production cows. Average calf crop = 97.5%. Losses pre-wolf were attributed to calving, open range cows, coyote predation, and winter weather. In 2008, the San Mateo Pack denned near calf core areas on Ranch B.

Pre-wolf losses: 5 head/year $3,300
2008 losses: 27 head $16,200
2009 losses: 58 head $34,800
Total losses 2008 and 2009 .... $51,000
Defenders of Wildlife Compensation $1,500

 Ranch C is located approximately 35 miles as the crow flies in a southerly direction from Ranches A and B. Records show that Ranch C had a 3% average annual pre-wolf-introduction loss. Total herd is 330 head. Herd makeup: 18 bulls, 0 steers, 30 replacement heifers (not expected to calve), and 282 production cattle. Average annual pre-wolf losses of 9 head per annum were noted. Losses were attributed to birthing, coyote depredations, open range cows, and winter weather. In 2005, the Luna Pack denned in calf core areas on Ranch C.

Pre-wolf losses: 9 head $5,076
2005 losses: 42 head $25,200
2006 losses: 82 head $49,200
2007 losses: 69 head $41,400
Defenders of Wildlife compensation 0
Ranch went out of business in 2007.

🐾 Ranch D is located to the west of Ranch C. When the
livestock were removed from Ranch C, the wolves
immediately left the vicinity of Ranch C and dispersed
to Ranch D where there were livestock. Records show
Ranch D had an 11% annual pre-wolf-introduction loss.
Total herd is 205 head. Herd makeup: 15 bulls, 0 steers,
10 replacement heifers (not expected to calve), and 180
production cattle. Average annual pre-wolf losses of 20
head per annum were noted. Losses were attributed to
birthing, coyote, bear depredations, open range cows, and
winter weather. In 2008, the Luna Pack denned in calf
core areas on Ranch D.

Pre-wolf losses: 20 head $8,100
2008 losses: 35 head $21,000
2009 losses: 23 head $13,800
Defenders of Wildlife compensation 0

🐾 Ranch E is located northeast of Ranch C and runs yearlings.
The allotment consisted of 3 pastures. There were 300
yearlings in excellent condition in Pasture A and B and
287 yearlings in Pasture C. In 2009, the Middle Fork Pack
denned in yearling core areas on Ranch E.

Pre-wolf annual losses: 5 head $2,828
2009 confirmed kills: 11 head $6,307
2009 carcasses: 14 head $7,917
2009 missing animals: 73 head $41, 281
Losses confirmed wolf kills carcasses missing.
Defenders of Wildlife compensation paid $6,307

The final analyses indicate that annual post-wolf-introduction livestock losses are higher than the average annual
pre-wolf losses for the five study ranches:

Total combined livestock losses .... 651 head
Total combined dollar value losses $382,199

Two of the five ranches in this study went out of business;
one sold the ranch, and the second is on the market. A
third ranch sold off its livestock in the fall of 2009 and
did not re-stock cattle in 2010. In addition, confirmed
and probable findings do not reflect the true number of
livestock losses. Wolf-caused stress disrupts a cow's
breeding cycle; the resulting calf loss must be measured
in monetary value as if the wolf depredated a calf.

The findings for these five ranches of confirmed and actual losses
and overall damages are consistent with other ranches across Catron
County where wolves den in calf and yearling core areas. Many ranchers
have cooperated with wolf recovery agencies, utilizing recommended
non-lethal schemes to prevent wolf-livestock interactions that result
in livestock depredation. These ranches have added additional range
riders, moved livestock to other pastures, penned livestock and fed
hay, and worked multiple additional hours to prevent wolves from
killing their livestock. All these measures add costs to operation. Still
the wolves depredate their livestock. The ongoing added effort, plus
stress and expense, represents a high loss cost factor far beyond pre-
wolf introduction.

Wolf-Human Interactions

The USFWS has changed its terminology from "supplemental
feeding" to "diversionary feeding" of wolves. This gives the public
the false sense that the wolves are able to hunt and make it on their
own. This diversionary feeding contributes to food conditioning,
which results in habituation. Habituated wolves are bold and fearless
and are a threat to children. Wolves come to homes and yards where
children play, and I am called in to investigate. Since the release of
Mexican wolves in 1998, unusual wolf behavior has been documented
by Catron County residents numerous times including:

- 🐾 Territorial scrapes at one residence where wolves were
  documented at the home 23 times.

- 🐾 Mexican wolves urinating on vehicle tires and on an ice
  chest located outside an occupied camp trailer.

212

- 🐾 A wolf defecating on the front of an ATV vehicle located in a front yard.

- 🐾 Wolves defecating on porches and yards at door entrances of occupied homes.

- 🐾 Territorial scrapes at occupied residences where the wolves were claiming the residence as part of their territory.

- 🐾 Wolves playing and/or mating with domestic dogs.

- 🐾 Wolves following young children as they walked to and from home from the school bus stop.

Just like cattle that are harassed by wolves, causing stress, we have documented cases of psychological trauma (post-traumatic stress disorder) in children and families of Catron County that were subject to fearless wolves approaching them. One case really sticks out in my mind. Wolves were regularly coming to the home of a family with a 14-year-old daughter. Wolves were documented at this home a total of 23 times. The county was in the process of trapping the wolves. As I sat in the living room with the family, the father had the ground telemetry unit on. Little was said as the signal began to beep, first faintly and then a little louder and louder and louder, as a wolf made its way to the residence from the nearby hills. No one said a word. Breathing was shallow for some time. I remember the look on their faces. It reminded me of when "Jaws" was coming to the boat with the harpoon tracking device putting out a steadily increasing signal.

Conclusion

From April 2006 to February 23, 2012, there were 420 reported wolf-animal or wolf-human interactions and 46 information reports; 214 were on private property, and 206 were on non-private property. The fact that approximately 50% of all wolf interactions occurred on private property clearly indicates the severe habituation of wolves towards humans and human-use areas. Mexican wolves seek out humans and human-use areas due to their habituation and their lack of an avoidance response towards humans. Therefore, designation of critical wolf habitat is useless when the wolf's definition of critical wolf habitat includes homes, communities, and people.

To alleviate the taking of private property (livestock) without compensation by the federal government, confirmation standards and the compensation scheme as a whole must be re-evaluated. In-depth studies must be conducted to evaluate the negative impacts of wolves' denning in calf and yearling core areas and the effects of wolf-related stress on livestock. Evaluation of data must include the wide spectrum of negative impacts to livestock and livestock producers, rather than the current focus solely on benefits to wolves. Recommended areas of study include the following:

- 🐾 Pre-wolf-introduction historic annual losses;
- 🐾 Post-wolf-introduction annual livestock losses;
- 🐾 Wolves denning in calf and yearling core areas;
- 🐾 Wolves denning near calf and yearling core areas;
- 🐾 Wolf rendezvous sites located in calf and yearling core areas;
- 🐾 Wolf-claimed territory overlapping livestock core areas; and
- 🐾 Wolf-caused chronic stress and effects on livestock.

Defenders of Wildlife, a pro-wolf organization, used to have a compensation fund to reimburse ranchers for livestock lost to wolf-confirmed kills; however, several ranches received no compensation on livestock depredation investigations conducted by Wildlife Services for documented, confirmed, or probable losses. Defenders of Wildlife no longer pays compensation.

Wolf recovery causes negative effects that can only be described as collateral damage. The mentality of the USFWS wolf recovery team to date is this: So what if our wolves are habituated and come to homes and communities, confront children, and kill family pets, farm animals, and livestock. You need to accept the Mexican wolf for the way he is and coexist by changing your daily lives to accommodate the true "New Age Mexican Wolf" that lacks wild wolf characteristics and is flawed to the extreme by habituation.

🐾 🐾 🐾 🐾 🐾 🐾 🐾 🐾

[1] http://www.fws.gov/southwest/es/mexicanwolf/cap_manage.shtml
[2] http://www.krqe.com/dpp/news/politics/new-mexico-abandoning-wolf-program
[3] http://www.azgfd.gov/w_c/es/mexican_wolf.shtml
[4] http://www.fws.gov/news/NewsReleases/showNews.cfm?newsId=3878CF69-1F55-4B23-AF326A3A9BA7F930

Chapter 13

THE WOLF IN THE GREAT LAKES REGION

By Ted B. Lyon

LittleMiss /Shutterstock.com

Editors' Statement of the Issues

Editors' Statement of the Issues

Historically, there is no question that, as wolf populations increase, so too do animal depredation and dangerous wolf/human encounters. Unless wolf populations are reduced and strictly controlled by hunting and trapping, more conflicts will occur.

Editors' Statement of the Facts

Wolf management programs must be designed to protect innocent lives—human, domestic livestock, and ungulate—as a priority before increasing or maintaining a large wolf population. Human/wolf encounters must be made available as well, so that people can learn to be wary. Livestock losses and aggressive behaviors are unacceptable and signal a specific need to reduce wolf populations.

> *Protecting wolves in the Midwest will cease to be practical*
> *or even possible, according to the Endangered Species Act,*
> *as species integrity is disappearing.*
> Ted B. Lyon

Before 1974

The nationwide campaign to eradicate wolves did just that by the 1960s or earlier, except for the upper Midwest where a small group of wolves retreated into wilderness. Wolves were declared extinct and given the equivalent of endangered species protection in Wisconsin in 1957 and Michigan in 1965. Until the mid-1970s, gray wolves in the Great Lakes region were almost entirely found in northern Minnesota in the Superior National Forest. This was the largest population in the United States, numbering about 1,000; it was kept in check by hunting and trapping. In Minnesota, there was a bounty on all predators, including wolves, until 1965. Between 1965 and 1974, Minnesota had an open season on wolves and a Directed Predator Control Program that removed about 250 wolves per year. Anti-wolf sentiment in that region ran fairly high.[1]

After 1974

Following the passage of the Endangered Species Act in 1973, wolves all across the U.S. were listed as endangered in 1974. Almost immediately, the wolf population in the Midwest began to increase and expand, aided by a succession of mild winters and a large

216

whitetail deer population. According to the Michigan Department of Natural Resources, wolves were breeding in Wisconsin by 1975 and in Michigan's Upper Peninsula by 1989.[2] There was also a small population of wolves on Isle Royale that got to the island by crossing the ice in winter; incidentally, these wolves, which fed on moose, were perhaps the most studied wolves in the world.

Due to their increase in population, wolves in Minnesota were reclassified to "threatened" in 1978. As the numbers of wolves in the Great Lakes area began to rise, people became more aware of them, and pro-wolf campaigns were launched by environmental and animal rights groups. Early public opinion surveys, after listing wolves under the Endangered Species Act, were largely positive. Naming Minnesota's pro basketball team the "Timberwolves" no doubt aided early positive public opinion of wolves.

Rising Population

Protection aided wolf population growth. According to former Michigan Department of Natural Resources wildlife biologist James Hammill, "Michigan and Wisconsin have typically had a 15% annual rate of increase in the number of wolves since 1977. The wolves in Minnesota have also continued to increase but at a slower rate of roughly 4% annually."[3]

Just how many wolves there are in this region is controversial. The U.S. Fish and Wildlife Service estimates that the current Midwest wolf population is about 4,400.[4] This is a conservative number. The Michigan Department of Natural Resources (DNR) says there are nearly 700 wolves currently in Michigan's Upper Peninsula.[5] Wisconsin DNR says it has at least as many.[6] And Minnesota DNR says that it has over 3,000 wolves.[7]

Hammill, who lives in Michigan's Upper Peninsula and is active in studying wolf behavior, believes that the Midwest wolf population of 2012 will actually approach 6,000 animals, about 4,000 of which are in Minnesota.[8] Wolves also have been reported in North and South Dakota, Iowa, and northern Illinois. And, of course, wolves are found in neighboring Canadian provinces.

There have been multiple reports of wolves in Michigan's Lower Peninsula.[9] Presumably, these wolves reached Michigan's Lower Peninsula by crossing Lake Michigan in the winter, but as in other states, some people raise wolves and wolf-dogs in captivity, and some escape or people purposefully release them into the wild. In 2000, Michigan took steps to reduce this problem by passing the

Wolf-Dog Cross Act, which requires wolf-dog owners to prove the hybrid has been sterilized, to keep the animal in a fenced area, and to display a sign reading: "A potentially dangerous wolf-dog cross is kept on this property."[10]

As Dr. Matthew Cronin states in his chapter on wolf genetics (Chapter 20), the identification of specific species and/or subspecies of wolf is not uniformly agreed upon among scientists. Basically, there are two wolf species: the gray wolf (*Canis lupus*) and the red wolf (*Canis rufus*). At one time, there were as many as 24 subspecies of wolf recognized in the U.S. Today, most scientists agree that there are five subspecies of gray wolf in North America: the Arctic wolf *Canis lupus arctos*, the Eastern Timber wolf *Canis lupus lycaon*, the Great Plains wolf *Canis lupus nubilis*, the Rocky Mountain wolf *Canis lupus occidentalis*, and the Mexican wolf *Canis lupus baileyi*. Of these five, at least two subspecies—the Great Plains wolf and the Eastern Timber wolf—interbreed in the Great Lakes region, and Ontario also considers the Arctic wolf territory along Hudson Bay to overlap with Eastern Timber wolf territory.

Today there are only 90-100 red wolves, a population which was bred in captivity and released into the wild, and they all live in the mountains of North Carolina. For this discussion of wolves in the Midwest, the red wolf needs to be noted as, at one time, it may have been found as far west as Texas and Missouri and northward to southern Ontario. Therefore, some Midwestern wolves may well also contain DNA of red wolves.[11]

The point simply is that, sooner or later, if it is not already the case, all wolves in the Midwest will be hybrids—either mixtures of wolf sub-species or wolf-coyote-dog hybrids.

> *I can tell you this: The adage that a wolf only kills the old animal or the sick or wounded is total bull. I had them kill 12 sheep in one night.[12]*
> Tony Cornish, retired MN game warden

Wolves and People in the Great Lakes States

Gray wolves in Minnesota have normal ranges of 25-150 square miles, which is much smaller than wolves in the Northern Rockies. As food availability influences the territorial behavior of wolves, presumably the smaller range area in Minnesota is due to the abundance of whitetail deer. These wolves are also smaller than the Canadian wolves that were transplanted into the Northern Rockies. Minnesota gray wolf females run 50-85 pounds, and males, 70-110 pounds.

Reviewing each state's website about wolf management, one finds fascinating differences of opinion about the number of deer per year that an adult wolf consumes. The Michigan Department of Natural Resources estimates that an adult wolf kills 30-50 deer per year.[13] The Minnesota Department of Natural Resources estimates only about 20 deer per year are taken by an adult wolf.[14] The two states share a border and are so close in habitat that it's hard to imagine that the Minnesota estimation is lower because of some dramatic difference in habitat.

As the wolf population in Minnesota, Wisconsin, and Michigan has grown, wolf hunting territory expanded, both to find new land not already occupied by a wolf pack and to find food. The Midwest has national forests, but extensive wilderness habitat like in the Northern Rockies is not found in the Great Lakes region, except in the far north. So Midwest wolves increasingly approach farms and urban areas where hunting has not been permitted. As wolf populations increase, they move closer to population centers including Minneapolis-St. Paul and Duluth, Minnesota; Green Bay, Wisconsin; and even Chicago, Illinois—and wolf-human contact increases. According to James Hammill, wolves have attacked and killed pets in the immediate vicinity of homes and within city limits in all three states.[15]

As predation on pets and livestock has increased, according to the USDA Wildlife Services, over 3,000 wolves in the Great Lakes area have been killed due to conflicts with human settlements. Wolf attacks on pets and livestock hit record levels in 2010. In Minnesota, 15 dogs were killed by wolves, up from an average of just two dogs per year from 2006 to 2008, according to the federal agencies. Minnesota officials verified 130 of 272 complaints—both records—involving 139 livestock and poultry and 23 dogs. The verified complaints were 31% above the five-year average. One person's safety was threatened by a wolf.[16]

In Wisconsin in 2010, wolves attacked livestock on 47 farms, 15 more than the previous high, killing 69 animals.[17] Also, wolves killed 24 dogs and injured 14 more, the most ever.[18]

In May 2012, the Michigan DNR killed eight wolves that had moved into the Upper Peninsula town of Ironwood, where they hunted pets and consumed garbage.[19]

> *Rapid City Journal*, Rapid City, South Dakota, March 12, 2012:
> In January of 2012, a large canine was killed near Custer, South Dakota. After six weeks of testing, it was determined by genetics that it was a gray wolf from the upper Great Lakes population. "Basically, there's no place for wolves in South Dakota," said Tony Leif of Pierre, Wildlife Division Director for the South Dakota Game, Fish & Parks Department. "There's no place biologically. There's no place socially."

Minnesota, Wisconsin, and Michigan have a livestock compensation program for animals killed or injured by wolf predation, and Wisconsin has a pet owners' compensation program. In 2011, Minnesota paid out $154,136 compensation to farmers for livestock—mostly for cattle but also for sheep, turkeys, pet dogs, a horse, and a llama.[20] The state paid 128 claims totaling $102,230 in fiscal 2011 and 104 claims for $106,615 in fiscal 2010. Contrast this with $72,895 paid for 71 claims in 2006.

In Minnesota, trappers with the USDA's Wildlife Services killed 192 problem wolves in 2011, down slightly from the 196 killed in 2010. The state Department of Agriculture paid about $96,000 in 2010 to people who lost livestock to wolves.[21]

Wolf Politics

Like the Northern Rockies and the Southwest wolf population centers, attempts to manage wolf numbers by hunting and trapping in the Midwest have been met by legal action from pro-wolf organizations. In 2003, the U.S. Fish and Wildlife Service (USFWS) reclassified Midwestern wolves from endangered to threatened, but this was judicially overturned. In April 2005, the USFWS issued subpermits to allow the Michigan and Wisconsin Departments of Natural Resources (DNR) to kill depredating wolves. But in September of the same year, the subpermits were stopped by Federal District Court in Washington, D.C., due to inadequate public notice of the states' applications.

In May 2006, a permit to kill depredating wolves was issued to the Wisconsin DNR and the Michigan DNR by the USFWS. The Humane Society of the U.S. (HSUS) and others filed suit against the Department of the Interior (DOI) and USFWS for issuance of those permits. On August 9, 2006, a U.S. District Court judge ruled against DOI, and the permits were no longer in effect.

In 2007, the USFWS de-listed the wolf in the Midwest, but again HSUS sued to block it. Finally, Endangered Species Act protection for gray wolves in the Western Great Lakes Distinct Population Segment (DPS) was published in the *Federal Register* on December 28, 2011.[22] However, in 2012, suits were filed to prevent hunting wolves with dogs in Wisconsin and to block the Minnesota management wolf hunt. These lawsuits have cost the federal government considerable money to defend.

Ontario: A Model for Wolf Management?

On the eastern side of the Great Lakes, the U.S. gray wolf population comes into contact with gray wolves from Ontario, Canada. Ontario presently has 8,000 to 10,000 wolves, about the same as Alaska. Provincial parks, such as Algonquin Provincial Park, are known to be wolf population centers. As the landscape is heavily forested and viewing wolves is less easy than in the more open western U.S., "howl-ins" are a popular tourist activity in some Ontario parks.[23]

The Ontario Ministry of Natural Resources (MNR) considers there to be two species of wolf in Ontario: the Gray wolf in northern boreal and tundra areas and the Eastern wolf in the central and northern hardwoods and conifers. MNR also states:

> *Coyotes, a close cousin to the wolf, co-exist well with humans and are common in the developed and agricultural areas of southern and northern Ontario. Where the ranges of Eastern wolves and coyotes overlap, interbreeding makes it difficult to distinguish between Eastern wolves and coyotes, and they can easily be confused.*

Other researchers say there are four subspecies of wolf in Ontario: 1) *Canis lupus hudsonicus* inhabiting the subarctic tundra, 2) a race (Ontario type) of the Eastern Timber wolf (*Canis lupus lycaon*) that inhabits the boreal forests, 3) a second race (Algonquin type) of *Canis lupus lycaon* that inhabit the deciduous forests of the upper Great Lakes, and 4) a small wolf (Tweed type) in central Ontario that may be a hybrid between the Algonquin-type wolf and expanding coyotes, *Canis latrans*.[24] The hybridization between wolves and coyotes is a major reason why some scientists are opposed to establishing a wolf population in the Adirondack Mountains, as it would not be able to be protected under the Endangered Species Act.[25]

Regardless of the species or subspecies of wolf, Ontario has some important lessons for U.S. wolf management. For example, Ontario has always had wolf hunting and trapping, and their wolf population is robust—more wolves than in the entire U.S., and the population has been stable for about 30 years.[26] The size of the Ontario wolf population in itself should answer concerns about hunting and trapping alone driving wolves to extinction. Science-based management is the key.

The Ontario Ministry of Natural Resources has an Enhanced Wolf Management Plan (EWMP) that seeks to insure that wolves occupy 85% of their historic range in the province. It states that hunting and trapping should play an integral role in controlling wolf populations within the province.[27]

What we can also learn from Ontario's experience is that, as wolf populations grow and expand, more and more conflicts between wolves and people will occur. This has translated into several recent attacks on people:

🐾 In Algonquin Provincial Park, there have been five attacks on humans, mostly on children, between 1987 and 1998 by four different wolves.[31]

🐾 September 2006. A lone black wolf attacked and seriously injured six people, including three children, at the popular Katherine's Cove Beach in Lake Superior Provincial Park. The wolf was killed by park staff and tested negative for rabies.[28]

🐾 December 31, 2011. A lone wolf attacked two ice fishermen and their Border Collie on a lake adjacent to Algonquin National Park, where hunting and trapping wolves had been curtailed.[29]

🐾 Summer 2012. There were two wolf attacks on people with dogs in the Thunder Bay, Ontario, area.[30]

Conclusion

Before wolves were placed on the Endangered Species Act protected list, public opinion was not too favorable to wolves. As soon as they were listed, environmental groups sought to exploit wolves for fund-raising, and people who did not have daily contact with wolves became more positive about them. A 1990 study by Steven Kellert found that 80% of Michigan's deer hunters favored Michigan's wolf population being replenished.[32] At that time, there were a handful of wolves in the Upper Peninsula. Fifteen years later, when the wolf population had reached about 400, 812 people attended a series of meetings about wolves in the Upper Peninsula—22% said they wanted no wolves in the Upper Peninsula, and 36% said they wanted fewer than existed at that time.[33] Another study conducted in 2002 found that residents of the Lower Peninsula (where there were no wolves) were much more supportive of having wolves than residents of the Upper Peninsula.[34] Elsewhere, as Midwestern wolf populations have increased, conflicts with farmers and dog owners have risen. As conflicts have increased, public acceptance of wolves has declined.[35]

Reviewing these studies and many others, wildlife biologist James Hammill concluded: "The current trajectory of public attitudes, especially in Michigan and Wisconsin, is not favorable to sustaining wolves in those states."[36] This conclusion leads Hammill to suggest that—in addition to the biological carrying capacity of an area determining the desired population for a species, especially species like wolves where contact with humans may lead to conflicts—a "Social Carrying Capacity" should also be developed for a realistic management plan.

As the Midwest states grapple with wolf management, much could be learned from Ontario about the practical sensibility of managing wolves in modern times.

🐾 🐾 🐾 🐾 🐾 🐾 🐾 🐾

[1] Weiss, et. al., "An Experimental Translocation of the Eastern Timber Wolf," *Audubon Conservation Report 5*, 1975, Twin Cities, MN: NAS.
[2] Michigan Department of Natural Resources, *Michigan Gray Wolf Recovery and Management Plan*, Lansing, MI, 1997.
[3] James Hammill, "Policy Issues Regarding Wolves in the Great Lakes Region," Transactions of the 72nd North American Wildlife and Natural Resources Conference, Wildlife Management Institute, 2007, pg. 378.
[4] http://www.fws.gov/midwest/wolf/aboutwolves/WolfPopUS.htm
[5] http://www.michigan.gov/dnr/0,1607,7-153-10370_12145_12205-32569--,00.html#recovery
[6] http://www.jsonline.com/news/wisconsin/45452492.html
[7] http://www.dnr.state.mn.us/rsg/profile.html?action=elementDetail&selectedElement=AMAJA01030
[8] Hammill, *Op. Cit.*, p. 387.
[9] http://greatlakesecho.org/2010/02/20/wolf-count-raises-questions-about-michigan-range-threats/
[10] http://www.animallaw.info/statutes/stusmi287_1001.htm
[11] http://www.wolf.org/wolves/learn/wow/regions/United_States/North_Carolina.asp
[12] http://www.startribune.com/sports/outdoors/169262076.html
[13] http://www.dnr.state.mn.us/mammals/wolves/mgmt.html
[14] http://www.dnr.state.mn.us/mammals/wolves/mgmt.html

[15] Hammill, *Op. Cit.*, p. 383.

[16] http://www.dnr.state.mn.us/mammals/wolves/mgmt.html

[17] http://www.startribune.com/sports/outdoors/116787898.html; http://dnr.wi.gov/topic/WildlifeHabitat/wolf/

[18] http://dnr.wi.gov/topic/wildlifehabitat/wolf/dogdeps.html

[19] http://www.mlive.com/outdoors/index.ssf/2012/05/government_killing_of_8_wolves.html

[20] http://www.twincities.com/localnews/ci_21518716/minnesota-wolfs-recovery-seen-higher-livestock-loss-payouts

[21] http://www.dnr.state.mn.us/mammals/wolves/mgmt.html

[22] http://www.thewildlifenews.com/category/wolves/wisconsin-wolves/

[23] http://www.mnr.gov.on.ca/en/Business/SORR/2ColumnSubPage/STEL02_163450.html

[24] http://jhered.oxfordjournals.org/content/100/suppl_1/S80.abstract

[25] http://consbio.org/

[26] http://www.ofah.org/news/Research-needed-on-Algonquin-wolves

[27] http://www.ofah.org/news/Government-announces-Enhanced-Wolf-Management-Plan

[28] http://www.propertyrightsresearch.org/2006/articles09/six_injured_in_rare_wolf_attack.htm

[29] http://www.ontariooutofdoors.com/News/?ID=203&a=read

[30] http://www.ontariooutofdoors.com/news/?ID=153&a=read

[31] http://www.fws.gov/mountain-prairie/species/mammals/wolf/FinalWolfHabPlanforemail.pdf

[32] Kellert, S.R., *Public Attitudes and Beliefs about the Wolf and its Restoration in Michigan,* Wisconsin: HBRS, Inc., 1990.

[33] Hammill, *Op. Cit.*, p. 383.

[34] Mertig, A.G., "Attitudes about Wolves in Michigan," *Report to the Michigan Department of Natural Resources-Wildlife Division,* 2004, East Lansing, MI.

[35] Mech, L. D. 2001. *Managing Minnesota's Recovered Wolves,* Wildlife Society Bulletin 29:70-77.

[36] Hammill, *Op. Cit.*, p. 387.

critterbiz / Shutterstock.com

Chapter 14

THE WOLF AS A DISEASE CARRIER

By Will N. Graves

William Larrison / Shutterstock.com

Editors' Statement of the Issues

Wolf proponents stated that wolf recovery was unlikely to have a measurable impact on disease or parasite transmission, but research confirms that wolves are carriers of up to 50 diseases and parasites that can infect hitherto uninfected wildlife, livestock, and people.

Editors' Statement of the Facts

In order to properly avoid or mitigate wolf-transmitted diseases, it is critical that specific information be made available to the general public regarding health safeguards necessary to protect humans and livestock, as well as requiring notification of disease control centers when encountered. Education regarding Hydatid disease in particular is critical for prevention.

> ### *The only way the E. Granulosus parasite will be eliminated is elimination of the wolf.*
> Will N. Graves

Wolves in Finland

There is a long history of wolves attacking domestic livestock, semi-wild reindeer, and people in Finland that traces back before Rome occupied the Scandinavian northland. In response, the Finns would periodically launch campaigns to reduce wolf numbers. In 1734, for example, while some game species were only hunted by noblemen, anyone could kill a wolf on any land, including the private property of another person. Additionally, at least one male from every family had to participate in hunts to control predators.[1]

In the 1960s, Finland conducted a war on wolves in northern Lapland. Planes were sent into the air with soldiers carrying sub-machine guns to kill the enemy. Wildlife biologist Dr. Karloo Nygren remembers the time well. He says there was "a radio program where the local Game Chief (we have 15 such in Finland) gave instructions on how to shoot a wolf through the window—what gun and cartridges one should use and how close to the glass one should keep the barrel when the beast is watching through it."[2]

The True Concern

The war against wolves in the 1960s and 1970s in northern Finland was not so much about predation but rather a tapeworm carried by wolves that was infecting reindeer and the people who herd them (*Echinococcus granulosus*) and the horrible illness (*Hydatidosis* or Hydatid Disease) created by the tapeworms.

Russian scientists have identified 50 different diseases that wolves may carry—including rabies, hoof and mouth disease, anthrax, and brucellosis—that can affect livestock and/or people. All canids, wild and domestic, can carry many diseases, including those caused by tapeworms. Every dog owner knows that his/her dogs should be periodically wormed, but relatively few realize that some tapeworms carried by canines can infect humans, as well as livestock and wild game, and that these diseases can be deadly.

After the wolf culling period in the 1960s and 1970s in Lapland, when wolf numbers were severely reduced, shooting and trapping wolves were greatly reduced. Soon, their populations began to rebound. Dr. Nygren wrote in 2010 that, as wolf populations increase, hydatid disease is again on the rise:

> ... It appears to be spreading in my own home area, Karelia, on both sides of the Fenno-Russian border. I am afraid it will not only affect our staple food and essential part of our heritage, moose, but also us directly. Hunters, dog owners, forest workers, and berry and mushroom pickers will indeed be in danger... none of it is exaggeration.

Wolf Disease Problem in U.S.

The possibility that wolves transplanted into the U.S. from Canada by the 1995-96 relocations might carry *Echinococcus granulosus* was recognized by the U.S. Fish and Wildlife Service. When the 1993 Draft Wolf Environmental Impact Statement to Congress on wolf relocation was made public, it stated, on Page 1-20: "Wolf recovery is unlikely to have any measurable impact on disease or parasite transmission."[3]

Veterinarian Jacob Wustner, who developed and coordinated the veterinary part of the 1995-96 wolf relocation program, stated: "We treated every wolf (at least twice) with three paraciticides, including Droncit, which is essentially 100% effective against *Echinococcus* in wolves with a single treatment. It is extremely unlikely that any reintroduced wolf from Canada could have carried *Echinococcus* tapeworms into the U.S." However, Wustner then conceded that *Echinococcus granulosus* was already present in the Northern Rockies, and so the wolves could have quickly picked it up as they became established.[4]

There is considerable disagreement with this statement from experts in North America and around the world. Parasitologist Dr. Delane C. Kritsky, Professor emeritus, Idaho State University, writes:

... I conducted research for seven years on E. multilocularis *in North Dakota during the 1970s, and it is a very dangerous parasite to human beings. However, the species of* Echinococcus *occurring in wolves and ungulates in Idaho is* Echinococcus granulosus, *a close relative of* E. multilocularis.

E. granulosus *is, in my opinion, more dangerous than the strain of* E. multilocularis *that occurs in the upper Midwest (North Dakota, eastern Montana, South Dakota, and points southeast). The strain of* E. multilocularis *in the north-central states appears to be relatively non-infective to human beings. However,* E. granulosus *is more dangerous because it is highly infective to man and, as a parasite of sheep and domestic dogs, it is much more easily brought into homes in Idaho, Montana, and Wyoming where human beings can be exposed.*

Utah had a focus of E. granulosus *during the 1970s and 1980s during which time people were dying or undergoing dangerous surgery for the parasite cyst. The Utah focus occurred primarily in rural areas where sheep were raised. My friend and colleague, Dr. Ferron Anderson at Brigham Young University, was conducting research on* E. granulosus *in Utah. He developed an educational program that primarily endorsed burying sheep carcasses and deworming dogs, which eventually eliminated the parasite in central Utah.*

The parasite in Idaho will not be dealt with as easily—and I doubt that it can ever be eliminated as long as wolves are present—because wolves and ungulates (deer and elk) will maintain a sylvatic (wild) cycle, which did not occur in Utah during the 1970s and 1980s. Thus, elimination of the parasite from sheep and dogs, as occurred in Utah, will not be successful as it was in Utah because the wild cycle will continuously provide eggs of the parasite for infection of man and his domestic animals in the future.

228

The only way that the parasite will be eliminated from our area is elimination of the wolf. I have examined coyotes, which can carry both species of Echinococcus, *and foxes from southeastern Idaho since 1974 and never found either* Echinococcus multilocularis *or* E. granulosus. *Ferron Anderson never found the latter species in Idaho either when he examined canines in Idaho during the 1970s and 1980s; that is, the* E. granulosus *was never in Idaho until the introduction of the wolf.*

Finally, I asked the Fish and Wildlife, during one of their public meetings concerning introduction of the wolf (prior to wolf introduction), and was "brushed off" by their "promise" that the wolves introduced to Idaho would be "dewormed," which action everyone (and especially they) should have known is never 100% effective.[5]

Hydatid Disease

The illness called Hydatid Disease or *Hydatidosis* is caused by the *Echinococcus granulosus* tapeworm. According to the U.S. Army, there are six species of *Echinococcus* that are currently recognized, and all are capable of infecting man. However, *E. granulosus* is the mostly likely of the six species to cause serious or lethal illness.[6] *Echinococcus granulosus* is a three-millimeter tapeworm that requires two hosts to complete its life cycle: the final host, which harbors the adult parasite, and the intermediate host, which harbors the larval stage. The adult parasite occurs in the small intestine of carnivores, canines (wolves, coyotes, dogs, and foxes); felines (lion, puma, and jaguarundi cat); raccoons; and their domestic counterparts (dogs and cats). Ungulates—deer, elk, moose, sheep, goats, caribou, reindeer, and antelope in the wild and domestic sheep, cattle, buffalo, and goats—are the main intermediate host-group. Humans also can be an intermediate host.

According to the Center for Disease Control, the larval stage of *Echinococcus granulosus* is transmitted as dogs and other canines eat the organs of animals that contain hydatid cysts. Canines pass the parasite to sheep, cattle, goats, and pigs from eggs released in canine feces. *Echinococcus* eggs can stay viable for up to a year.[7]

Humans typically are exposed to eggs through handling carcasses and skins, touching infected animals, and contact with feces of infected animals. Those photos you may have seen of people hugging or kissing wolf-dogs or tame wolves potentially spell trouble, unless the animals have been wormed, as wolves are the most common wild carrier of *Echinococcus granulosus*. All around the world, the people most likely to contract hydatid disease are those who have close contact with dogs that have had contact with either livestock that have the intermediate larval stage or wild, infected canines. This is why the incidence of infection in many areas is highest among women, who clean house, and children, who play with dogs. Anyone who tends to sheep and other livestock and hunters or ranchers who butcher animals also are at risk from contact with the hair of the animals. Thoroughly sanitizing your hands after contact with dogs that may have rolled in wolf scat or are infected themselves is imperative if you live in areas with wolves.

When eggs are ingested through contaminated food or water, they hatch in the small intestine, releasing a small larva (known as an oncosphere) that then penetrates the gut wall and enters the circulatory system. The oncosphere is then carried via the circulatory system first to the liver. Typically, 60 to 75% localize there as cysts. *Echinoccocus granulosus* is the most common cause of liver cysts in the world.[8]

If the larvae pass through the liver, they are carried into the right auricle of the heart and then to the lungs, where they develop into the larval cyst stage called a hydatid cyst. Cysts may become so large and sometimes numerous that organ failures may occur. In wild elk, deer, sheep, goats, caribou, reindeer, and moose, nearly all larvae are found in the lungs. Predators that catch and eat infected animals contract the disease from eating the cysts.

While the lungs and liver are the most common places to find hydatid cysts in ungulates, in human beings, it is not uncommon to find them in other organs of the body, including the brain and reproductive organs. Human beings are not the natural host for *E. granulosus*; therefore, the oncosphere (the larva of the tapeworm), in a sense, becomes lost when entering the blood stream of a human being and can lodge in almost any organ in its body.

After the larvae are established in an organ, growth begins quickly, resulting in a solid mass in which a central cavity soon appears—the beginnings of the hydatid cyst. A membrane soon lines the entire cavity. This membrane is in direct contact with a thin connective tissue capsule separating it from normal host tissue. The hydatid cyst swells and fills with a milky fluid. The larvae grow steadily and are produced in great numbers. Secondary germinal vesicles may develop inside the first cyst, increasing the infected area and the overall size of the cyst.

Cysts in the liver may not be detected until discovered accidentally, depending on their size. In the lungs, difficulty breathing and inflammation are common. If a cyst breaks open, more and more cysts are formed, and they occupy more space and draw vital fluids from the body, resulting in a wide variety of symptoms as they spread throughout the organs and body cavity.

Hydatid disease in the human kidney, spleen, or brain is generally serious. Seizures and paralysis may occur. Osteoporosis of the skull bones is possible. If larvae reach bone marrow, the bones are weakened, resulting in fractures. Spontaneous or traumatic (including surgical) rupture of the cysts may cause severe anaphylactic shock, which can be fatal. Treatment options include medication and surgery. Surgical removal of cysts must be done carefully, without cutting the cyst open. If a cyst is cut, the liquids can spread the disease throughout the entire body.

Detection of *hydatidosis* is not easy. While the parasites quickly become established in the body, actual symptoms may not appear for some time, as much as 10 to 20 years. Often the cysts are not recognized until other illness or body trauma occurs. New ultrasound and serology tests aid detection, but these are not always able to detect hydatid disease. Often infections are ultimately found in autopsies of people who died of other causes.

Prevalence

There is no question that hydatid disease can be serious and deadly. It is also well established that, among wildlife, wolves are a major carrier. They may ingest larvae by feeding on infected wild or domesticated ungulates, rolling in feces, and licking each other or themselves. The real question is just how common is hydatid disease among people?

An international survey of hydatid disease published in 1977 in the Bulletin of the World Health Organization (WHO) entitled "Hydatidosis: A Global Problem of Increasing Importance" finds that hydatid disease is global and increasing not only where it is endemic but also in countries where it was not previously found.[9] The authors—professors of bacteriology and parasitology from Lebanon, Australia, and England—report that, during the 19th century, Iceland had the highest prevalence of human *hydatidosis* in the world. In 1900, it was 25%. They attribute this to the prevalence of sheep and dogs, and they note that an aggressive public health effort to control the disease resulted in a significant decline in human cases during the 20th century.

Other areas with high levels of *hydatidosis* were northern Scandinavia; throughout the Middle East (Israel once reported an average of 100 deaths a year from human *hydatidosis*); the Indian subcontinent, where hydatid disease was the most important helminth[10] health problem in Afghanistan, and where 22% of the adult humans in one village in India tested positive for hydatid disease; the northern part of Russia; North Africa; Alaska and Canada where the disease is frequently found in Indians and Eskimos; southern South America, especially Argentina, Chile, Peru, and Uruguay; and Australia, where the disease is most prevalent in the south and west.

The authors conclude: "Prevention is better than cure... Unless urgent action is taken to redress this situation, the natural pattern of the disease is likely to change for the worse in the next few years." Some may argue that, as the WHO article was published in 1977, it does not reflect contemporary times. In response, following are some brief quotes from other articles:

China. "According to Weiping Wu of China's Institute of Parasitic Diseases, quoted in *Scientific American* in 2005, at least 600,000 Chinese are currently infected by the deadly disease, and an additional 60 million are at risk. "It is an epidemic," he says."[11]

East Africa. In 1988, 18,565 nomadic pastoralists, from 12 different groups—living in the vast, semi-desert regions of Kenya, Sudan, Ethiopia, and Tanzania—were screened for hydatid cysts using a portable ultrasound scanner. High prevalence of *hydatidosis* were recorded among Die northwestern (5.6%) and northeastern (2.1%); the Turkana of northwest Kenya; the Toposa (3.2%) of southern Sudan; the Nyangatom (2.2%), Hamar (0.5%), and Boran (1.8%) of southwest Ethiopia and northern Kenya; and the Maasai (1.0%) of Tanzania.[12]

Uruguay. A study of 1,149 people in the village of LaPaloma in central Uruguay in 1998 found 5.6% had *hydatidosis*. Almost 20% of the dogs in the village were infected.[13]

Wales, UK. *Echinococcus granulosus* in sheep and dogs has been known to be endemic for many decades. An analysis of national hospital records for 1974-1983 found that the incidence of human cystic *echinococcosis* was 0.2 cases per million in England, 2 cases per million in Wales, and the highest rates (5.6 cases per million) occurring in southern Powys County.[14]

Australia. Scientists believe that hydatid disease came to Australia with sheep and dogs, and today it is most prevalent in sheepherding areas. It was most common in the late 1800s, but between 1987 and 1992, 321 people were diagnosed with hydatid disease in New South Wales and the Australian Capital Territory.[15]

Nepal. Hydatid disease is of considerable economic and public health significance in this neighbor of China. One study found that 25% of the residents of one region showed symptoms of *hydatidosis*. Testing of blood samples of patients admitted to different hospitals of Kathmandu Valley showed that the disease had slightly higher prevalence among the males (53%) and considerably higher prevalence among the 35 or older age group (76%).[16]

Finland. In northeastern Finland, 25% of the wolves today are carrying hydatid disease. The incidence of the disease is increasing, although it is not yet commonly found among dogs.[17]

Kazakhstan. Kazakhstan has the most wolves per capita of any country in the world. In 2002, researchers reported that human cystic *Echinococcus* rates increased four-fold within 10 years after post-Soviet independence due to dismantling collectives and changes in organized livestock and farming practices.[18]

Russia. Over 500 cases of hydatid disease are reported in Russia every year, and the number is increasing. Hydatid disease is on the rise in the Bashkiria region of Russia where 53 cases were identified in 2008, which is 1.7 times the number of cases reported the year before.[19]

Hydatid Disease in Northern Rockies

The relocated wolves came from Canada, where hydatid disease is well known. For example, between 1991 and 2001 in a clinic in Edmonton, Alberta, Canada, 42 cases of hydatid disease were identified in people ranging in age from 5-87 years—77% were female, and

Idaho To Take Proactive Measure
To Investigate Hydatid Disease

Senator Tim Corder
Senate Agricultural Affairs Committee
Senator Gary Schroeder
Senate Resources and Environment Committee

Idaho State Senate
State Capitol
P.O. Box 83720
Boise, Idaho 83720-0081

February 16, 2010

Recently a great deal of information has become available, via the internet, about a little known parasite, Echinococcus, and a disease caused by E. granulosus called Hydatid disease. Because of the concern indicated in the information and the assertions that factual data is being purposefully withheld, it would seem prudent to adopt a plan for discovery and action. The actions and discoveries, outlined below, will assist in determining the nature of the risk to human health directly or indirectly through pets, game animals, or livestock. Science-based discoveries will dictate the actions to be taken to protect the public health, game herds, pets, and livestock. Discoveries and actions to be taken:

The Department of Health and Welfare's Division of Public Health will request for the human medical provider community to notify state epidemiologist Dr. Hahn of suspicious or irregular findings consistent with echinococcal disease. They will also get information to family physicians, radiologists, and state disease specialists and survey them about possible cases they may have investigated in Idaho. In addition, they will work to raise awareness in the physician community of the potential for echinococcal disease transmission in Idaho and they will also:

1. Evaluate the value of explicitly adding human echinococcal disease to the state's reportable diseases and conditions (it currently could be reported as an "extraordinary occurrence of illness" under the Reportable Disease Rules).
2. Department of Agriculture will evaluate placement of the disease on the "Notifiable" list of livestock diseases.
3. Request for wildlife biologists that have worked with canids to be checked for the presence of Hydatid cysts.
4. Request federal inspection reports from slaughter houses processing livestock.
5. The Department of Health and Welfare will update their web site to place a link for hunters and those handling wild game. The site will offer safe handling techniques and precautions that will minimize risk for echinococcal and other disease transmission.
6. The Department of Agriculture will offer similar information to livestock producers.
7. State medical authorities Dr. Hahn, Dr. Barton, and Dr. Drew will contact the individuals whose research has been cited on the internet to ensure they have a full awareness of those researchers' positions on echinococcal disease and transmission risk.
8. Drs. Hahn and Tengelsen from the Department of Health and Welfare will make the CDC aware of the changing epidemiology and concerns about potential risk and enlist their assistance in gathering information that will help clarify the epidemiology of echinococcal disease in Idaho.
9. Develop a plan for expanding the examination of canid, cervid, ungulate and other carcasses and the accumulation of data.
10. Use examinations to determine whether more than one species of echinococcus is present.
11. Designate a resident location for the compilation of data and available resources.

Source: http://www.skinnymoose.com/bbb/2010/02/16/idaho-to-take-proactive-measure-to-investigate-hydatid-disease/

41% were native (women and children tend to most often contract the disease as women clean house and children play with dogs); 40% of the patients had cysts in their lungs, and 55% had cysts in their liver. The researchers report that the most common intermediate hosts are barren ground caribou and then moose. Studies have shown that 50% of the moose in Ontario and British Columbia have hydatid disease.[20]

Between 2006 and 2008, a team of federal and state veterinarians and biologists—Foreyt, Atkinson, and McCauley—evaluated the small intestines of 123 gray wolves collected in Idaho and Montana, looking for *Echinococcus granulosus* tapeworms. Reporting their findings in the *Journal of Wildlife Diseases* in 2009, they stated, "The tapeworm was detected in 39 of 63 wolves (62%) in Idaho, and 38 of 60 wolves (63%) in Montana. The detection of thousands of tapeworms per wolf was a common finding."

Intermediate-form hydatid cysts were found in Idaho elk, mule deer, and mountain goats and in Montana elk. The researchers concluded, "To our knowledge, this is the first report of adult *E. granulosus* in Idaho or Montana… Based on our results, the parasite is now well established in wolves in these states and is documented in elk, mule deer, and a mountain goat as intermediate hosts."[21]

The extent of human infection in the United States may not be known for years; however, in 2011 a woman in Idaho, who lives in an area where wolves are common, was operated on to remove a large hydatid cyst on her liver.[22]

What Can Be Done?

Dr. Valerius Geist, who had a relative die of hydatid disease, advises hunters and those with domestic dogs where wolves are present to take the following steps to protect themselves and their pets:

 To prevent infection of dogs, do not consume or allow your dog to consume uncooked meat or organs of wild or domestic ungulates. If your dog does have access to carcasses, talk to your veterinarian about an appropriate deworming treatment. Deworming your dog and preventing access to carcasses of dead sheep, which should be buried immediately, are important preventive steps.

 For hunters: do not touch or disturb wolf, coyote, or fox scat. Wear gloves when field dressing a canid carcass, and

wash any body part that may have been exposed to feces or contaminated fur.

Neospora Caninum

Wolves also may carry a prozoan parasitic disease, *Neospora caninum*, which so far has not been shown to affect people, but it does cause cattle, sheep, and goats to abort, resulting in considerable economic loss to ranchers and farmers. *N. caninum* was not identified until 1988. It has since been found worldwide.[23] It is a major cause of abortions in ruminants. Adult infected cattle appear healthy, but they will abort at least 20% of the time during their lifetime,[24] causing millions of dollars in damages every year.[25]

Like *Echinococcus, N. caninum* has a life cycle with two hosts—canids as adult hosts and ruminants as the intermediate host. *Neospora caninum* may be transmitted by dogs, wolves, coyotes, and foxes. Wild, free-ranging wolves make it very difficult to control the disease in areas where there are cattle, sheep, and goats. Ruminants contract the infectious eggs when they feed on grass where there are wolf, fox, and coyote feces.[26]

When introduced into a herd, up to 90% of the unborn calves could be infected. When the brain and nervous system of cows, as well as sheep and goats, are infected, it results in abortions. When canines eat the aborted fetuses and placenta, the disease is spread, as their feces contain oocysts (hardy, thick-walled spores able to survive for lengthy periods outside a host). When ruminants feed where the canines have defecated, they ingest the oocysts, which then infect the dogs, wolves, foxes, and coyotes, and the life cycle begins again. *Neospora caninum* is now recognized as being worldwide and one of the most damaging of all parasites transmitted to cattle.[27] *Neospora caninum* does not appear to be infectious to people, but in dogs, it has been found that infection can lead to neurological damage.[28]

In 2011, a team of researchers from the USDA Animal Parasitic Research Laboratory found *Neospora* oocysts in three of 73 necropsied wolves in Minnesota.[29] One study in 2004 found 39% of 164 wolves from Minnesota were infected with *Neospora caninum*.[30] A study of wolves in Yellowstone National Park published in 2009 found that 50% of the wolves tested positive for *Neospora caninum,* with the oldest wolves most likely to be infected.[31]

Researchers in Wisconsin also report wolves testing positive for *Neospora caninum.* As both Minnesota and Wisconsin have a significant dairy industry and wolf populations are increasing in both states, one would expect that *Neospora caninum* would also be increasing.[32] In a 2011 study reported in the *Journal of Veterinary Parasitology*, *Neospora*-like oocysts were found in the feces of three of 73 wolves in Minnesota examined by necropsy.[33]

N. caninum has been shown to be a large economic loss to the dairy and beef industry with infected animals being three to 13 times more likely to abort than non-infected cattle. The researchers also note that the presence of wolves near cattle and other livestock causes the livestock to run about in fear, which adds further stress that also increases abortions.[34]

Considering there are wild and domestic intermediate hosts in areas where wolves are now found, eradication is very difficult. However, a vaccine has recently been developed that yields partial protection to ruminants.[35]

Rabies

Rabies is a viral disease normally spread by the saliva of infected mammals that bite others. After being bitten by a rabid animal, the disease attacks the central nervous system, resulting in swelling, which progresses to the brain. Symptoms may appear as early as seven days after a bite to as long as several years. The average incubation time is three to seven weeks. Early-stage symptoms of rabies include malaise and fever, progressing to acute pain, violent movements, uncontrolled excitement, depression, hydrophobia, hallucinations, and delirium. As rabies progresses, violent mood swings may occur before the victim lapses into a coma. If not treated with a vaccine before symptoms begin, rabies is usually fatal. Death typically occurs two to 10 days after symptoms appear. Modern vaccines seldom have any side effects beyond those from common influenza shots.

Rabies vaccines for dogs have severely curtailed bites by dogs as a cause of rabies in the U.S. In underdeveloped countries abroad, however, dogs are a major vector for rabies. In North America, rabies is most often spread by wild animals, especially raccoons, bats, skunks, foxes, coyotes, and, where they are found, wolves. When animals become rabid, they seem to lose all fear. Thus, a rabid animal becomes easy prey for a predator, and conversely, rabid animals are not afraid of anything. Rabies is present in almost all countries of the world, and especially in Asia, attacks on humans by rabid wolves are not uncommon. There are an estimated 55,000 human deaths from rabies worldwide every year—about 31,000 in Asia, and 24,000 in Africa.[36] There were 45 cases of rabies in the U.S. between 1995-

2010; of these, nine are thought to have been acquired abroad. Bats are the most common carriers in North America.[37]

Conclusion

Wolves are wild ancestors of dogs, but unlike pet dogs, wild wolves do not receive treatment from veterinarians. As a result, the 50-some diseases they may carry double their negative impact on human settlements. This reality amplifies the need to keep a safe distance between man and *Canis lupus*.

🐾 🐾 🐾 🐾 🐾 🐾 🐾 🐾

[1] http://www.ekoi.lt/info/azl/2003/AZL%2013_15-20.pdf

[2] http://www.savewesternwildlife.org/killer-wolves.html

[3] http://huntingnewsdaily.com/2011/07/08/idaho-fg-perpetuates-ignorance-with-misinformation/

[4] http://missoulian.com/mobile/article_cfd5615e-77ce-11df-bcde-001cc4c03286.html

[5] Personal correspondence with Dr. Delane C. Kritsky, Professor Emeritus, Idaho State University, 2011.

[6] http://tmcr.usuhs.mil/tmcr/chapter3/intro.htm

[7] http://www.cdc.gov/parasites/echinococcosis/epi.html; for detailed life-cycle information, see: www.dpd.cdc.gov/DPDx/html/Echinococcosis.htm

[8] http://www.ispub.com/journal/the-internet-journal-of-pulmonary-medicine/volume-10-number-1/unusual-presentation-of-hydatid-disease.html

[9] Matossian, R.M., M.D. Rickard, and J.D. Smythe, "Hydatidosis: A Global Problem of Increasing Importance," *Bulletin of the World Health Organization*, 1977, 55 (4) 499-507.

[10] Pertaining to intestinal worms, such as the tapeworm and the roundworm.

[11] *Scientific American*, July 2005 issue, page 22; http://www.khamaid.org/about_kham/news/hydatidosis.htm

[12] http://www.tropicalmedandhygienejrnl.net/article/0035-9203%2889%2990664-0/abstract

[13] http://www.ncbi.nlm.nih.gov/pubmed/9790441

[14] http://findarticles.com/p/articles/mi_m0GVK/is_4_11/ai_n13609429/

[15] http://www.science.org.au/nova/056/056print.htm

[16] http://homepage.usask.ca/~shb292/hydatid.pdf

[17] http://www.sciencedirect.com/science/article/pii/S0304401702003813

[18] Togerson, P.R., B.S. Shaikenov, K.K. Baitursinov, and A.M. Abdybekova, "The Emergent Epidemic of Echinococcus in Kazakhstan," Translated, *Society of Tropical Medical Hygiene*, 2002; 96: 124-8.

[19] http://www.wormsandgermsblog.com/2009/01/articles/animals/dogs/echinococcus-on-the-rise-in-bashkiria-russia/

[20] http://www.biomedcentral.com/1471-2334/5/34/prepub

21 Foreyt, William J., Mark L. Drew, Mark Atkinson, and Deborah McCauley, "*Echinococcus Granulosus* in Gray Wolves and Ungulates in Idaho and Montana, USA," *Journal of Wildlife Diseases*, 2009: 1208-1212.

22 http://www.sciencedirect.com/science/article/pii/S0304401711003566

23 Dubey, J.P., "Neosporosis—The First Decade of Research," *International Journal for Parasitology*. 29 (10):1485–8.

24 http://www.wvdl.wisc.edu/PDF%5CWVDL.Info.Recognizing_and_Preventing_Neosporosis_Infections.pdf

25 http://cmr.asm.org/content/20/2/323.abstract

26 http://www.ars.usda.gov/Main/docs.htm?docid=11007

27 http://www.merck-animal-health-usa.com/binaries/NeoGuard_Fact_Sheet_tcm130-126564.pdf

28 Barber, J.S., C.E. Payne-Johnson, and A.J. Trees, "Distribution of Neospora Caninum Within the Central Nervous System and Other Tissues of Six Dogs with Clinical Neosporosis," *Journal of Small Animal Practice* 37 (12): 568–74.

29 "I am a Woman with a Story to Tell," *The Outdoorsman*, March-May, 2011: 7-9.

30 http://dnr.wi.gov/org/land/er/publications/pdfs/wolf_impact.pdf

31 http://www.ncbi.nlm.nih.gov/pmc/articles/PMC2738425/

32 http://www.wvdl.wisc.edu/PDF%5CWVDL.Info.Recognizing_and_Preventing_Neosporosis_Infections.pdf

33 J.P. Dubey, M.C. Jenkins, C. Rajendran, K. Miska, L.R. Ferreira, J. Martins, O.C.H. Kwok, and S. Choudhary, "Gray Wolf (*Canis lupus*) Is A Natural Definitive Host for *Neospora Caninum*," Veterinary Parasitology, 2011.

34 http://dnr.wi.gov/org/land/er/publications/pdfs/wolf_impact.pdf

35 http://www.merck-animal-health-usa.com/binaries/NeoGuard_Fact_Sheet_tcm130-126564.pdf

36 "Rabies," World Health Organization, September 2011. Retrieved 31 December 2011.

37 "Rabies Surveillance Data in the United States," Centers for Disease Control and Prevention; http://www.cdc.gov/rabies/

Dennis Donahue / Shutterstock.com

Chapter 15

THE WOLF AS A CASH COW

By Karen Budd-Falen

Karen Budd-Falen grew up as a fifth-generation rancher on a family-owned ranch in Big Piney, Wyoming. She received her undergraduate degrees and her law degree from the University of Wyoming. She is admitted to practice law before the State Court of Wyoming; U.S. District Courts for Wyoming, Colorado, Nebraska, and District of Columbia; U.S. Courts of Appeals for the Seventh, Eighth, Ninth, Tenth, District of Columbia, and Federal Circuits; U.S. Court of Federal Claims; and U.S. Supreme Court. With her husband Frank Falen, Budd-Falen is the owner of the Budd-Falen Law Offices, located in Cheyenne, Wyoming. She served for three years in the Reagan Administration, U.S. Department of the Interior, Washington, D.C., as a Special Assistant to the Assistant Secretary for Land and Minerals Management. She later served as a law clerk to the Assistant Solicitor for Water and Power, and she has also worked as an attorney at Mountain States Legal Foundation. Karen has been featured in *Newsweek*'s "Who's Who: 20 for the Future" for her work on property rights issues, and she was awarded Wyoming's Outstanding Ag Citizen in 2001.

Editors' Statement of the Issues:
Although environmental groups will argue that the litigation they file against the federal government related to the wolf is primarily because of their altruistic concerns over protection of this predator, the amount of taxpayer dollars paid to these groups for these cases suggests another, much more financially beneficial reason to continue bringing wolf cases before the federal courts.

Editors' Statement of the Facts:
The massive amounts of litigation related to the wolf and other species either listed, or proposed to be listed, under the Endangered Species Act will not subside until Congress acts to close the loopholes in the attorney's fees payment statutes that financially reward environmental groups and their litigation counsel with American taxpayer dollars.

> *The environmentalists have taken one animal, the wolf,*
> *and used it and the ESA to develop*
> *a profit-generating business opportunity for its lawyers.*
> Karen Budd-Falen

The Economics of Litigation

Environmental groups have profited immensely by "defending wolves" in federal litigation. Interestingly, in their fundraising appeals, environmental groups explain their need for funds as essential to support education, administration, lobbying, and litigation. What they fail to explain is that the litigation itself is also a significant source of income. But the wolf is but one of many species. Even more troubling is that the attorney's fees paid to these groups by the federal government are not properly tracked, so the public cannot know for sure exactly how much money the environmental groups are taking in by their constant filing of litigation.

Additionally, the funding source of the attorney's fees payments made to the environmental groups is not always clear. Generally, payments are made either from the Department of Treasury's Judgment Fund (an open-ended budget item) by way of the Endangered Species Act of 1973's citizen-suit provision or from the responsible agency's budget pursuant to the Equal Access to Justice Act by way of the Administrative Procedure Act (APA). Oftentimes, environmental groups file their complaints mixing both ESA citizen-suit claims and APA claims; therefore, even a review of the claims cannot provide a clear answer.

242

Environmental Litigation Gravy Train

There are two major sources of attorney fees that can be paid to plaintiffs that "prevail" in litigation either by winning a case on the merits or by the Justice Department agreeing that the group "prevailed" in a settlement by achieving the purpose of the litigation.

One source is called the "Judgment Fund." The Judgment Fund is a Congressional line-item appropriation and is used for Endangered Species Act cases, Clean Water Act cases, and other statutes that directly allow a plaintiff to recover attorney fees. There is no central database for tracking the payment of these fees; thus neither the taxpayers nor members of Congress nor the federal government knows the total amount of taxpayer dollars spent from the Judgment Fund on individual cases.

The second major source of payments to "winning" litigants against the federal government is the Equal Access to Justice Act (EAJA). EAJA funds are taken from the "losing" federal agencies' budget. Within the federal government, there is no central data system or tracking of these payments from the agencies' budgets.

Karen Budd-Falen, 9/15/2009

Judgment Fund

Prior to 1956, most judgments against the United States could not be paid from existing appropriations, but required specific congressional appropriations. In 1956, Congress enacted a permanent, indefinite appropriation called the Judgment Fund for the payment of final judgments which were "not otherwise provided for" by another source of funds.

EAJA

In 1980, Congress embarked upon an experiment in "curbing excessive regulation and the unreasonable exercise of government authority" by directing that attorney fees be awarded in favor of private parties who resist unjustifiable government conduct in litigation. Signed into law by President Carter on October 21, 1980, the EAJA had a sunset provision and expired by its own terms on October 1, 1984. In 1985, Congress passed new legislation that reinstated EAJA retroactively to October 1, 1984. On August 5, 1985, President Reagan signed the bill, which made the EAJA a permanent statute.

The EAJA awards costs of litigation and attorney fees (up to $125/hour) when a citizen, non-profit organization, or small business wins a case involving the federal government and can show that the federal government's position was not "substantially justified." The EAJA's purpose is to help those who had been locked out of the decision-making process by virtue of their income, their race, their economic scale, or their educational limitations.

Thirty Years of Profits

While the time span of wolf litigation actually covers almost 30 years of profits, the ridiculousness of the attorney's fees awards ballooned in the last 10 years, when the average case garnered over a quarter million dollars in attorney's fees. The following is a summary of major cases where environmentalists have won awards of attorney's fees by representing the interests of wolves over the interests of those people living nearby.

> ***Sierra Club v. Clark***. In 1985, the Sierra Club secured an appeals court win in the Eighth Circuit Court of Appeals in *Sierra Club v. Clark*.[1] In *Sierra Club*, the environmentalists challenged rules concerning the Eastern Timber wolves (gray wolves living in Minnesota). At the time, the wolves were listed as threatened species under the Endangered Species Act of 1973 (ESA).[2] As the population of wolves in Minnesota spiraled out of control, the state requested the federal government transfer control of the wolves to the state so it could create a public sport season to control wolf numbers.
>
> While the U.S. Fish and Wildlife Service (USFWS) did not transfer control of the wolves to the state, in 1983 the USFWS promulgated regulations permitting trapping of the wolves in certain areas where repeat livestock predation had occurred. Shortly after publication of the new regulations, environmentalists filed suit. The Eighth Circuit held that the regulations violated the ESA because it found the ESA did not provide discretion to the Secretary of the Interior to allow for "taking" threatened species. The environmentalists won and were awarded $55,369.45 in attorney's fees. That number included a 30% upward adjustment of the fees of the two attorneys for the "extraordinary circumstances and the importance of the case to the management of all threatened species." Little did anyone know that, years later, we would look back on an award of $55,000 of taxpayer funds as a modest payment.
>
> ***Biodiversity Legal Foundation v. Babbitt***. Ten years passed before the environmental groups again sued over wolves in *Biodiversity Legal Foundation v. Babbitt*.[3] In *Biodiversity*

Legal, the environmentalists petitioned USFWS to list the Alexander Archipelago wolf, which is found in the Tongass National Forest in Alaska.[4] For several years, the U.S. Forest Service had been considering major revisions to the Tongass Land Management Plan (TLMP). Because the Forest Service had major changes planned for the TLMP, the Regional Director of USFWS found that listing the Alexander Archipelago wolf was not warranted, as the Forest Service's revisions would reverse any declining population of the wolf. The District Court in Washington, D.C., overruled the agency's action, holding that the ESA only allowed consideration of current regulations and their effects on species, not upcoming changes. The environmentalists won and were awarded attorney's fees.[5]

Defenders of Wildlife v. Department of the Interior. Another 10 years passed before the environmentalists recognized their cash cow: ESA litigation. In *Defenders of Wildlife v. Department of the Interior*,[6] the environmentalists cashed in on their first quarter-million-dollar wolf case. After numerous years of much higher than expected "recovery" of wolves throughout the country, the USFWS issued a rule creating three distinct population segments (DPSs) for the gray wolf and downlisting the wolf from endangered to threatened in the Eastern and Western DPSs (2003 Final Rule). Wolves in the Southwestern DPS remained listed as endangered, and the experimental populations in the Northern Rocky Mountains remained unchanged. The environmentalists succeeded in convincing the court that the USFWS had violated the ESA by creating the DPSs and downlisting wolves in major parts of the country. In doing so, the environmentalists found what they were looking for: attorney's fees paid by the federal government to the tune of $272,710.54.

National Wildlife Federation v. Norton. In the same year as *Defenders of Wildlife*, environmentalists double-dipped on the same issue. Nine months after the U.S. District Court for the District of Oregon held the 2003 Final Rule promulgated by the USFWS violated the ESA, another environmentalist group scored another quarter of a million dollars from challenging *the same rule*. In *National Wildlife Federation v. Norton*,[7] the environmentalists challenged USFWS' final rule because it had combined two DPSs listed in its proposed rule into one DPS in its final rule. The environmentalists won their argument and, even though the final rule had already been invalidated more than six months prior, secured more funds for their coffers: $255,500.

Humane Society of U.S. v. Kempthorne. After paying the environmentalists half a million dollars to litigate against its 2003 Final Rule, the USFWS issued a new rule in 2007 creating a Western Great Lakes DPS and simultaneously delisting that DPS, as the wolf numbers had over the years exploded in Minnesota. In *Humane Society of U.S. v. Kempthorne (HSUS)*,[8] the animal rights group challenged USFWS for creating a DPS based on geographical boundaries and at the same time removing that DPS from ESA protection by delisting the DPS. HSUS was able to successfully argue that USFWS was incorrect in considering the ESA clear on its face for providing authority to create a DPS and delist it at the same time. The court did not hold that USFWS' interpretation of the statute was unreasonable—only that the statute was not clear on its face. The court vacated the 2007 rule and remanded it back to the USFWS to explain its interpretation. For that, HSUS cashed a check for $280,000.

Defenders of Wildlife v. Hall. The final case that environ-mentalists used to make wolf litigation a million-dollar industry was *Defenders of Wildlife v. Hall (Defenders I)*.[9] In *Defenders I*, the environmentalists challenged a rule similar to that in *HSUS*, only the rule in *Defenders I* concerned wolves in the Northern Rocky Mountain DPS instead of in Minnesota. In 2008, USFWS created this DPS and delisted all the wolves in the DPS. Judge Molloy held that USFWS acted arbitrarily by delisting these wolves without proof of genetic exchange between the subpopulations and without explaining its reversal concerning Wyoming's wolf management plan.[10] Judge Molloy granted a preliminary injunction, finding that the environmentalists would likely succeed on the merits of their case. As a result, the wolves in Idaho, Montana, and Wyoming were relisted, and the environmentalists took in an additional $263,099.66. (The environmental groups had actually requested $388,370 in attorney's fees.)

USFWS attempted to comply with Judge Molloy's ruling in *Defenders I*. In 2009, USFWS issued another final rule, this time delisting the wolves in Montana and Idaho. With respect to the wolves in Wyoming, FWS stated that the proposed management plan for Wyoming was deficient, and therefore, the wolves in Wyoming would remain listed until the issues were sufficiently addressed.

Again, the environmentalists challenged this rule in *Defenders of Wildlife v. Salazar* ("*Defenders II*").[11] And again, Judge Molloy ruled in favor of the environmentalists, this time on the basis that ESA does not allow for designations smaller than DPSs, which the court held USFWS had done by delisting the wolves from the same DPS in Idaho and Montana, but not delisting them in Wyoming.

No Matter What

The United States Congress eventually took notice that, no matter what USFWS would do with wolves, the environmentalists would challenge those decisions in court. In response to Judge Molloy's decision in *Defenders II* and while that case was on appeal to the Ninth Circuit, Congress passed H.R. 1473, the Department of Defense and Full Year Continuing Appropriations Act of 2011. Congress, in §1713 of that Act, directed USFWS to reissue the 2009 rule. Moreover, Congress mandated that the 2009 rule could not be subject to judicial review, thus shielding it from the environmentalists' never-ending challenges.

Yet again, the environmentalists filed another challenge in court, this time to Congress' passage of §1713 in *Alliance for the Wild Rockies v. Salazar*.[12] The environmentalists claimed that Congress overstepped its authority and unconstitutionally violated the separation of powers doctrine. Judge Molloy held §1713 as constitutional. The environmentalists appealed that decision, but the Ninth Circuit affirmed the constitutionality of §1713 in *Alliance for the Wild Rockies v. Salazar*.[13]

Finally, more than 15 years after USFWS introduced Canadian wolves to the Rockies and after the public paid more than $1 million to the environmentalists' attorneys, wolves in Idaho and Montana have been removed from ESA protection.

Yet Another Law Suit

The final step concerns the experimental wolves in Wyoming. Since *Defenders I*, Wyoming government officials have worked to satisfy the issues with its management plan. On September 10, 2012, after finding Wyoming's latest management plan sufficient, FWS published a final rule delisting the wolves in Wyoming.[14] On the same day, four environmental groups (Defenders of Wildlife, Center for Biological Diversity, Natural Resources Defense Council, and the Sierra Club) gave notice of their intent to sue USFWS over the 2012 rule delisting the wolves in Wyoming.[15] Under the citizen-suit provision of the ESA, these environmental groups will once again sue the federal government in their attempt to "defend wolves" and cash in on big-dollar attorney's fees.

Tip of the Iceberg

The environmentalists have taken one animal, the wolf, and used it and the ESA to develop a profit-generating business opportunity for its lawyers. In a span of only a few years, the environmentalists have been able to promote the interests of wolves over the interests of those people living near the animals and other species of wildlife and, in doing so, cashed in on over $1 million of taxpayer funds. Those funds could have been used to conduct many other wildlife management programs, and the man-hours spent on the case by the federal government—both attorneys and staff—could also have been directed to other purposes.

As if this was not bad enough, the wolf is the tip of the iceberg in environmentalist and animal rights litigation against the federal government. According to Pulitzer Prize-winning *Sacramento Bee* journalist Tom Knudsen, during the 1990s, the U.S. government paid out $31.6 million in attorney fees for 434 cases brought against federal agencies by environmental organization attorneys.[16]

Numerous studies find ever-increasing sums of money being paid by the federal government to environmental and animal-rights groups. Moreover, even though the law calls for the federal government to keep track of money paid out for attorney's fees, the United States Governmental Accountability Office (GAO) has found that accounting stopped in 1995, so the government does not know how much money is actually being paid to environmental litigants.

The GAO report, April 12, 2012, finds: "Most USDA and Interior agencies did not have readily available information on attorney fee claims and payments made under EAJA and other fee-shifting statutes for fiscal years 2000 through 2010. As a result, there was no way to readily determine who made claims, the total amount each department paid or awarded in attorney fees, who received the payments, or the statutes under which the cases were brought for the claims over the 11-year period."[17]

When the GAO asked 75 bureaus and agencies at USDA and the Department of the Interior for records on payments, only *ten* could provide data on cases and attorney fee reimbursements. Moreover, the records provided were incomplete and unreliable, based on manual calculations from older files and the memory of career employees. As such, the GAO is not certain of its totals. Even with the data the GAO could find from those ten units that had any records at all, the GAO identified $4.4 million per year of payments to environmental groups during the period of 2000-2010.

The GAO's minimum numbers do not add up to totals available from public court records. Public federal court records from just 13 federal courts revealed $5.2 million in legal fees per year, compared to the GAO's estimate of $4.4 million. An analysis published in the Notre Dame Law School *Journal of Legislation* considered evidence from the public tax returns of just 20 environmental groups; in fiscal year 2010, those documents showed $9.1 million in attorney's fees.[18]

Similarly, according to data from the United States Department of Treasury, attorney's fees payments made from the Judgment Fund alone for ESA cases—not including payments made pursuant to EAJA—averaged more than $6.6 million per year for fiscal years 2009 through 2011. That figure was up from the prior three-year time span, when the average payments from the Judgment Fund alone for ESA cases were $1.6 million per year for fiscal years 2006 through 2008.

Additionally, a careful review of the cases finds that many are about challenging "minor procedural decisions"—red tape—rather than substantive issues. Thus, not only does the government not know just how much money is being spent on attorney's fees, but also the adversarial actions of the environmental and animal rights groups are causing more stress on the federal agencies, which is burying them in paperwork and making them more prone to overlook bureaucratic details, which in turn makes them easier targets for more legal actions.

Conclusion

In short, the wolf is a cash cow for environmental litigants, and the successes of wolf litigation cases have helped establish a sizeable source of revenue for environmental and animal rights groups, all at the expense of the taxpayer in particular and wildlife in general. And unfortunately, the wolf is only one example of how environmental and animal rights groups are using the ESA and the EAJA to drain economic resources from taxpayers and reduce the effectiveness of federal agencies in practicing effective wildlife management.

🐾 🐾 🐾 🐾 🐾 🐾 🐾 🐾

[1] 755 F. Supp. 2d 608, 8th Cir. 1985.

[2] 16 U.S.C. §§ 1531-1543.

[3] 943 F. Supp. 23, D.D.C. 1996.

[4] *Id.* at 23.

[5] The exact amount is not available on Public Access to Court Electronic Records (PACER), the public database for federal court records.

[6] 354 F. Supp. 2d 1156, D. Ore. 2005.

[7] 386 F. Supp. 2d 553, D. Ver. 2005.

[8] 579 F. Supp. 2d 7. D. D.C. 2008.

[9] 565 F. Supp. 2d 1160, D. Mont. 2008.

[10] Previously, USFWS rejected Wyoming's plan; Wyoming held its ground that wolves should not be protected throughout the entire state, much of which is not suitable habitat for wolves.

[11] 729 F. Supp. 2d 1207, D. Mont. 2010.

[12] 800 F. Supp. 2d 1123, D. Mont. 2011.

[13] 672 F. Supp. 3d 1170, 9th Cir. 2012.

[14] 77 Fed. Reg. 55530.

[15] http://earthjustice.org/sites/default/files/wy-delisting-noi.pdf

[16] http://icecap.us/images/uploads/ENVIRONMENTINC.doc

[17] http://www.gao.gov/products/GAO-12-417R

[18] http://www.boone-crockett.org/news/featured_story.asp?area=news&ID=138

John L. Absher / Shutterstock.com

Chapter 16

HOW THE FIRST POLITICAL WOLF WAR WAS WON

By Ted B. Lyon

Holly Kuchera / Shutterstock.com

Editors' Statement of the Issues

In the highly polarized political climate we find ourselves in recent years, it is all too common for valid issues to get sidelined by party lines. This we/they attitude breeds controversy and eliminates meaningful dialogue between legislators as well as constituencies.

Editors' Statement of the Facts

Both sides of the aisle are required to support legislative change, and educating influential political leaders is critical if they are to help influence important legislation such as the delisting of the wolves. But more important than education is the necessity that supporting documents are objective and factual—beyond reproach—such that, when presented, leaders have the confidence to move forward.

I knew that, if we could get the sportsmen groups together
and tell the truth about what was happening,
the law would eventually be changed.
Ted B. Lyon

Democracy

As Americans, we are used to the fairness of democracy. We are used to seeing the will of the people acknowledged and their choices made clear through legislation—slowly perhaps but inexorably. It became painfully clear that this matter of wolf management *had* to come to Washington, D.C., and sooner *not* later. It was also clear that an extraordinary effort was going to be needed to undo the damage done by the introduction of Canadian wolves, and I wanted to be part of that effort.

The real beauty of democracy is the potential for individual concerned citizens to reach out to their legislators to enact change, and I felt strongly that I should add my voice to the clamor of the stakeholders affected by the wolves. Because of my background in politics and my relationships with so many sportsmen and conservationists, I already knew many who might be willing to take some time to see what plans could be laid to either change the Endangered Species Act (ESA) or, failing that, simply find a way to get the wolves delisted and under individual state's management.

As a country, we are the victims of free speech as well as its beneficiaries. Just about every point of view is available to be heard, and it takes an educated and determined listener to winnow through the chaff to find the wheat. What we were dedicated to do was to provide very busy men and women the kernels of truth not currently available to the general public in such a way that they would be able to reassess their perceptions of the wolf management issue and then work together to make changes.

Once a group of us started the dialogue with legislators about getting things changed, it created an energy of its own, and we found stakeholders all across the country, angered by the continued listing of the wolves, who were contacting their congressmen for action. We started with small meetings—here and there across the country with concerned organizations, governors, senators, and representatives— to establish the basic groundwork of presenting the hard cold facts of the situation at hand. And those facts spoke for themselves. They spoke so well, in fact, that some of the legislators were stunned by what they learned.

Allies

The first step in any political battle is to find allies. In early 2010, I called Ray Anderson and asked him, as a member of the Rocky Mountain Elk Foundation (RMEF), to go to their headquarters in Missoula, Montana, and ask them to call me. RMEF is one of the premier wildlife conservation groups in North America. They have almost 200,000 members and have conserved over six million acres of land in conservation easements for wildlife. RMEF had never really taken a stand on wolf issues because some of their former executives thought it might be too controversial. Ray must have been effective because I was asked to call Rod Triepke, Chief Operating Officer, the next day.

Rod Triepke suggested that I talk to the RMEF President, David Allen. David is active on a national basis and knows many of the big name players in the conservation movement. Additionally, RMEF has tremendous credibility throughout the West and across the conservation community.

Networking

That was in March of 2010. No one involved in the wildlife conservation community really knew who I was. I was simply some guy from Texas, who had a home in Montana and who was complaining about wolves. I was certain that, if we could get the sportsmen groups together and tell the truth about what was happening, the law would eventually be changed.

Bombshells and Shock Waves

On March 30, 2010, Mike Leahy, Director of the Rocky Mountain Region of the Defenders of Wildlife, and Kirk Robinson, Executive Director of the Western Wildlife Conservancy in Salt Lake City, Utah, sent David Allen an adversarial letter in which they accused the Rocky Mountain Elk Foundation of polarizing the important conservation issue of wolves and elk. They claimed that the wolves introduced into Yellowstone were native (not true), and they then called for "a scientific review of wolf recovery criteria to incorporate the best available science, followed by a regional stakeholder process to guide development of state plans that meet wolves' biological needs while addressing the legitimate concerns of affected people and communities." In an effort to intimidate Allen, they sent a copy to every Rocky Mountain Elk Foundation board member and published it.

David sent me a copy of the letter. He responded to it immediately, and he released his letter to the press. His reply letter—covered throughout the West and dated April 8, 2010—was a bombshell. He pointed out that Defenders of Wildlife and the Western Wildlife Conservancy were "contributing to perhaps one of the worst wildlife management disasters since the destruction of bison herds in the 19th century."

David went on to tell them that—until the federal lawsuit they had brought, which stopped the states of Montana and Idaho from managing the wolves, was dropped—there really was no need to meet. David went on to point out that groups such as Defenders had continued to move the "goal line" of wolf introduction, which had gone from 300 total wolves in Montana, Idaho, and Wyoming under the original introduction to a current estimate of a minimum of 1,700 wolves with many respected biologists considering there to be many more. David pointed out that, therefore, the wolves were not endangered at all. He further pointed out that the states and the federal government had, through their wildlife agencies, already called for delisting of wolves but that the wolf support groups opposed those scientists. At the end, he offered to meet with groups to try to work the issues out. They eventually met, but Leahy could provide no new science to support the Defenders' position.

David's letter, which was widely published, sent shock waves throughout the political world. If an organization that was as well respected in wildlife conservation circles as RMEF felt so strongly about the Endangered Species Act and wolves, then surely something must be wrong with current policy.

254

Help from Congressman Edwards

In May 2010, I called Congressman Chet Edwards. Chet and I served together in the Texas Senate for 10 years, and he and I are extremely good friends. While serving in the Senate, Chet always looked to me for advice on hunting and fishing issues since I was such an avid outdoorsman and served for a number of years as the vice chairman of the Senate Natural Resources Committee. I asked Chet if he would introduce a bill that would change the Endangered Species Act and take wolves out of the act. Chet knew that I had a home in Montana, and after I explained to him what was happening, he agreed that something needed to be done.

Chet directed his staff to go to legislative counsel and draft a bill. The first bill that was sent to me was so complex that I could see it being litigated for years—just like the Endangered Species Act. I called Chet's aide up and dictated to him over the phone how I thought the bill should be drafted. The next draft was simple. Chet filed the bill, and from there, things started to happen very fast. This gained me the advantage of instant credibility with many of the sportsmen groups.

I met with David Allen, RMEF President, in Missoula, Montana, and outlined my plan to get all of the other sportsmen groups to support Chet's bill to take the wolves out of the Endangered Species Act and let the states manage them.

Politics

Also in May of 2010, I reached out to Don Peay, Founder of Sportsmen for Fish and Wildlife (see Chapter 7). Don is from Utah and is truly a human dynamo. He has done more for wildlife conservation in Utah and the West than most other people. I asked him to fly to Dallas and meet Congressman Edwards, a Democrat. Don is a moderate, who supports both Democrats and Republicans that support wildlife. Don is very close to a number of Republican senators and understands politics better than anyone I know in the conservation movement.

Don, Chet, and I met that evening and formulated a plan. Don would work with the Republicans, and I would work with the Democrats. We knew going in that the end result would not be the bill that Chet had introduced, but a bill that would probably have to be phased in. Chet was able, within a few weeks, to get Congressman Denny Rehberg, a Republican from Montana, to sign on as a co-sponsor of the bill, and by that fall we ended up with close to 60 co-sponsors in the House, both Democrats and Republicans.

In June of 2010, I drove to Missoula, Montana, and met with David Allen and Montana Governor Brian Schweitzer. A rancher himself, Governor Schweitzer was fed up with wolves and what they were doing to the wildlife and people of Montana, Idaho, and Wyoming. He agreed to support the concept of the bill and, over the next few months, was quite helpful in lobbying Secretary of the Interior Ken Salazar, who comes from a ranching family in Colorado, to help get the wolf delisted.

The summer and fall of 2010, I attended a small event with President Obama and other events in Dallas with Senators Claire McCaskill, Sherrod Brown, and Charles Schumer. At every one of those events, I talked about wolves to the Senators and their staff and to close advisors to the President. While the initial reaction was skepticism when I talked about the loss of elk in Yellowstone, the other elk herds that had been lost, the livestock killed by wolves, and the way the federal government treated the people that suffered livestock losses, these political professionals came to realize the situation was critical. The clincher for most of the Democrats was when I told them of polling that showed that over 80% of the people in the United States wanted wolves managed by the states and not the federal government.

In the meantime, I also had lunch in Dallas with my long-time friend, U.S. Trade Representative Ron Kirk, and relayed to him the problem of wolves. He was well aware of the problem.

Politics and Intrigue

As time went on that summer, more and more people became aware of Chet's bill, but the pro-wolf groups did not believe it had a prayer of passing.

Don Peay, who has worked for years with Senator Orrin Hatch, was told very clearly by Senator Hatch that, in order to pass the bill, it would be necessary to get the support of the two Democratic senators from Montana—Max Baucus and Jon Tester. The offices of Senators Tester and Baucus were flooded with emails, phone calls, and personal visits as dozens of private citizens across the West began to pressure their senators and congressmen to get something done.

I personally talked on the phone with Idaho Governor Butch Otter after my old friend, Brent Hill, called Governor Otter and asked him to support Chet Edwards' legislation.

One Republican senator called me personally and told me that, while I may have passed hundreds of bills in the Texas Legislature, passage of legislation of this magnitude changing the Endangered Species Act in these partisan times was simply not going to happen. While thinking to myself, not so, I was extremely polite and told him that, if he could get the Republicans to cooperate, I would somehow get the Democrats.

Don Peay told me that, with my Democratic connections, he felt strongly that we could get wolves delisted and asked me to fly to Washington, D.C., to meet with industry leaders about Chet Edwards' wolf delisting bill.

I got on a plane and flew to D.C. the next day and attended a luncheon meeting with a number of professional D.C. folks that worked for Safari Club International, the National Rifle Association, the Congressional Sportsmen's Foundation, and several other organizations. The consensus was that Congress would not tackle anything that controversial in an election year. This meeting occurred on September 15, 2010. One participant was particularly adamant that nothing could get done. I told him, quite frankly, that we didn't have a choice, that elk were being slaughtered by the thousands as the records showed, that this was a wedge issue, and that, if it wasn't taken care of, congressmen and senators from both parties would be beaten in the next election.

L to R: Ted Lyon, Senator Harry Reid, Don Peay, Clint Bentley, and Miles Moretti.

Breakfast Meeting

In September of 2010, I was able to set up a breakfast meeting with Senator Reid in Washington, D.C. In attendance at the meeting were Don Peay with Sportsmen for Fish and Wildlife; Miles Moretti, President, Mule Deer Foundation; and Clint Bentley, former President of Nevada Bighorns Unlimited and former member of the Nevada

Wildlife Commission. Clint, who had just had open heart surgery two weeks before, was there on behalf of the Wild Sheep Foundation at the request of Don Peay and had come because he was close to Senator Reid.

As we sat down to breakfast, we had a plan. We had met the night before, and it was decided that no one else would talk until I made my presentation to Senator Reid. This 15-minute presentation was one of the most important presentations of my life. I knew that, if we could persuade Senator Reid, undoubtedly the most powerful person in the U.S. Senate, then other Democratic senators would follow. The whole ballgame to me was the U.S. Senate. I had stayed up half the night writing and rehearsing what I was about to say. I knew that, if I could convince the Senator, then we would have a good chance of passing the legislation in the Senate. The House, I always felt, with 60 co-sponsors, was doable.

At the beginning of the meeting, I asked Senator Reid if he wanted the science or the politics first. Without hesitation, he replied that he wanted the science first so I gave him the facts. Almost all of the information came from the 2009 Montana Wolf-Ungulate Report and from RMEF biologists. (What is so sad is that, even as we go forward in 2013 as I am finishing my book, the elk herds have dropped even more in number.)

Miles Moretti and Don Peay both made very brief but heartfelt pitches on behalf of their groups. The 15-minute meeting stretched to 45 minutes. I finally wrapped it up by telling the Senator that—under the law as it existed in Montana, Idaho, and Wyoming at that time—he could be walking his dog down a country road and, if a wolf attacked it, there was nothing he could legally do to stop the attack or to defend his dog. The Senator could not believe it. He was visibly upset and, without consulting his staff, said that he would help us. He also said to me, "That was a hell of a presentation, lawyer."

Senator Reid had known in a minute what the right thing to do was, and he had committed to doing it: the states needed the right to be able to manage the wolves, and the endless stream of lawsuits by groups simply using the wolf as a fundraising tool had to stop.

Montana's Senator Tester

We still did not have the support for our bill from the Montana Senators whose state was arguably hit the hardest, but at least we were making progress. I had reached out to Senator Tester's office through my good friend, former Lieutenant Governor Ben Barnes of Texas, and he gained me entry to Tester's Chief of Staff. Senator Tester's staff was well aware that Representative Denny Rehberg, Tester's Republican opponent in 2012 for the Senate, was holding town hall meetings all across Montana about delisting the wolves. This was a smart move on Rehberg's part and enabled him to get 100 or more people at several meetings across the state. Representative Rehberg's move to talk about the wolves put a lot of pressure on the two Montana Senators, Tester and Baucus.

On October 13, 2010, Senator Tester and I had a meeting in Missoula, Montana, with his staff. Tester made it clear to me that he had no love for wolves and would do everything he could to pass legislation to delist them. My goal was to get total delisting of the wolves across the U.S. His goal was to explain to me that a compromise was the only way we could get anything through the Senate. He was right. Interestingly enough, in 2013, the USFWS has proposed delisting gray wolves in all states with the exception of the Mexican gray wolf in Arizona and New Mexico.

Senator Tester filed Senate Bill 3864, which was drafted by the Senate's legislative counsel and Senator Tester's and Senator Baucus' staff. It was an extremely complicated bill that would have resulted, in my opinion, in an endless series of lawsuits. Senator Tester asked for my opinion, and I gave it to him on November 30, 2010, along with a legal memorandum addressing the consequences of S. 3864 as written.

Wins and Losses

At the same time this was going on, Senator Hatch was working on a rider to the Omnibus Spending bill that Congress had to pass by the end of 2010. Senator Hatch's staff was on the phone with me on a regular basis to talk about the rider, and Don Peay and I talked every day. I was also on the phone on a regular basis with Senator Tester's staff.

As the year 2010 came to a close, the Republican and Democratic troops were fighting, and our rider never got put on the bill. Many of our troops were truly demoralized, but I pointed out to them that we had come a long way in a little less than a year. We had on our side all the U.S. Senators from Montana, Idaho, and Wyoming; the governors of Wyoming, Montana, and Idaho; the Majority Leader in the Senate; the Republican Vice Chairman of the Senate Finance Committee; and the Democratic Chairman of the Senate Finance Committee. Additionally, we had over 60 co-sponsors in the House of Representatives, and numerous other people in the wildlife conservation community supporting our cause.

Online Support

The wolf supporters, who still did not understand the forces that they were up against, sat idly by, not knowing that they were faced with folks who, armed with the truth, were making tremendous headway in Washington, D.C. The issue, with the help of a group of bloggers on the internet, caught the imagination of thousands of sportsmen across America. George Dovel of Idaho, who publishes the Outdoorsman magazine; Toby Bridges of Missoula, Montana, publisher of the LOBO Watch blog; and Tom Remington of the Black Bear Blog, on an almost daily basis, were on the internet giving the U.S. Fish and Wildlife Service, the Idaho Fish and Game, and the Montana Fish, Wildlife and Parks, as well as any number of people in political office, hell for what was happening to the elk, mule deer, and moose across the West. Steve Alder and Scott Rockholm from Idaho, along with Robert Fanning with the Friends of the Northern Yellowstone Elk Herd from Montana, were also active in criticizing the actions of the U.S. Fish and Wildlife Service. They pointed out—with facts and statistics—what was really going on. Their combined efforts had a tremendous effect upon the issue coming to the forefront of people's minds.

Non-Partisan Support

The effort was non-partisan, and we needed every sportsman's group united. The National Rifle Association (NRA) lobbyist had committed at our earlier meeting in D.C., and having the NRA on our side was huge. With them would come most Republicans as well as many Democrats. Don and I went to the Congressional Sportsmen's Foundation meeting the night that Montana Senator Jon Tester was sworn in as its new President.

By mid-November, Montana Senators Tester and Baucus had filed a bill that took care of the delisting of the wolf in Montana and Idaho. Idaho Senators Risch and Crapo were also very involved and filed their own legislation. Wyoming Senators Enzi and Barrasso were also very involved as was Wyoming Representative Lummis.

Win Some, Lose Some

The elections came, the Democrats lost many seats, and the rider to be attached to the spending bill failed because it was going to require unanimous consent, and Maryland Senator Ben Cardin objected. The message I gave to every Democratic senator that I talked to was that, in the West, senators were in deep trouble in 2012 unless the wolf problem was solved. Again, a budget resolution was needed to keep the government running, and again our plan was to put a rider on the budget resolution that would delist the wolves.

Looking forward to the 2012 election, we shared an interesting data point with senators and congressmen: although Utah is a very Republican state, by a slim margin, traditional Republican voters in Utah had re-elected Democratic Congressman Jim Matheson, who had been a vocal and early supporter of returning wolf management to the states. Conversely in the Democratic state of New Mexico, Democratic Congressman Harry Teague was a big supporter of wolves, and the pro-wolf groups had poured hundreds of thousands of dollars against Teague's challenger, Steve Pierce. However, Pierce, a pro-wolf management Republican was elected.

We also pointed out that we have a political data point in the Western states where hunters and ranchers will not vote party lines; instead, they will vote for those who support the "right" wildlife management policies, and this group, when motivated, can be a difference maker in congressional elections.

Language, Lawsuits, and Law

The final language that came out from the group that included Senators Reid, Hatch, Tester, Crapo, and Risch delisted wolves in Montana, Idaho, and Wyoming while still preserving at least 150 wolves and 10 breeding pairs in Montana and Idaho.

In March of 2011, the U.S. Department of Interior dropped its appeal in Wyoming Wolf Coalition v. U.S. Department of Interior, Case No. 09-CV-118J, and Park County v. U.S. Department of Interior, Case No. 09-CV-138J, which meant that Judge Alan Johnson's decision that Wyoming's proposal to designate wolves as trophy game animals in certain areas around the national parks and as predators in the rest of Wyoming was now law.

> *The absence of wolf breeding pairs in the 88% of Wyoming where the wolf is designated as a predatory animal has no impact on Wyoming's ability to maintain its share of the greater Yellowstone area metapopulations segment because that area is largely unsuitable wolf habitat and it is not located between the core recovery areas so it cannot affect the rate of natural dispersal between the three core recovery areas.*
>
> *AR 2009-032170*

Testimony by Ed Bangs, the top wolf biologist for the U.S. Fish and Wildlife Service, stated for the record:

The 2007 Wyoming wolf plan is a solid, science-based conservation plan that will adequately conserve Wyoming's share of the greater Yellowstone area population so that the Northern Rocky Mountain wolf population will never be threatened again.

Actions and Reactions

Harriet Hageman—the Cheyenne, Wyoming, lawyer who won the lawsuit for Wyoming, which is the first lawsuit anyone ever won for a group not supporting the proliferation of wolves—was highly upset in March of 2011 when she found out about an amendment sponsored by U.S. Representative Mike Simpson, a Republican from Idaho, and Senators Jon Tester and Max Baucus, Democrats from Montana. The amendment, as originally proposed, would have delisted wolves in Idaho and Montana and in parts of Oregon, Utah, and Washington, while Wyoming and the rest of the states would be left to fend for themselves. In a letter to her clients called "A Call To Action!" Hageman blasted the original amendment by Simpson, Tester, and Baucus that was also supported, as her letter notes, by the NRA, Safari Club International (SCI), Congressional Sportsmen's Foundation (CSF), and Boone and Crockett.

I am absolutely certain that neither Senators Tester or Baucus nor Representative Simpson knew that the amendment was flawed; nor do I believe that the NRA, SCI, CSF, or Boone and Crockett would have supported that kind of bill. But once Hageman sent her letter to her clients, all hell broke loose, and there were harsh words exchanged between the NRA, Big Game Forever, and Sportsmen for Fish and Wildlife, creating bad feelings that still exist to this day.

Success At Last

Needless to say, after a flurry of action, Representative Simpson, to his full credit, amended the language in the budget rider to fix the problem. The key language was that the 2009 wolf delisting rules adopted by the U.S. Fish and Wildlife Service "shall not be subject to judicial review." Ryan Benson of Big Game Forever, who is a Harvard-educated lawyer, helped draft the final language that concerned Wyoming. Congress also stated in that rider that the bill "… shall not abrogate or otherwise have any effect on the order and judgment issued ..." in the Wyoming delisting lawsuit, which Wyoming had won in Judge Johnson's federal court earlier in 2011. Don Peay called me to see if we should accept the language that was eventually agreed upon by all of the Western Senators, and I told him not only *yes*, but *hell yes*.

The appropriations bill with the rider passed the Senate by a vote of 81 to 19 on April 14, 2011. "This wolf fix isn't about one party's agenda. It's about what's right for Montana and the West," Senator Tester said in a statement. President Obama signed the bill into law the next day, and the Endangered Species Act, for one of only two times in its history, was amended. Sportsmen groups had all worked together in a nonpartisan way, and as a result, wolves will now be managed by the states up to certain levels. Wyoming will have a predator zone outside of the national parks where wolves are treated the same as coyotes, and Oregon, Washington, and a small part of Utah will be managed by the individual states.

Was this legislation everything we wanted? Absolutely not, but it was a start. As Confucius once said, "The journey of a thousand miles begins with a single step."

Conclusion

Although much more needs to be done, this first and critical victory in wolf management proves the effectiveness of sportsmen's groups working together. When stakeholders make their voices heard to their legislators, with hard evidence supporting their claims, laws can be changed.

All across the Western states of New Mexico, Arizona, Washington, and Oregon, as well as states where wolves are just beginning to show up—California, Utah, Colorado, Texas, and North and South Dakota, wolf management issues remain to be resolved, and stakeholders need to vigilantly continue their efforts. Additionally, there are growing wolf populations in Michigan and Wisconsin, dispersals into neighboring states, and proposed plans for wolves in New York, and all these populations will need proper management—and soon.

In order to insure that wolves are managed responsibly, people need to know the truth, and stakeholders need to be willing to do what it takes to communicate with their legislators. This first victory in wolf management sets the stage for the future.

Ted Lyon's Note: From my viewpoint, the legislative heroes who stood up for their states include Senators Orrin Hatch, Jon Tester, Harry

Reid, Mike Crapo, James E. Reich, Mark Pryor, Mike Enzi, and John Barrasso along with Representatives Chet Edwards, Bob Simpson, Denny Rehberg, Cynthia Lummis, and Jim Matheson. There are many people and groups that deserve credit for their involvement in the Montana, Wyoming, and Idaho delisting of the wolves under the Endangered Species Act. It took all of our efforts—individually and as groups—to achieve this first step in wolf management, and it will take the same nationwide involvement to achieve further delisting. This is one person's perspective of how it happened.

Holly Kuchera / Shutterstock.com

Chapter 17

HOW THE SECOND POLITICAL WOLF WAR CAN BE WON

By Ted B. Lyon

Stayer//Shutterstock.com

Editors' Statement of the Issues

For progress to be made in any area of legislation, it is critical that the dialogue remain (or become) civil. When important, even life- and livelihood-threatening, issues are at stake, however, it is too easy to wage war on individuals and parties instead of issues.

Editors' Statement of the Facts

We are all in this together, and participants must practice restraint when communicating with each other in order to avoid unnecessary and unmanageable strife. Facts, objective and well presented, regarding wolf management and delisting must be made available in a palatable manner to make it obvious to legislators and the public alike that current wolf populations must be managed responsibly.

> *The biggest culprit that I found in my research*
> *was not the human beings*
> *trying to implement government policy*
> *because it was their job,*
> *but the law they were charged to implement.*
> Ted B. Lyon

Just Doing Their Jobs

I want to conclude by stating that, in the 1990s, the governmental employees for the U.S. Fish and Wildlife Service (USFWS) and all of the state wildlife agencies were doing their jobs, and that was the selling and introduction of the Canadian gray wolf throughout the West. There are many people whose lives have been impacted by this introduction who dislike or even despise those government employees, but I'm not one of them. All the people that I interviewed for this book who worked for wolf introduction were extremely nice people who love wildlife and who thought then, and in many cases still believe, that wolves are a positive force in nature. I think they are wrong, and I believe that history will show them to be wrong, but I still respect their opinion. I can't really believe that people who spend their careers working for fish and wildlife agencies want to destroy the wildlife that they have spent their lives protecting.

266

Nolodymyr Burdiak / Shutterstock.com

Declining Moose Populations

The biologists sometimes, however, cannot see the forest for the trees. For example, as this book goes to print in 2013, we are seeing alarming rates of declines for moose wherever wolves exist. In 2013, Wyoming Game and Fish Biologist Doug Brimeyer counted 239 moose, the second lowest count in 28 years. Wildlife Conservation Society biologist Joel Berger was quoted in the *Jackson Hole Wyoming News & Guide* on March 2, 2013, as stating that wolf predation had played a role in the moose's decline.

Minnesota, which has approximately 4,000 wolves, announced on February 5, 2013, that it had canceled the state's 2013 and future moose hunting seasons, citing a "precipitous decline in moose population" as reported by the Associated Press. "The population has dropped 35% over the past year and 52% from 2010 to an estimated 2,760 moose left in northeastern Minnesota, according to the Department of Natural Resources in January 2013." The Associated Press story noted that some scientists suspected global warming, parasites, disease, and changes in vegetation. The story did not, however, mention that 4,000 wolves eat a lot of meat and that the likely culprits, if one listens to the hunting community in Minnesota, are the wolves.

From Yellowstone National Park to Minnesota, Montana, Idaho, and Wyoming, the moose are disappearing, and the one common ingredient that all four states and Yellowstone National Park have is the presence of wolves.

The Role of the ESA

The biggest culprit that I found in my research was *not* the human beings trying to implement government policy because it was their job, but the law they were charged to implement. The Endangered Species Act (ESA) itself is a large part of the reason wolves were introduced into Yellowstone National Park and Idaho and also why the states were unable to control wolves until 2011.

To that end, I asked a number of experts who deal with the ESA to convene in Bozeman, Montana, on January 12, 2013, to come up with ideas on what could be done to improve the ESA and to come up with legislative solutions that would make the ESA more effective in achieving its mission of recovering species threatened with extinction. The attendees were as follows: Arnie Dood, endangered species biologist for Montana Fish, Wildlife, and Parks; Miles Moretti, President and CEO of the Mule Deer Foundation, who has a degree in biology and formerly served as acting director of the Utah Division of Wildlife Resources; Karen Budd-Falen, an attorney who is recognized as an expert on environmental policy issues; Dr. Matthew Cronin, Associate Professor of Animal Genetics, School of Natural Resources and Agricultural Sciences at the University of Alaska; Ryan Benson, attorney, graduate of Harvard Law School with a degree in political science, and co-founder and National Director of Big Game Forever; and Ben Barmore, an associate at my law firm who graduated *Cum Laude* in 2011 from SMU Dedman School of Law and who worked closely with me in defending the amendment to the Endangered Species Act in the U.S. Court of Appeals for the Ninth Circuit.

After a full day of debate, this high-powered group came up with the following problems and proposed solutions:

Problems Identified

🐾 The failure to include state wildlife agencies and local authorities in listing and wildlife management decisions undermines support for recovery efforts under the Endangered Species Act (ESA).

🐾 The ESA lacks a clear methodology and practice for the delisting of recovered species.

🐾 The governmental funds spent on wildlife recovery efforts under the ESA are disproportionately spent on "charismatic" species that are not threatened at a global level.

🐾 "Animals of greatest concern" are determined based on litigation from non-governmental organizations instead of on objective, science-based analyses.

Proposed Resolutions

After vigorous discussion and debate, the meeting's participants agreed that legislative changes are needed to enable the ESA to achieve its original purpose of efficiently and effectively recovering species threatened with extinction. Consistent with this belief, the participants agree that the following legislative changes are necessary:

🐾 The states affected by a listing determination should be statutorily authorized as equal partners in management decisions affecting wildlife populations within their states. States possess broad trustee, police powers and primacy over fish and wildlife within their borders and will ultimately be responsible for management of the recovered species populations upon a successful delisting of the species. As such, they should be partners in wildlife management to ensure support for recovery at a local level and a smooth transition to state-only management upon delisting of the species. The cooperation between federal, state, and local governmental agencies—in the Memorandum of Understanding on the management of the Mexican gray wolf experimental population that was recently dissolved in response to attacks from the Defenders of Wildlife—is a good example of how an equal partnership between federal and state agencies can be achieved.

🐾 State wildlife agencies must have an opportunity to participate in listing decisions on species with populations that are located in the state or that, with reasonable probability, will be located in the state after recovery. At a minimum, this opportunity should include a formal "vote" from the affected state on whether listing is necessary to preserve the species and consideration of any scientific data provided by the state's wildlife agencies without any special deference to the scientific evidence produced by the U.S. Fish and Wildlife Service (USFWS). If USFWS determines that listing is necessary against the state's recommendations, a reasoned opinion supporting the decision should be required to be published by the USFWS.

🐾 The Recovery Plan for a listed species should be required to include quantifiable recovery goals that are developed in coordination with the wildlife agencies of the affected states. Additionally, there should be statutorily mandated deadlines for preparing recovery plans, and states should be partners in development of those plans. The statute should also provide the necessary funding for state participation. The ESA should then provide a regulatory mechanism for states and non-governmental organizations to initiate delisting once the published recovery goals have been met.

🐾 Listing determinations should be made based on threat to the species on a global level. The current distinction between "endangered" and "threatened" is meaningless and allows for the listing of species that are not in danger of extinction. The ESA authorizes the federal

270

government to usurp the states' inherent primacy in the management of wildlife resources within their borders and so should be used sparingly in a manner consistent with our nation's founding principles. To more effectively distinguish the level of threat to a species, so as to justify listing under the ESA, there should be additional categories of "threatened" to more accurately reflect the species' danger of extinction.

Note: Our discussion centered around the IUCN (International Union for Conservation of Nature) categories, which include:

🐾 Extinct: No free-living, natural population.

🐾 Extinct in the wild: Captive individuals survive, but there is no free-living, natural population.

🐾 Critically endangered: Faces an extremely high risk of extinction in the immediate future.

🐾 Endangered: Faces a very high risk of extinction in the near future.

🐾 Vulnerable: Faces a high risk of extinction in the medium-term.

🐾 Near threatened: May be considered threatened in the near future.

🐾 Least concern: No immediate threat to the survival of the species.

🐾 When developing the above-described categories of "threatened," the vitality of the species should be assessed on a global level without consideration of international borders unless there is an affirmative finding that the foreign nation's policy actually threatens the species with the danger of extinction.

🐾 A species that is not currently threatened with extinction should not be listed based solely on predictive models indicating that the species *may* become threatened or endangered in the future. Listing decisions should be based solely on the current status of a species.

🐾 Listings under the ESA should be made exclusively at a species level. The determination of "subspecies" and "distinct population segments" is too subjective and controversial to be a meaningful, science-based determination.

🐾 The U.S. Fish and Wildlife Service should be required to provide an annual accounting to Congress detailing the amount of spending on recovery efforts for each species listed under the ESA.

🐾 There should be an annual accounting of attorneys' fees awarded for actions brought under alleged violations of the ESA or NEPA (National Environmental Policy Act). This accounting should be available, and it should list each entity that received an award and the total amount of fees awarded in the calendar year. This should be reported to the GAO (Government Accounting Office).

🐾 A litigant who wins a legal proceeding based solely on "process" under the ESA or NEPA should not be entitled to attorney's fees.

🐾 It should be specifically provided that private property may not be listed as "critical habitat" under the ESA; instead, the protections for private landowners that enter into Conservation Reserve Program (CRP) contracts for the purpose of wildlife conservation should be strengthened. This is necessary to provide incentives for private landowners to voluntarily cooperate in conservation efforts.

Conclusion

There are times where critical habitat designation may appropriately include private lands. There are significant numbers of endangered species that are only found on private lands. I feel there certainly needs to be additional discussion about critical habitat and how it is applied. It has always been my feeling that the statement "habitat is critical" is true and accurate; however, "critical habitat" as currently designated often rewards those who get out of the various conservation programs early and penalizes those who have stayed in the programs, done a good job, and maintained species on their lands.

As far as protections for private landowners, there should probably be additional discussion about validating programs such as the Safe Harbor Program, the Candidate Conservation Agreements with Assurances, and so on by securing needed flexibility within the statute.

Critterbiz / Shutterstock.com

Debbie Steinhausser / Shutterstock.com

Chapter 18

CANIS STEW

By Ted B. Lyon

Holly Kuchera / Shutterstock.com

Much of the wolf introduction proponents' dialogue has centered on preserving individual wolf populations, subspecies, and species, although these categories are scientifically arbitrary (see Chapter 19 by Cronin). Because wolves travel great distances, populations and subspecies will mix, and because wolves can interbreed with coyotes and dogs, maintaining "pure" wolves is also not certain.

Editors' Statement of the Facts
With the determined efforts of wolf activists, the Canadian wolf was introduced right inside the ranges of our native American wolves in Montana, Wyoming, and Idaho. Arguments about subspecies and locally adapted populations are undermined by the USFWS importing a non-native Canadian subspecies into the northern U.S. Rockies. All wolves are determined and effective predators, which is the primary fact relevant to wildlife management.

> *I was much struck how entirely vague and arbitrary is the distinction between species and varieties.*
> Charles Darwin[1]

> *The domestic dog is an extremely close relative of the gray wolf, differing from it by, at most, 0.2% of mtDNA sequence....*
> *In comparison, the gray wolf differs from its closest wild relative, the coyote, by about 4% of mitochondrial DNA sequence.*
> Robert Wayne, Ph.D.[2]

The Disappearance of Purebred Wolves

In 2011, a lone gray wolf, OR7, left his pack in northeastern Oregon and journeyed 1,062 miles south, crossing into northern California. His path was monitored by a radio collar. Becoming an overnight media celebrity, OR7 roamed about 900 miles in northern California, and then in early March of 2012, he turned north and headed back into Oregon, seemingly because he was unable to find any wolves to unite with in California. If only he would have gone a little farther south, he might not have returned to Oregon.

In 2009, a sheep rancher contacted a USDA Wildlife Control Specialist for APHIS (Animal and Plant Health Inspection Service)

276

in northern California as his livestock were being attacked and killed by an animal that was presumed to be a large coyote by the size of the tracks—a large Western coyote can weigh 30 pounds. The APHIS trapper put out snares and caught "two animals that weighed over 100 pounds each that looked like wolves." A subsequent examination of the animals showed that they were wolf-dog hybrids. The trapper said that, during the last 10 years, he had trapped several dozen similar animals. He said that people who grow marijuana prefer to use wolf dogs for guard dogs and that there is evidence that some people in northern California are raising and releasing wolves and wolf dogs to create a California population.

A U.S. Fish and Wildlife Service (USFWS) Special Agent commented, "If they were 100% wolves, catching and killing them without a proper permit would have been breaking the law. Since they were wolf-dog hybrids, the trapper was following the local law that says that any dog or coyote attacking livestock can be shot on sight without needing a hunting license."

This incident foreshadows a problem that, in the long run, seems inevitable. Wild gray wolves will become ineligible for protection under the Endangered Species Act, regardless of their numbers, because they can breed with coyotes and dogs, so purebred wolves may disappear in many areas.

Tracing the Origins of Canis lupus

Wolf-like canids—wild dogs—began to appear about three million years ago. Originally, there were three species in North America—the gray wolf, *Canis lupus*; the Dire wolf, *Canis dirus,* a stocky carnivore with large teeth and jaws for crunching bones of megafauna (large-bodied mammals weighing more than 100 pounds) and nearly twice the body size of the gray wolf; and the coyote, *Canis latrans*. For reasons we do not understand, the gray wolf, *Canis lupus*, migrated to Asia over the Bering Straits land bridge and spread throughout Eurasia. Ultimately, they became the most widely distributed predatory mammal on earth. Later in the Paleolithic period,[3] gray wolves migrated back over the land bridge into North America. The three wild dogs—*Canis lupus*, *Canis dirus*, and *Canis latrans*—lived together for about 10,000 years. The Dire wolf ultimately went extinct about 10,000 years ago, seemingly along with the extinction of the megafauna.

After the recession of the Wisconsin Glacier some 10,000 years ago, smaller wild game became plentiful, and gray wolves proliferated.

There may have been as many as two million gray wolves in North America when the Europeans arrived in the 1500s. The gray wolf numbers were so large because of the abundance of big game animals, especially 60 million buffalo. Lewis and Clark reported large packs of wolves following the buffalo herds of the Great Plains—"We scarcely see a gang of buffalo without observing a pack of these faithful shepherds on their skirts... ."[4] They also reported that wolves were relatively unafraid of people and that, on occasion, wolves attacked members of their expedition.[5]

As waves of settlers pressed westward, the buffalo was nearly driven to extinction by market hunters and as a planned mass killing by the military to force Indians onto reservations. The European settlers replaced the dwindling herds of native buffalo, elk, and deer with sheep and cattle. Wolves switched their predation to domestic livestock, which resulted in a robust campaign to eradicate wolves, ultimately leading to the virtual extinction of wolves by the 1940s in the lower 48 states by hunting, trapping, and poisoning. It should be noted that, while the wolves were nearly removed completely, their little brother, the coyote, prospered.

Canis latrans, the coyote, which is normally one-third or less the size of an adult gray wolf, was originally found only west of the Rocky Mountains. Despite the eradication campaign that killed off wolves and equal efforts to control coyotes, the coyotes proliferated, spreading eastward. They are now found in 49 of the 50 states (excluding Hawaii) and all of the Canadian provinces. Their population is estimated to be somewhere between 10 million and 100 million, even though as many as 400,000 coyotes are shot, trapped, and poisoned every year.

When wolves meet coyotes, one of two things normally happens. The wolves attack the coyotes and drive them away. Or, some wolves, especially beta wolves without mates, will breed with coyotes, creating "coy-wolves."

As gray wolf populations are being restored in the Northern Rockies, Pacific Northwest, Great Lakes area, New England, Southeast, and Southwest, this generation of wolves is encountering unprecedented numbers of coyotes as well as tens of millions of domestic dogs. There are somewhere between 70 and 80 million domestic dogs in the U.S.; 39% of American households own at least one.

The word *Canis* means "dog" in Latin. The *Canis* genus contains seven to 10 living species, including dogs, wolves, coyotes, and jackals. They are all capable of interbreeding and producing fertile offspring. Interspecies hybridization may result in wolf populations that are not "pure" wolf.

The Evolution of Dogs from Wolves

The wolf has 78 pairs of chromosomes, but there can be millions of variations. There is not a standard allele.[6] Normally, through successive generations of breeding, mutations occur, which—combined with adaptations for changing environmental conditions—result in new self-replicating species or sub-species forming. Hybridization between wolves and dogs or coyotes is occurring today in some areas of the lower 48 states.

Domestic dogs are descended from wild wolves, *Canis lupus.* The Latin name for the dog is *Canis lupus familiaris, or Canis familiaris,* depending on a taxonomist's tendency to consider dogs as a subspecies of wolf or a different species (see Chapter 19 by Cronin). The domestication process began somewhere between 60,000 and 135,000 years ago. Humans first used wolves for food and fur. Domestication most likely began when humans raised very young wolf puppies. Some animals were friendlier than others and could be tamed. This resulted in them performing tasks such as hauling loads, serving as watchdogs, and becoming hunting allies. Animals with desirable behavioral and morphological[7] traits were bred with others with similar traits to create new unique breeds. This process of selective breeding has been demonstrated by Russian scientist Dmitry Belyaev with foxes in Siberia to form a new subspecies.[8]

What is a "species" is hotly debated in scientific circles. John Wilkins has listed 26 different "species concepts" that scientists may use to determine what is a species of animal or plant.[9] Originally, morphology was the key, but now with DNA analysis, one must also look at genetics, and the geneticists cannot agree on species identification either. The problem with canid species identification is that all canines can interbreed with other species of canines. Thus, the concept of preserving a "species" becomes difficult when one cannot agree on what it is, and its genetics are constantly changing from hybridization.

The USFWS is currently considering subspecies of wolves in an inconsistent and scientifically subjective fashion (see Chapter 19 by Cronin). The Mexican wolf (*Canis lupus baileyi*) is considered a subspecies, and there is disagreement on whether the eastern gray wolf and red wolf are species or subspecies. In 1978, USFWS classified the gray wolf as an endangered population at the species level *(C. lupus)* throughout the conterminous[10] 48 states and Mexico, except for the Minnesota gray wolf population, which was classified as threatened.[11]

Species Conservation and the ESA

The Endangered Species Act of 1973 (ESA) is the third in a series of acts designed to preserve wildlife species in danger of extinction.[12] Its predecessors were the 1966 Endangered Species Preservation Act and the 1969 Endangered Species Conservation Act.

Richard Nixon signed the ESA into law on December 28, 1973. Its stated purpose is to protect species and "the ecosystems upon which they depend." ESA's primary goal is to prevent the extinction of imperiled plant and animal life and to recover and maintain those populations by removing or lessening threats to their survival.[13]

According to ESA, "The term 'species' includes any subspecies of fish or wildlife or plants and any distinct population segment of any species or vertebrate fish or wildlife which interbreeds when mature." The term "species" as used in the ESA has "vernacular, legal, and biological meanings."[14] The law defines "species" vaguely as follows: "Species: includes any subspecies of fish or wildlife or plants, and any distinct population segment of any species of vertebrate fish or wildlife which interbreeds when mature ... based solely on the best scientific and commercial data available."[15]

An "endangered" species is one that is "in danger of extinction" throughout all or a significant portion of its range. A "threatened" species is one that is "likely to become endangered" within the foreseeable future.

To be considered for listing, the species must meet one of five criteria:

🐾 *There is the present or threatened destruction, modification, or curtailment of its habitat or range.*

🐾 *There is an over-utilization for commercial, recreational, scientific, or educational purposes.*

🐾 *The species is declining due to disease or predation.*

🐾 *There is an inadequacy of existing regulatory mechanisms.*

🐾 *There are other natural or man-made factors affecting its continued existence.*

The Secretary may treat an unlisted species as listed if: it so closely resembles a listed species that enforcement personnel would have substantial difficulty in attempting to differentiate between the species; the effect of this difficulty is an additional threat to the listed species; this treatment will substantially facilitate enforcement and further the Act's policy. Regulations must be issued to provide for the conservation of threatened species.

One of the chief concerns of species preservation is whether the population number is too small, which will result in in-breeding, which will lead to loss of health and ultimately extinction of the species. The gray wolf was placed on the Endangered Species List in 1974, and aside from a brief delisting in 2009 in Montana and Idaho, it remained there until delisting was done by amending the Endangered Species Act in 2011, despite numerous efforts to delist it in the Northern Rockies and the Midwest. Some wolf advocates worry about species extinction due to wolves being killed or dying from disease. The current Northern Rockies gray wolf population is about five and a half times higher than the minimum population recovery goal and about three and a half times higher than the breeding pair recovery goal, and the Eastern timber wolf population is no longer listed, according to the USFWS.[16]

Canid Hybridization

Increasingly, it is difficult to determine just exactly what a "true wolf" is, as wolves can and do breed with coyotes and domestic dogs. The results are called coy-wolves, "dogotes," and wolf-dog hybrids. There are an estimated 300,000 wolf-dogs in the U.S. in captivity. They are banned completely in 40 states, and anything less than a 100% wolf is considered a wolf-dog.[17] Wolf dogs are considered dangerous. They have the size and strength of wolves, but seem to have less wild instinct. A number have been released into the wild by people, who find that, after two or three years, the animal cannot be controlled.

The hybridization problem is further complicated by feral dogs, a growing problem in many areas of the United States. When domestic dogs become wild, they lose their domestic behavior, form packs, and become secretive, surviving as opportunistic predators as well as scavengers, much like wolves and coyotes.[18]

In February of 2012, a rancher in Valencia County, New Mexico, reported that he and his employees had killed more than 300 dogs over the past 18 months. The feral dogs were attacking young calves and weak cattle for sport.[19]

The New Mexico rancher is not alone. In 1999, the National Agricultural Statistics Service reported that feral dogs were partly responsible for killing cows, sheep, and goats worth about $37 million a year.[20]

In 1997, Colorado was faced with the task of how to handle a growing number of wolf-dogs in the state. The State Veterinarian convened a special committee to study the problem and come up with suggestions. Based on advice from three men—Dr. Ray Pierotti, a geneticist at University of Kansas; Dr. Nick Federoff, a wildlife biologist; and Dr. Erick Klinghammer, an ethologist from Wolf Park in Indiana—the report concluded:

> *All forms of wolf-dog identification are problematic. There is no genotype (the genetic constitution of an animal) or phenotype (the observable appearance of an animal) to distinguish between a dog, a wolf-dog cross, and a wolf. All DNA tests to differentiate wolf-hybrids from domestic dogs are subject to challenge. There are no known DNA markers uniquely distinguishable in wolves that are not present in dogs. Blood tests, skull measurements, and skeletal measurements all have some merit but have not withstood legal challenge.[21]*

Many wildlife forensic scientists agree that morphology alone cannot be used to determine whether an animal is a purebred or a hybrid. The only way that one can tell the species of an animal is with DNA analysis, which today has become sophisticated, but as the Colorado report states, with that complexity comes disagreement among scientists about what is a species.[22]

It has been known for well over two decades that Rocky Mountain gray wolves have interbred with domestic and wild dogs.[23] The DNA of Rocky Mountain gray wolves is currently being tested, and the scientific community is very aware that hybridization has biological, forensic, legal, and social implications.[24]

Genetic research from Stanford University and UCLA has already revealed that wolves obtained the genes for black pelts through hybridization with domestic dogs, perhaps 10,000 years ago.[25]

Wolves normally kill coyotes, but they also breed with them, creating coy-wolves.[26] Coy-wolves were responsible for the death

282

of Canadian folk singer Taylor Mitchell in Nova Scotia in 2009. Geneticists say that "Eastern timber wolves" that are found in Ontario, Quebec, and Nova Scotia are primarily coy-wolves.[27]

The "Eastern coyote" of Massachusetts contains the DNA of Western coyotes and Eastern wolves and not domestic dogs or the gray wolf. According to researchers Kay, Rutledge, Wheeldon, and White, the "Eastern coyote" should be called a "coywolf."[28]

Some "wolves" in Oregon also have been found to be hybrids with coyotes and dogs.[29] The "Red wolf" of the Southeast, *Canis rufus*, is thought to be a coy-wolf by many scientists.[30] Similar research in the Northern Rockies also finds hybridization taking place.[31] The same hybridization issue is also found in Europe.[32]

"Wolves" and the ESA

The Endangered Species Act (ESA) is meant to protect species that can self-replicate. When one species breeds with another species, the resulting hybrid becomes problematic for ESA protection. The "Hybrid Policy" was originally a rigid standard that said: "Protection of hybrids would not serve to recover listed species and would likely jeopardize that species' continued existence."[33] As DNA analysis has become more sophisticated and species identification has become more complex, the USFWS has rescinded this position. There is currently no official U.S. policy to provide guidance for dealing with hybrids.[34]

The answer to the question, "What really is a wolf?" can only be determined by DNA analysis, and DNA analysis finds that 100% purebred wolves are growing increasingly rare due to hybridization. A harbinger of the future, hybrid wild canids in general seem to be growing in size and number. In 2004, a wolf-dog was killed in Pennsylvania that weighed 105 pounds.[35] In 1996, an 86-pound male wolf-like canid was killed in Maine that was later confirmed to be a wolf-coyote hybrid. In 1997, a 72-pound coy-wolf was killed in Vermont.[36] And in 2010, a 104-pound "coyote" was killed in Missouri.[37]

As you learned in an earlier chapter on Mexican wolves, if zoos and other small enclosures are used to preserve purebred breeding stock, it is not likely that these wolves could be released into the wild, as they would be habituated. Their capacity to survive without preying on livestock and interacting with ranches and farms and their dogs, as well as feral dogs, is questionable.

When wolves co-exist with millions of dogs and coyotes, as they do today throughout North America, inter-species breeding will inevitably occur.[38] And spontaneous hybridization of wolves will most certainly increase.[39] It seems practical to not list wolves as an endangered species, for the only way to tell what is a wolf requires expensive and time-consuming DNA analysis that can only be done when the animal is dead or captured. The number of "questionable wolves" being sent to the USFWS Forensic Laboratory in Ashland, Oregon, is growing. And as the Colorado report concluded, even with the best methods and equipment, the test may not always be completely conclusive.

Conclusion

It seems inevitable that the "pure" wolf will become increasingly rare, especially as it spreads throughout the lower 48 states, and management ultimately will shift from protecting "endangered species wolves" to controlling wolf-like, apex predator, hybrid wild dogs. Currently, such hybrids can be shot on sight if they are chasing wildlife, threatening human safety, and/or preying on domestic livestock.

If purebred wolves do exist, and some scientists question their existence, it seems that the only way to preserve the genome of *Canis lupus* is to impede the contact of purebred wolves with dogs and coyotes, which is probably impossible.

🐾 🐾 🐾 🐾 🐾 🐾 🐾 🐾

[1] Darwin, C., *On the Origin of Species by Means of Natural Selection or the Preservation of Favored Races in the Struggle for Life,* London: Murray, 1859.
[2] From Wayne's article, "Molecular Evolution of the Dog Family."
[3] Prehistoric period of human history, which extends from the earliest use of human tools to the end of the Pleistocene period.
[4] From Meriwether Lewis' journal, May 5, 1805.
[5] http://fwp.mt.gov/mtoutdoors/HTML/articles/2006/LCmisadventures.htm
[6] A different form of a gene that is located in a specific place on a chromosome that controls a specific trait and is responsible for genetic variation.
[7] Morphology is the study of how things are put together, like the makeup of animals and plants.
[8] http://www.abc.net.au/animals/program1/factsheet5.htm
[9] http://scienceblogs.com/evolvingthoughts/2006/10/a_list_of_26_species_concepts.php
[10] All the states except Alaska and Hawaii.
[11] http://ecos.fws.gov/speciesProfile/profile/speciesProfile.action?spcode=A00D
[12] http://www.fws.gov/endangered/laws-policies/index.html
[13] http://www.nmfs.noaa.gov/pr/pdfs/laws/esa_section3.pdf

[14] Natural Research Council, "Science and the Endangered Species Act," *National Academy Press*, 1995:5.

[15] http://wildlifelaw.unm.edu/fedbook/esa.html

[16] http://ecos.fws.gov/docs/five_year_review/doc3978.%20lupus%205-YR%20review%20PDF.pdf

[17] http://www.wolfdogalliance.org/legislation/statelaws.html

[18] http://icwdm.org/handbook/carnivor/FeralDog.asp

[19] http://www.cattlenetwork.com/cattle-news/New-Mexico-ranchers-protect-cattle-from-wild-dogs-139392968.html

[20] http://icwdm.org/handbook/carnivor/FeralDog.asp

[21] http://www.naiaonline.org/articles/archives/coloradotask.htm

[22] http://onlinelibrary.wiley.com/doi/10.1046/j.1523-1739.1992.06040559.x/abstract; http://www.springerlink.com/content/kw1132542l0p2mgn/

[23] http://www.rmrs.nau.edu/publications/pilgrim_et_al_1998/pilgrim_et_al_1998.pdf

[24] http://www.jstor.org/pss/3783983

[25] http://www.nytimes.com/2009/02/06/science/06wolves.html?_r=1&ref=todayspaper

[26] http://www.jstor.org/pss/2409486

[27] Kyle, C.J., A.R. Johnson, B.R. Patterson, P.J. Wilson, K. Shami, S.K. Grewal, and B.N. White, "Genetic Nature of Eastern Wolves: Past Present and Future," *Conservation Genetics*, 2006, 7:273-287.

[28] http://easterncoyoteresearch.com/downloads/GeneticsOfEasternCoywolfFinalInPrint.pdf

[29] http://www.defenders.org/publications/northern_california southwestern_oregon_gray_wolf_dps.pdf

[30] http://www.scientificamerican.com/article.cfm?id=what-is-a-species; http://www.jstor.org/pss/2386371and http://www.nal.usda.gov/awic/newsletters/v5n4/5n4wille.htm

[31] http://www.jstor.org/pss/3802344http://www.jstor.org/pss/3783983

[32] http://www.nature.com/hdy/journal/v90/n1/full/6800175a.html

[33] O'Brien, S.J., and E. Mayer, "Bureaucratic Mischief: Recognizing Endangered Species and Subspecies," *Science*, March 8, 1991.

[34] Ellstrand, Norman C., David Biggs, Andrea Kaus, Pesach Lubinsky, Lucinda A. McDade, Kristine Preston, Linda M. Prince, Helen M. Regan, Veronique Roriver, Oliver A. Ryde, and Kristina A. Schierenbeck, "Got Hybridization? A Multidisciplinary Approach for Informing Science Policy," *BioScience,* May 2010, 60:5.

[35] http://www.snopes.com/photos/animals/coyote.asp

[36] Adkins, Collette L., "The Big Bad Wolf Hybrid: How Molecular Genetics Research May Undermine Protection for Gray Wolves Under the Endangered Species Act," *Minnesota Journal of Science and Technology,* 2004-2005, 6:2.

[37] http://mdc.mo.gov/newsroom/hunter-shoots-unusually-large-coyote-northwest-missouri

[38] Mallett, James, "Hybridization as an Invasion of the Genome," *Trends in Ecology & Evolution,* May 2005, 20:5.

[39] Ellstrand, et. al., *Op.Cit.*

Holly Kuchera / Shutterstock.com

Chapter 19

WHAT IS A WOLF?

By Matthew A. Cronin, Ph.D.

Matthew A. Cronin is a research professor of Animal Genetics at the University of Alaska Fairbanks, School of Natural Resources and Agricultural Sciences. He received a Ph.D. in Biology from Yale University in 1989, an M.S. in Biology from Montana State University in 1986, and a B.S. in Forest Biology from State University New York, College of Environmental Science and Forestry in 1976. Dr. Cronin was a U.S. Coast Guard officer from 1981 to 1984 and worked for the U.S. Fish and Wildlife Service as a geneticist from 1989 to 1992. From 1992 to 2004, he worked in the private sector with a focus on wildlife research and impact assessments for the oil, timber, and mining industries. In 2004, Dr. Cronin was appointed to the university faculty and focused his research on population genetics of wildlife and livestock. His work emphasizes proper application of science to natural resource management, and his experience with private industry, academia, and government provides insights for achieving multiple use objectives.

Editors' Statement of the Issues

There has been a great deal of effort devoted to the questions centering around the classification of wolf species and subspecies, not to mention the use of good taxpayer money, for what appears to the non-scientist to be just another way to list more wolves.

Editors' Statement of the Facts

Efforts to separate wolf species and subspecies simply provide more opportunities for wolf activists to attempt to force the protection of additional wolves on the tax-paying public. What is needed is recognition that wolves interbreed—not only with other wolves, but also with dogs and coyotes, and that their offspring in whatever form are large, dangerous predators that needs delisting.

> *This concept of the subspecies is fallacious....*
> *The better the geographic variation of a species is known,*
> *the more difficult it becomes to delimit subspecies*
> *and the more obvious it becomes*
> *that many such delimitations are quite arbitrary.*
> Ernst Mayr[1]

Taxonomic Issues

The wolf was classified as a species, *Canis lupus*, by the Swedish naturalist Carl Linneaus in 1758.[2] This included all of the wolves worldwide with a range that takes in most of the northern hemisphere north of 30^0 latitude and south into India, Pakistan, the Middle East, and North Africa. Since Linneaus, classifications of wolves have included numerous subspecies and, recently, new species and populations in North America. These classifications are important because species, subspecies, and populations are all considered in the definition of "species" in the U.S. Endangered Species Act (ESA). The U.S. Fish and Wildlife Service (USFWS) and the National Marine Fisheries Service (NMFS) implement the ESA and determine if a group qualifies as a species, subspecies, or population and if it is threatened or endangered with extinction. The National Research Council has addressed taxonomic (i.e., classification) issues related to the ESA[3] and notes that there is considerable debate among scientists as to what qualifies as a species, subspecies, or distinct population segment (DPS).

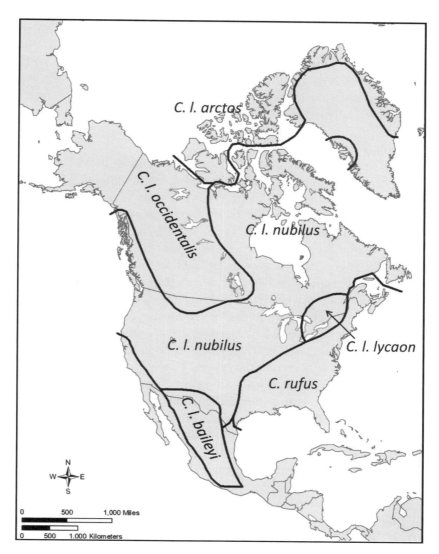

Figure 1.
This map shows the ranges of the North American gray wolf (*Canis lupus*) and red wolf (*Canis rufus*) recognized by R.M. Nowak in "Another Look at Wolf Taxonomy." The subspecies are Mexican wolf (*Canis lupus baileyi*), Northern Timber wolf (*Canis lupus occidentalis*), Plains wolf (*Canis lupus nubilus*), Eastern wolf (*Canis lupus lyacon*), and Arctic wolf (*Canis lupus arctos*). The Eastern wolf is recognized as a species (*Canis lyacon*) by some authors. This map is Figure 3 in Chambers et al., "An Account of the Taxonomy of North American Wolves from Morphological and Genetic Analyses."

In this chapter, I will review the current taxonomic status of wolves in North America. I will begin with a description of how scientists classify species, subspecies, and populations. A key point I will emphasize is that noted in the quote of Ernst Mayr above: that designation of species, populations, and subspecies, in particular, is not scientifically definitive.

Taxonomy

To understand wolf classification, we must understand the science of biological classification known as taxonomy. The basis of taxonomy is that organisms (plants and animals) are grouped according to phylogeny, which is simply the genealogy or ancestry of a group. In simple terms, we classify animals according to common ancestry, so two groups with recent common ancestry are classified together, and groups with more distant common ancestry are classified as different groups. This has been recognized since before Darwin.[4] Groups in taxonomy are called *taxa* (singular: *taxon*). The basic groups of taxonomy are kingdom, phylum, class, order, family, genus, and species. For example, wolves are in the animal kingdom, chordate phylum, mammal class, carnivore order, Canidae family, *Canis* genus, and *lupus* species. There are sub-levels of each of these levels as well, including subspecies, so a wolf subspecies, such as Mexican wolves, is further classified with a third Latin name (*Canis lupus baileyi*).

There are other methods of classifying animals besides using phylogeny. For example, one might classify animals as aquatic (living in water) or terrestrial (living on land). In this case, aquatic fish and whales would be classified together separate from land mammals. However, we know that all mammals, including whales, share a more recent common ancestry than any mammal does with fish, so in biology, aquatic whales are classified as mammals along with terrestrial mammals. Another example is aquatic sea otters and river otters, which are classified as mustelids—along with the terrestrial weasels, mink, fisher, marten, skunks, badger, and wolverine—and not with the aquatic, but distantly related, seals, sea lions, and walruses.

Another non-taxonomic classification is the designation of "ecotypes," which are groups of populations of a species with similar characteristics due to local adaptations or environmental influence, not shared genealogies. Hence, ecotypes are not a part of taxonomic classification. An example is caribou (*Rangifer tarandus*) that includes high arctic, mountain, forest, and barren ground tundra ecotypes. These ecotypes have different characteristics in different ecological conditions, but members of an ecotype do not necessarily share common ancestry and should not be classified as taxonomic groups.[5]

Species

A very important level of taxonomy is the species. Species are classified in two related, but subtly different, ways: by phylogeny (i.e., ancestry) and by reproductive isolation. There is a huge literature on what is known as the "species problem" in biology. The problem is that, because closely related groups may interbreed, identification of species as distinct groups is not always possible. To identify species, it must be determined if two populations in different geographic areas share common ancestry and can interbreed. This is complicated because populations may interbreed, but the offspring may be inviable or infertile or have reduced fitness. For example, horses (*Equus caballus*) and donkeys (*Equus asinus*) can interbreed but produce sterile mules. Horses and donkeys are considered different species because of this incompatibility. In some cases, wild populations have non-overlapping ranges, and interbreeding is not testable. This leads to uncertain species designations. For example, if there are two wolf populations in different geographic areas that differ in size or coat color, should they be one species or two?

Species is an important taxonomic level because species are groups that are separate and will remain so. If two populations are isolated and then reconnect and interbreed freely, then they are not species but simply populations of the same species. Different species concepts are used by scientists and include the "biological species concept" (BSC) in which reproductive isolation defines species. The BSC defines species as follows: Groups of interbreeding natural populations that are reproductively isolated from other such groups.[6] The BSC is the most commonly used species concept.

Another species concept is known as the "phylogenetic species concept" (PSC) and considers a species to be a group with an independent genealogy, without consideration of reproductive isolation. Under the PSC, groups that share a most recent common ancestor are considered members of the same species. The PSC is most useful for very distinct groups (e.g., wolves and foxes) and not for closely related groups (e.g., wolves and coyotes). Regardless of how they are designated, species are considered groups that are now, and will remain, separate. Clear identification of species definition is important with regard to wolves because of the designations of the eastern wolf and red wolf as species in a recent review by USFWS biologists.[7]

Subspecies

As noted above, subspecies are groups classified below the species level, and two definitions are commonly used:

Mayr[8]: "A subspecies is an aggregate of phenotypically similar populations of a species inhabiting a geographic subdivision of the range of the species and differing taxonomically from other populations of the species."

Avise and Ball[9]: "Subspecies are groups of actually or potentially interbreeding populations, phylogenetically distinguishable from, but reproductively compatible with, other such groups. Importantly, the evidence for phylogenetic distinction must normally come from the concordant distributions of multiple, independent, genetically-based traits."

These technical definitions simply mean that there must be genetically-based traits that differ between populations for them to qualify as subspecies. However, the fact that subspecies are members of the same species and can interbreed results in subspecies that share genes and, hence, are not definitive groups. The scientific community recognizes the subjectivity of subspecies as exemplified in Mayr's quote at the beginning of this chapter and the following:

Vanzolini[10] noted that "... present applications of the subspecies concept are uneven, frequently undocumented, and lead to no improvement of either evolutionary theory or practical taxonomy."

Futuyma[11] noted that there is so much variation among populations of most species that some combination of characters will distinguish each population from others, and consequently, there is no clear limit to the number of subspecies that can be recognized.

Ehrlich[12] stated: "Widespread species thus can be divided into any number of different sets of 'subspecies' simply by selecting different characteristics on which to base them. ... As is the case with other species, geographic variation in human beings does not allow *Homo sapiens* to be divided into natural evolutionary units. That basic point ... has subsequently been demonstrated in a variety of organisms... and use of the subspecies (or race) concept has essentially disappeared from the mainstream evolutionary literature."

O'Gara[13]: "Classification below the species level often has been subjective because there are no standard criteria for naming subspecies or populations. ... Subspecies designations should be based on phylogenetic relationships, the same for species. In practice, they seldom are."

Zink[14]: "Mitochondrial DNA sequence data reveal that 97% of ... avian subspecies lack the population genetic structure indicative of a distinct evolutionary unit. ... A massive reorganization of classifications is required so that the lowest ranks, be they species or subspecies, reflect evolutionary diversity. Until such reorganization is accomplished, the subspecies rank will continue to hinder progress in taxonomy, evolutionary studies, and ... conservation."

Haig et al.[15]: "Among taxonomists, definitions of subspecies are a source of considerable disagreement. ... In an extensive literature review, we found no universally accepted subspecies definition within or across taxa... The scientific community has some level of comfort with the subjective nature of subspecies classification..."

Avise[16]: "In intermediate situations (and also in hybrid settings), educated nomenclatural judgments will remain necessary at species and subspecies levels."

I provide these quotes to emphasize that the scientific community acknowledges that subspecies is not a rigorous scientific category. However, the subspecies concept can be useful to describe populations in geographic areas that differ from others, as long as the inherent subjectivity of the category is acknowledged. In addition, subspecies can be listed as "endangered species" under the ESA, so there is a practical need to deal with the category. Because subspecies are subjective, it is apparent that neither side in debates over subspecies designations is right or wrong in an absolute sense.[17] **However, one of the primary problems with the ESA is the designation of subspecies by the USFWS, without open acknowledgement of the inherent subjectivity of the category. This has resulted in many management and legal controversies.**

Populations

The problems with taxonomic uncertainty and the ESA become even more apparent at the population level, which is also included in the Endangered Species Act definition of species. Scientifically, a population is considered a group of interbreeding animals in a specific geographic area. It is a general term and can be applied to a local area (e.g., the Yellowstone National Park population of wolves) or an entire continent (e.g., the North American population of wolves). However, the ESA includes in its species definition "distinct population segments" (DPS), which are defined as populations that are discrete from other populations and significant to the species to which they belong. There are no empirical criteria on what constitutes a discrete or significant population so, like subspecies, DPS designations are quite arbitrary and subjective.[18]

Genetics, DNA, and Taxonomic Inflation

Traditionally, species and subspecies have been identified with morphology (i.e., anatomy), including body size, skull and bone proportions, coat color, or other characteristics. During the last half of the 20th century, genetics technology (i.e., biotechnology) advanced to where DNA could be compared between animals. This is highly technical work, but the basic idea is that similarity of DNA reflects recent common ancestry and, hence, can shed light on taxonomic relationships.

There are many technical approaches for using DNA to compare species, subspecies, and populations. One approach is to compare the proportions of gene variants (called alleles) in populations. For example, if there are alleles for a gene-controlling black and white coat color in wolf populations, the proportion of black and white alleles in each population can be compared, and the level of interbreeding between the populations can be estimated. Another approach involves comparing actual DNA molecules between groups. In this approach, the number of differences in the DNA sequence (which is composed of a series of four molecules, designated A, G, C, and T) can be counted, and the relationship of the DNA sequences determined. This can then be used to infer the relationship and even the time of divergence, of the groups. This is straightforward for very divergent groups (e.g., dogs and cats), but for related groups, it is complicated because they often share the same DNA sequences. Both of these approaches have been used in wolf taxonomy.

The use of genetics and DNA over the last 50 years has allowed better resolution of variation in wildlife populations. However, it has also been accompanied by a decided tendency to split groups into more species and subspecies. This has been termed "taxonomic inflation," and its impetus is often to elevate the status of groups for conservation. For example, there has been an increase in elevating subspecies of wildlife to species status to enhance their chances of legal protection with laws such as the ESA.[19] The designation of the eastern gray wolf as a distinct species, *Canis lycaon* described below, is an example of taxonomic inflation, and it remains to be seen if advocates attempt to designate these wolves an endangered species in northeastern North America.

Taxonomic inflation reflects the current tendency of extensive splitting of animals into species, subspecies, and populations. I believe

294

this tendency springs from three primary factors: the recent availability of DNA technology that allows detailed comparison of populations, people's inherent tendency to classify animals into discrete groups, and the current focus on conservation and the use of the ESA to protect wildlife. This has led to acceptance of many new species, subspecies, and DPS without full acknowledgement of the scientific uncertainty of such designations and has allowed science to be influenced by political and legal maneuvering.[20]

Wolf Species

In addition to wolves, the canid family (called the Canidae) includes foxes, coyotes, jackals, dingoes, African wild dogs, and domestic dogs (*Canis lupus familiaris*). There are wild canids on every continent except Antarctica. Both wolves and jackals were thought to be the progenitors of domestic dogs, but genetics has now shown that dogs were domesticated exclusively from wolves in Europe and Asia within the last 15,000 years or so.

Fossils suggest that the genus *Canis* arose in North America more than three million years ago during the Pliocene Epoch. They spread to Eurasia and Africa and differentiated into jackals (side-striped jackal (*Canis adjustus*), golden jackal (*Canis aureus*), black-backed jackal (*Canis mesomelas*), gray wolves (*Canis lupus*), and the Ethiopian wolf (*Canis simensis*). Another idea is that gray wolves emerged in Africa and then spread to the northern hemisphere. Coyotes (*Canis latrans*) developed from the original *Canis* and have only occurred in North America.

Gray wolves migrated from Eurasia to North America during the Pleistocene Epoch over the last two million years. This was possible because the sea level dropped during glacial periods, exposing the Bering land bridge between Asia and Alaska. It has been suggested that gray wolves entered North America at least three different times, with the new migrants mixing with or displacing the existing populations.

Gray wolves (*Canis lupus*) now occur across Eurasia and North America. Until recently, wolves in North America have been classified as two species: the gray wolf that ranges across most of the continent, and the red wolf (*Canis rufus*) in the southeastern U.S.[21] Some taxonomists consider the red wolf a subspecies of the gray wolf (*Canis lupus rufus*) and not a full species. Red wolves were extirpated in the wild and now exist only in captive populations and an introduced population on an island in North Carolina.

Recently, a third species has been designated as the Eastern wolf (*Canis lycaon*) in the Canadian provinces of southern Quebec and

Ontario and in the northeastern U.S. as far west as Minnesota.[22] The Eastern wolf is also considered a subspecies of the gray wolf (*Canis lupus lycaon*) by some taxonomists.[23]

The species designations of the eastern wolf and the red wolf are based on their apparent common ancestry with coyotes in North America, separate from gray wolves in Eurasia.[24] The assessment is uncertain because of the extirpation of wolves from much of their North American range and recent interbreeding between gray wolves, the proposed eastern wolf, and coyotes and interbreeding between red wolves and coyotes. This greatly complicates the genetic patterns of living populations. Also, the species designations are based on a limited number of genes and an uncertain number of samples from pure red wolves and eastern wolves. In my opinion, this makes the species designation of an eastern wolf premature and unwarranted.

The eastern wolf and red wolf species designations are based on the theory that these two groups and coyotes all evolved in North America from a common ancestor. It is proposed that gray wolves evolved independently in Eurasia and then entered North America where they came into contact with the existing coyote, eastern wolf, and red wolf. The fact that these four groups (gray wolf, eastern wolf, red wolf, and coyote) interbreed makes the historical reconstruction with DNA data very complex and speculative. Gray wolves and coyotes don't typically interbreed where they co-occur, but they may in areas where coyotes expanded their range and where wolves are rare (like the northeast U.S. and southeast Canada). In these areas, it is believed that coyotes have hybridized with wolves (red, gray, and eastern) within the last 100 years, making the genetic composition of each group mixed to varying degrees.

It is important to note that the red wolf and eastern wolf species designations are based on limited data.[25] The primary supporting data is from skull measurements and short sequences of mitochondrial DNA (mtDNA) and Y-chromosome DNA. These data are not in agreement with an extensive DNA assessment (48,000 single nucleotide polymorphisms, called SNP) by vonHoldt et al.[26] that shows the eastern wolf is genetically more similar to gray wolves than coyotes, and it does not support the eastern wolf as a distinct species. A reanalysis of vonHoldt et al.'s data suggests that the eastern wolf may still be a legitimate species,[27] but there is not definitive support for the eastern wolf as a species. The analyses of both Chambers et al. and vonHoldt et al. are consistent in showing that red wolves are more similar to coyotes than to gray wolves. Because the genetic patterns of both red wolves and eastern wolves reflect recent interbreeding between various groups of wolves and coyotes, the taxonomy of the groups remains unclear.

The uncertain species status of wolves and coyotes in North America indicates there is a need for a taxonomic assessment with full consideration of species concepts and definitions, additional genetic data, and the study of fossils. Most importantly, the subjectivity of species classifications in situations with extinct and mixed populations, and the possibility that rigorous designations may not be possible, should be emphasized. In addition, comparisons should be made with other species groups for which DNA data are not concordant with species designations.[28] This includes whitetail deer (*Odocoileus virginianus*), mule deer (*Odocoileus hemionus hemionus*), and black-tailed deer (*Odocoileus hemionus sitkensis*) as well as brown bears (*Ursus arctos*) and polar bears (*Ursus maritimus*). In the case of the deer, mtDNA is more similar between whitetail deer and mule deer (different species) than between mule deer and black-tailed deer (same species). In the case of the bears, mtDNA is more similar between the southeast Alaskan brown bears and polar bears (different species) than between the southeast Alaskan brown bears and other brown bears (same species).

It is well known among scientists that mtDNA relationships may not reflect species relationships.[29] In both of these cases, other DNA data show concordance with the species designations (i.e., whitetail deer are different from mule deer and black-tailed deer, and all brown bears are different from polar bears). Comparisons such as these will provide insights for taxonomic appraisal of wolves.

Wolf Subspecies

Many subspecies of wolves were recognized during the 1900s based on morphology, particularly skull measurements. The primary classification included recognition of 23 subspecies of gray wolves in North America,[30] although up to 27 subspecies were recognized.[31] This was later reduced to five subspecies,[32] including the Mexican wolf (*Canis lupus baileyi*), the northern timber wolf (*Canis lupus occidentalis*), the plains wolf (*Canis lupus nubilus*), the eastern wolf (*Canis lupus lycaon*), and the Arctic wolf (*Canis lupus arctos*) whose ranges are shown in Figure 1 (page 289). The red wolf was designated a different species in these analyses. Chambers et al.[33] reviewed these five gray wolf subspecies with morphological and genetic data. They found the existing data support designation of the Mexican wolf, northern timber wolf, and plains wolf subspecies. They also found that the data for the Arctic wolf is not definitive, and that the Eastern wolf is a full species, not a subspecies.

The primary support of the three subspecies recognized by Chambers et al. (Mexican wolf, northern timber wolf, and plains wolf) include the skull measurements of Goldman[34] and Nowak[35] and genetic data showing limited interbreeding among them. These three subspecies and the possible Arctic wolf subspecies have a degree of DNA similarity with Eurasian wolves and are thus classified as subspecies, while the eastern wolf and the red wolf share DNA similarity with coyotes and are thus classified as species (although the eastern wolf also has gray wolf DNA). The data do indicate a degree of differentiation of the proposed subspecies, but the assessment is complex because data come from different studies with different sets of samples. Also, the subspecies range boundaries are uncertain so there is likely interbreeding across the original ranges of the different subspecies. The near extinction of the plains wolf and Mexican wolf subspecies makes attempts to characterize their genetic makeup particularly problematic because the living populations from which samples can be obtained may not represent the original genetic patterns.

The subspecies classification of Goldman[36] and Hall[37] includes another subspecies of gray wolf in southeast Alaska and coastal British Columbia, the Alexander Archipelago wolf (*Canis lupus ligoni*). However, both Nowak[38] and Chambers et al.[39] consider these wolves to be part of the plains wolf subspecies (*Canis lupus nubilus*). Figure 1 (page 289) shows the range of the plains wolf subspecies including the coastal region of British Columbia and southeast Alaska. Indeed, wolves from this area share DNA with wolves from the late 1800s from Kansas and Nebraska (sampled from museum specimens). This group of wolves is thought to have colonized from the western U.S. northward to coastal British Columbia and southeast Alaska following the melting of the continental glaciers within the last 15,000 years. This is important because the USFWS is considering the *ligoni* subspecies for listing as an endangered species under the ESA. The finding that *ligoni* is not a legitimate subspecies by Chambers et al., who are USFWS biologists,[40] makes this an unwarranted action.

Another interesting facet of the subspecies designations is the introduction of wolves from the range of the northern timber wolf subspecies (*Canis lupus occidentalis*) into the northern U.S. Rocky Mountains including Yellowstone National Park, which is within the original range of the plains wolf (*Canis lupus nubilus*)—see Chapter 5 by Cat Urbigkit. These wolves were introduced (and listed a part of the lower 48 states as a distinct population segment under the ESA), apparently without concern for the subspecies of origin. This situation may become more complex if the wolves in the northern U.S. Rockies expand southward and come into contact and interbreed with the Mexican wolf subspecies in New Mexico and Arizona.

The northern timber wolf subspecies is a large wolf from western Canada and Alaska. Skull and body sizes indicate that the northern timber wolf is larger than the plains wolf. Plains wolf males weigh 95–99 pounds, and northern wolf males weigh 85 to 115 pounds and occasionally reach 130 pounds. A male northern wolf caught in Alaska in 1939 weighed 175 pounds. The data showing the northern timber wolf subspecies is generally larger than the plains wolf subspecies is consistent with reports that the introduced wolves in the northern U.S. Rockies are considerably larger than the native plains wolves. In Yellowstone National Park, one male wolf has been caught that weighed 143 pounds, with a full stomach, and another weighed 147 pounds, with an empty stomach. However, it is important to recognize that we do not know if the size differences of wolves in these areas is due to heritable (i.e., genetic) or environmental (e.g., good nutrition) causes.

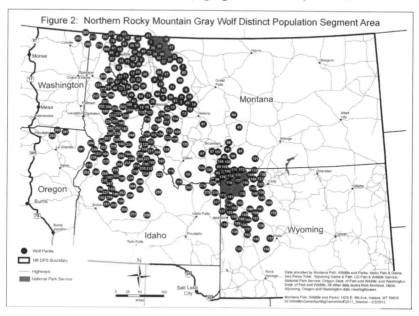

This map—from the U.S. Fish and Wildlife website www.fws.gov (August 15, 2013)—shows the wolf distribution in the Northern Rocky Mountains.

In summary, the subspecies designations of the northern timber wolf, plains wolf, and Mexican wolf endorsed by Chambers et al.[41] are not definitive because of indistinct boundaries, interbreeding between these groups prior to the near-extermination of Mexican wolves and plains wolves, and of course, the inherent subjectivity of the subspecies category.

Wolf Populations and the ESA

As noted above, "population" is a general term, and wolf populations can be designated at any scale. For our discussion, the population category used in the ESA, "distinct population segment" (DPS), is relevant. A population must be discrete and significant to be considered a DPS. Discreteness can be simple geographic separation or a combination of factors, including genetic differentiation.

Gray Wolf - Western Great Lakes Distinct Population Segment

Figure 3.
This graph—from the U.S. Fish and Wildlife website (August 15, 2013) www.fws.gov—shows the wolf distribution in the Great Lakes states.

300

Significance reflects the importance of the population to the "taxon to which it belongs," which can be the entire species or a subspecies. Significance is usually a subjective judgment by USFWS or National Marine Fisheries Service (NMFS). Despite Congress' order that DPS be used "sparingly" under the ESA, the USFWS and NMFS have listed many DPS,[42] including wolf populations.

In the case of wolves, the introduced population in the northern U.S. Rocky Mountains (Montana, Wyoming, and Idaho) and the population in the western Great Lakes region (Minnesota, Wisconsin, and Michigan) were listed as a DPS under the ESA. Both are currently delisted following recovery of the populations to large numbers of wolves. The gray wolf in the northern Rocky Mountains was delisted in 2011 in Montana and Idaho and in 2012 in Wyoming. Wolves in the northern Rockies have spread into Oregon and Washington (see Figure 2 on page 299). The gray wolf in the Great Lakes states was delisted in 2012. That population has now spread beyond Minnesota, Wisconsin, and Michigan to include North Dakota, South Dakota, Iowa, Illinois, and Ontario (see Figure 3 on page 300).

This review indicates that the current state of North American wolf taxonomy is confusing and not definitive. However, the management of wolves in the U.S., first under the federal ESA and then by states after ESA-delisting, can be influenced by taxonomic assessments. Both populations in the northern U.S. Rocky Mountains and western Great Lakes region have been managed with considerable population increases. However, both populations are of questionable taxonomic status. As noted above, the wolves in the northern U.S. Rocky Mountains were transplants of the northern timber wolf subspecies into the historic range of the plains wolf subspecies. The wolves in the western Great Lakes states are within the range of the plains wolf subspecies,[43] but are also considered the eastern wolf species,[44] and there has been mixing of the wolf subspecies in this area.[45] There are also inconsistencies in which subspecies are considered important in some cases (e.g., the Mexican wolf), but subspecies designations are ignored in others (e.g., introducing the non-native northern timber wolf subspecies into the northern U.S. Rocky Mountains).

Ecological Role

I contend that for wildlife management it is most important to recognize that the wolves in the northern U.S. Rocky Mountains and western Great Lakes regions fill the ecological role of a large predator, regardless of opinions regarding species, subspecies, and DPS.

Extensive effort and resources can be wasted debating wolf species and subspecies status and saving "pure" species or subspecies, instead of focusing on practical wildlife management.[46]

It is perhaps most important that taxonomy should be practiced as a science by taxonomists and not by wildlife biologists and that the inherent uncertainty associated with subspecies, DPS, and some species designations be openly and widely acknowledged. In particular, the USFWS should not favor one taxonomic opinion on species, subspecies, or DPS designation and ignore equally valid differing opinions.

Conclusion

Considering all of these points, my conclusion is that wolf management and wildlife management in general, including implementation of the ESA, should focus on populations in specific geographic areas, without being shackled by unresolvable debate over species, subspecies, and DPS taxonomy.

🐾 🐾 🐾 🐾 🐾 🐾 🐾 🐾

[1] Mayr, E., *Populations, Species, and Evolution,* Cambridge, MA: Belknap Press, Harvard University Press, 1970.
[2] Linnaeus, C., *Systema Naturae*, 10th Edition, Stockholm, 1758.
[3] National Research Council, Committee on Scientific Issues in the Endangered Species Act, *Science and the Endangered Species Act,* Washington, D.C.: National Academy Press, 1995.
[4] Darwin, C., *On the Origin of Species by Means of Natural Selection or the Preservation of Favored Races in the Struggle for Life,* London: Murray, 1859.
[5] Cronin, M.A., M.D. MacNeil, and J.C. Patton, "Variation in Mitochondrial DNA and Microsatellite DNA in Caribou (*Rangifer tarandus*) in North America," *Journal of Mammalogy* 86, 2005, 495-505.
[6] Mayr, E., *Op. Cit.*
[7] Chambers, S.M, S.R. Fain, B. Fazio, and M. Amaral, "An Account of the Taxonomy of North American Wolves from Morphological and Genetic Analyses," *North American Fauna* 77, 2012, 1-67.
[8] Mayr, E., *Op. Cit.*
[9] Avise, J.C., and R.M. Ball, Jr., "Principles of Genealogical Concordance in Species Concepts and Biological Taxonomy," *Oxford Survey of Evolutionary Biology 7,* 1990, 45–67.
[10] Vanzolini, P.E., *Systematics, Ecology, and the Biodiversity Crisis,* New York, New York: Columbia University Press, 1992, 185–198.
[11] Futuyma, D.J., *Evolutionary Biology,* Sunderland, Massachusetts: Sinauer Associates, 1986.

[12] Ehrlich, P.R., *Human Natures,* Washington, D.C.: Island Press, Shearwater Books, 2000.

[13] O'Gara, B.W., *Taxonomy,* Washington, D.C.: Smithsonian Institution Press, 2002, 3-65 [Chapter 1, North American Elk Ecology and Management, compiled and edited by D.E. Toweill and J. W. Thomas].

[14] Zink, R.M., "The Role of Subspecies in Obscuring Avian Biological Diversity and Misleading Conservation Policy," *Proceeding of the Royal Society of London 271,* 2004, 561-564.

[15] Haig, S.M., E.A. Beever, S.M. Chambers, H.M. Draheim, B.D. Dugger, S. Dunham, E. Elliott-Smith, J.B. Fontaine, D.C. Kesler, B.J. Knaus, I.F. Lopes, P. Loschi, T.D. Mullins, and L.M. Sheffield, "Taxonomic Considerations in Listing Subspecies under the U.S. Endangered Species Act," *Conservation Biology 20,* 2006, 1584-1594.

[16] Avise, J.C., *Phylogeography: The History and Formation of Species,* Harvard University Press, Cambridge, Massachusetts, 2000.

[17] Cronin, M.A., "The Preble's Meadow Jumping Mouse: Subjective Subspecies, Advocacy and Management," *Animal Conservation 10,* 2007, 159-161.

[18] Cronin, M.A., "A Proposal to Eliminate Redundant Terminology for Intra-species Groups," *Wildlife Society Bulletin 34,* 2006, 237-241.

[19] Marris, E., "The Species and the Specious," *Nature, 2007,* 446:250-253,

[20] Crandall, K.A., " Advocacy Dressed Up as Scientific Critique," *Animal Conservation 9,* 2006, 250-251.

[21] Nowak, R.M., "Another Look at Wolf Taxonomy," 1995, Pages 375–397 in L.N. Carbyn, S.H. Fritts, and D.T. Seip, editors, *Proceedings of the Second North American Symposium on Wolves,* Edmonton, Alberta: Canadian Circumpolar Institute, University of Alberta; Nowak, R.M., "The Original Status of Wolves in Eastern North America," *Southeastern Naturalist,* 2002, 1:95–130.

[22] Chambers, et. al., *Op. Cit.*

[23] *Ibid.;* Nowak, 1995, *Op. Cit.*; Nowak, 2002, *Op. Cit.*

[24] Chambers, et. al., *Op. Cit.*

[25] *Ibid.*

[26] vonHoldt, B.M., J.P. Pollinger, D.A. Earl, J.C. Knowles, A.R. Boyko, H. Parker, E. Geffen, M. Pilot, W. Jedrzejewski, B. Jedrzejewski, V. Sidorovich, C. Creco, R. Ettore, M. Musiani, R. Kays, C.D. Bustamante, E.A. Ostrander, J. Novembre, and R.K. Wayne, "A Genome-Wide Perspective on the Evolutionary History of Enigmatic Wolf-Like Canids," *Genome Research 21,* 2011, 1294–1305.

[27] Rutledge, L.Y., P. J. Wilson, C.F.C. Klütsch, B.R. Patterson, and B.N. White, "Conservation Genomics in Perspective: A Holistic Approach to Understanding *Canis* Evolution in North America," *Biological Conservation,* 2012, 155:186-192.

[28] Cronin, M.A., "Mitochondrial DNA in Wildlife Taxonomy and Conservation Biology: Cautionary Notes," *Wildlife Society Bulletin 21*, 1993, 339-348.

[29] *Ibid.;* Cronin, 2006, *Op. Cit.*; Zink, *Op. Cit.*

[30] Goldman E.A., "Classification of Wolves: Part II," Pages 389–636 in S.P. Young and E.A. Goldman, editors, *The Wolves of North America*, The American Wildlife Institute, Washington, D.C., 1944; E.R Hall, *The Mammals of North America, 2nd. Ed.*, John Wiley and Sons, New York, 1981.

[31] Chambers et. al., *Op. Cit.*

[32] Nowak, 1995, *Op. Cit.*; Nowak, 2002, *Op. Cit.*

[33] Chambers et. al., *Op. Cit.*

[34] Goldman, *Op. Cit.*

[35] Nowak, 1995, *Op. Cit.*; Nowak, 2002, *Op. Cit.*

[36] Goldman, *Op. Cit.*

[37] Hall, *Op. Cit.*

[38] Nowak, 1995, *Op. Cit.*; Nowak, 2002, *Op. Cit.*

[39] Chambers et. al., *Op. Cit.*

[40] *Ibid.*

[41] *Ibid.*

[42] Cronin, 2006, *Op. Cit.*

[43] Nowak, 1995, *Op. Cit.*; Nowak, 2002, *Op. Cit.*

[44] Chambers et. al., *Op. Cit.*

[45] Cronin, M.A. and L.D. Mech, "Problems with the Claim of Ecotype and Taxon Status of the Wolf in the Great Lakes Region," *Molecular Ecology 18*, 2009, 4991-4993.

[46] *Ibid.*; Cronin, 1993, *Op. Cit.*; Cronin, 2007, *Op. Cit.*; R.R. Ramey, "On the Origin of Specious Species," *Institutions and Incentives in Regulatory Science,* J. Johnston, editor, Lexington Books, 2012.

Arto Hakola / Shutterstock.com

Chapter 20

OF WHAT VALUE ARE IMPORTED CANADIAN WOLVES?

By Rob Arnaud and Ted B. Lyon

Dennis Donahue / Shutterstock.com

Editors' Statement of the Issues

One of the promised benefits to the Canadian wolf introduction was all revenue the wolves would bring in, revenue critical to offset the incredible costs associated with both the introduction and the management of these wolves. This benefit has failed to appear.

Editors' Statement of the Facts

Revenue critical to the management of wildlife has historically been a key component of wildlife population management. The promised wolf-related revenue has not materialized, and the wolves have decimated those herds of wildlife that used to bring in abundant revenue through hunting and related businesses, thus losing both revenue and treasured wildlife in one throw.

> ***Wolves are elusive, difficult to see, and occur in
> low population densities—
> all characteristics that inevitably confound the ecotourist's
> desire to see or get close to a wild wolf.***
> Matthew Wilson

Lost Revenue

Two economic techniques that can be used for quick reference of an environmental program or project are as follows:

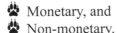 Monetary, and
🐾 Non-monetary.

Historically, hunters and fishermen generate annual (monetary) revenues used to maintain and manage America's prized big game herds, game birds, and fish stocks. Contributions from sportsmen and outdoor organizations have added to the revenues generated by license sales. For decades, this money was used successfully to balance and manage America's wildlife, contributing billions[1] directly into America's economy annually, including the following examples:

🐾 When the wood duck population began to decline, duck hunters and their organization, Ducks Unlimited, built predator-resistant wood duck boxes and placed them in swamps; that helped bring America's wood duck population back to near three million, enough to support annual harvests in the hundreds of thousands.[2]

🐾 When hunters became aware that the number of wild turkeys was declining, they helped restore them to their original range. Today, wild turkeys are found in virtually every state in the nation.[3]

🐾 When Gulf Coast fishermen saw that the red fish were being decimated by netters, they formed the Gulf Coast Conservation Association and stopped it.[4] (Incidentally, Ted Lyon helped pass the first legislation in the nation banning netting of red fish on Texas coasts.)

🐾 There are more rare African antelope preserved on Texas ranches than there are in Africa.[5]

Hunters, sportsmen, property owners, ranchers, farmers, community businesses, outdoor organizations, and state wildlife agencies alike have successfully managed and maintained America's abundant wildlife for decades. Now, huge numbers of Canadian wolves imported by the U.S. Fish and Wildlife Service (USFWS) and released in Yellowstone Park in 1995 and 1996 have multiplied and destroyed America's prized big game herds, decimating moose, elk, mule deer, and other wildlife—along with thousands of head of livestock throughout the West, forcing outfitters and other industry-related businesses to close or to lay off employees and bring to an end what thousands of citizens have worked for decades to establish.

With the disappearance of this wildlife have gone huge sums of hunting revenue generated historically to maintain these herds and the businesses connected to them—outfitters, motels, cafes, gas stations, sporting goods stores, and groceries. State wildlife agency budgets where these big game herds were traditionally located have experienced plummeting hunting license sales. Ranchers and farmers whose cattle, sheep, horses, and dogs have been slaughtered like the nation's wildlife are likewise losing thousands of dollars' worth of livestock to these wolves. To be reimbursed for these losses, ranchers must be sure that each head of livestock killed by wolves is independently verified by the government. T.R. Mader, Research Director of the Abundant Wildlife Society of North America, said that as many as 90% of livestock wolf kills could be classified as "unconfirmed" based on Animal Damage Control Officer reports.[6] He pointed out that:

🐾 Virtually all of the carcass is usually consumed by the wolves,

🐾 Scavengers and decay, especially in hot weather, rapidly eliminate evidence confirming cause of death. Coyotes, eagles, fox, skunks, crows, ravens, and magpies—all are usually waiting to feed on what's left of a carcass once the wolves finish, and

🐾 Dense undergrowth and vegetation, especially in heavy timber, often make it difficult to even find a kill once a rancher notices livestock are missing.

Ranchers or farmers have to make up these losses out-of-pocket, which cuts deep into their annual operating revenue and profit margins and directly affects their livelihood and ability to stay in business. Rocky Barker with the *Idaho Statesman* captured this financial reality in an article he wrote following the slaughter of 176 of a rancher's prize rams caused by wolves[7]:

> ...*176 sheep died on the west slope of the Tetons in Idaho when wolves panicked them and they ran over a ridgeline and were trampled or suffocated. The sheep were part of a 2,400-head band owned by the Siddoway Sheep Co, grazing in the area of Fogg Hill near Tetonia. Wildlife Services confirmed that wolves were the culprits, based on eyewitnesses, bite marks, and tracks. At about $200 a head (for reimbursement), the Siddoway's loss is $35,000.*

It doesn't take many slaughters like this to put a ranch or farm family out of business.

Anyone familiar with wolves—never in danger of biological extinction—had to know what would happen once their numbers exploded. Benjamin Corbin, a trapper and hunter, gave as good a firsthand description of what thousands of wolves running rampant across America's Western states looked like in 1900 (while pitching his bosses for a raise) and the tens of thousands of livestock they were killing.[8]

This domestic livestock and wildlife tragedy began with the introduction into Montana and Idaho by the USFWS of these Canadian wolves, which multiplied quickly into thousands. Their slaughter of

America's prized big game animals, if not stopped, has the potential to rank alongside the carrier pigeon and buffalo exterminations.

Wolves as Revenue Generators

Cheerleaders for the introduction of the Canadian wolves into the U.S. have tried to put a counter "value" on these predators, saying they generate interstate commerce and significant revenue through:

- 🐾 "Wolf howlings" (monetary),

- 🐾 "Wolf watchings" (monetary), or

- 🐾 The notion that wolves are "priceless" and that you cannot put a value on them (a non-monetary opinion and assumption).

Wolf Howlings

The "wolf howlings" gambit to establish "commercial value" of wolves first appeared in the 1970s when the USFWS trapped what red wolves they could find roaming between Texas and Louisiana, put them in a captive breeding program, and declared the species "extinct in the wild." When they checked the red wolves' DNA, however, it was found that feral dogs and coyotes had polluted the gene pool, so technically, the species may have already been "extinct" when they were captured and, thus, ineligible for any kind of "protection" under the ESA.[9]

Time passed, the red wolves multiplied in captivity, and the USFWS proposed introducing them in Kentucky and Tennessee. State officials, stockmen, and others in those states met them at the gate and told them they didn't want the wolves. So the USFWS shifted its wolf release to federal land in the Alligator River National Wildlife Refuge in North Carolina and nearby Pocosin Lake Refuge, assuring the public that, if one of their collared wolves strayed off federal land, the agency would simply "…recapture and return the animal to the refuge."

The wolves quickly exited these forests and multiplied, and new generations of collarless wolves began scouring private land for meals, some more than 100 miles away. A landowner who shot one threatening his livestock was fined and required to feed captive red wolves for a year. When North Carolina's Hyde and Washington counties began having similar problems with the wolves, they passed local ordinances making it legal for ranchers and farmers to kill wolves they believed a threat to people or livestock on their land. The North Carolina General Assembly joined and supported them.

A lawsuit—*Gibbs vs. Babbitt*—challenged the USFWS' authority to prevent landowners from killing wolves that were killing their livestock.[10] This suit, filed in Federal District Court for the Eastern District of North Carolina, argued that citizens had the right under state law to protect their property from marauding wolves, the ESA notwithstanding. In court, USFWS lawyers argued, straight-faced, that ranchers and farmers who killed wolves that killed livestock and wildlife on their property were "... interfering with interstate commerce and thus violating the Commerce Clause."

Establishing a "fact" that wolves generated interstate commerce was critical for USFWS' long-range plans for the wolves because the Commerce Clause prohibited states from interfering with interstate commerce (the wolves), which was another shield, like the Endangered Species Act (ESA), for the USFWS to use for the release of imported Canadian wolves in Montana and Idaho. So, what kind of interstate commerce would wolves generate? The kind—argued contract lawyers for USFWS—that tourists generated when they came from places like New York to North Carolina to "hear the wolves howl."

Gibbs lawyers checked the logbook at the federal forest where the so-called "wolf howlings" were held and found approximately 200 people had attended these events over the previous decade. And no one knew if they came to hear the wolves howl or just happened to be camping there when the "howlings" were held.

Other Types of Commerce

What other type of "commerce" could be harmed if a red wolf was shot? Growth of a valuable wolf-pelt industry might be disrupted. *Gibbs* lawyers again checked the historical records and could find no value placed on wolf pelts in the last century in North Carolina. They did find a bounty on wolves in the late 1700s, however.

What other type of "commerce" might be affected? Researchers who crossed state lines to study red wolves would be deprived of a resource if red wolves were shot. The Federal District Court for the Eastern District of North Carolina agreed with USFWS lawyers, and the case was referred to the 4th U.S. Circuit Court of Appeals which, in a split decision, agreed with the lower court. The federal court's logic refuting the landowners' claim that they had a right to kill wolves killing their livestock went something like this:

 That wolves were "things of interstate commerce" because they moved across state lines and their movement was followed by "tourists, academics, and scientists."

🐾 That the killing of the wolves implicated this "variety of commercial activities associated with interstate commerce" and thus was constitutional under the Commerce Clause.

🐾 That landowners killing wolves on private lands would reduce the total number of wolves, which in turn "could reduce the number of wolves on federal land."

🐾 That the (ESA) regulation allowing (these wolves to roam unchecked) was an integral part of an overall federal scheme to "conserve valuable wildlife resources important to the welfare of the country," and

🐾 That, since the landowners were killing wolves to protect their livestock and since livestock were clearly part of interstate commerce, Congress had the authority to regulate the wolves because the law affected interstate commerce and was, therefore, allowed under the Commerce Clause, even if that effect was a negative one.

The Competitive Enterprise Institute subsequently filed a petition for certiorari with the Supreme Court, but on February 20, 2001, the Supreme Court refused to hear the case. Pro-wolfers were worried about *Gibbs* from the beginning. Had the Supreme Court heard the case and ruled in favor of *Gibbs*, it would have prevented the USFWS from using the Commerce Clause to let wolves wreak havoc on domestic livestock and private property throughout America, and it stopped the USFWS from coming onto private property to charge stockmen and property owners with interfering with interstate commerce for killing wolves once states passed laws to deal with red wolf depredations.

Wolf Watchings

Washington, D.C., and USFWS shifted from "wolf howlings" to "wolf watchings" as the next interstate commerce *raison d'être*. Yellowstone National Park became ground zero for the introduction of Canadian wolves and for "wolf watchings," which the USFWS said would increase tourism (interstate commerce) at the Park. Early economic forecasting by the USFWS predicted people hoping to see a wolf in Yellowstone would add an estimated $23 million annually to the greater area.

An often-quoted 2005 study of the value of Yellowstone's wolf ecotourism—written by John Duffield, a University of Montana economist, and funded by the Yellowstone Park Foundation, an official fund-raising partner of the Park—estimated that at least $35 million annually in economic benefits would flow from wolf ecotourism into Yellowstone[11] and could even double as the "multiplier effect" of wolf-ecotourism money circulated through the local economy.

Duffield began his study in 2004 with random samplings taken at Park entrances and from the Lamar Valley parking areas where wolves were most likely to be seen. About two-thirds of these visitors (64%) responded. Based on the survey, it was estimated that approximately 10% of Yellowstone's visitors came to see a wolf; this was when the wolf population was at its highest in the Park. Since then, Yellowstone's wolf population has decreased 50%. Of their favorite food sources—moose and elk—the moose herd that historically maintained itself at approximately 1,000 animals has been decimated, the majority of which were eaten by wolves along with 80% of the Park's historic elk herd. The *Associated Press* reported on March 13, 2013, that:

> ...*scientists from the Park and the Montana Department of Fish, Wildlife and Parks said the Northern Yellowstone elk herd is down 6% this winter to 3,915 animals. The herd peaked at about 20,000 animals in 1992. That was just a few years before gray wolves were reintroduced to the Yellowstone area from Canada after being absent from the region for decades...*[12]

Lack of adequate food, deadly territorial battles between wolf packs for dominance, and wolves contracting distemper and mange have accounted for declining wolf numbers in Yellowstone in recent years.

When the Wyoming Sierra Club also surveyed visitors to Yellowstone in 2005 and asked them why they came, their number one reason for visiting the Park was to view the scenery, and the number two reason was to watch the wildlife. Only 3.5% said they would not come if there was no opportunity to see a wolf.[13] Elk and moose were their two most popular ungulates, along with the grizzly.

Additional reports have found that the presence of large numbers of wolves in Yellowstone has driven the surviving elk away from frequenting open fields and into thicker timber and cover to have a better chance of evading wolves, that the chance of seeing elk in Yellowstone is declining, and that moose have virtually disappeared.

312

Outfitters that attempted to react to the disappearance of the big game herds in the Rockies by offering wolf-watching tours reported that, by 2005, tourists were more interested in seeing grizzlies and geysers than wolves. The 1994 federal environmental impact statement (EIS) of the Canadian wolf introduction into the U.S. predicted that the wolves' presence would result in a 5-10% increase in annual visitation to Yellowstone and generate $20 million annually for Idaho, Montana, and Wyoming.[14] Instead, from 1995 through 2009, Yellowstone visitation declined in 11 years and remained essentially unchanged in the others.[15]

People who study eco-tourism and wolf watching say that opportunities to see wolves without professional assistance are rare and limited to areas of open terrain.

> ...*Wolves are elusive, difficult to see, and occur in low population densities—all characteristics that inevitably confound the ecotourist's desire to see or get close to a wild wolf. ... Only with the aid of science, technology, and adaptations in recreational strategies have current opportunities for viewing and hearing wolves in the wild become possible.*[16]

In the upper Midwest, wolves are even more difficult to see due to dense forest cover and different predation patterns. "Wolf watching" in the Midwest is focused at the International Wolf Center in Ely, Minnesota (which reports that it draws about 50,000 people a year[17]) and at Wolf Park in Battle Ground, Indiana, where captive wolves in enclosures ensure that people get to look at a wolf.

In the Southwest, more than $26 million has been spent trying to reintroduce Mexican wolves, excluding the additional costs associated with wolves in the mountain states. It's estimated that only about 75 of these animals exist in the wild after a 30-year attempt to reintroduce them. USFWS appears to be using a similar reintroduction strategy tried with the red wolves with exception—they are either inadvertently or deliberately opening the door to a Mexican wolf-Canadian wolf cross as the Canadian wolves drift into New Mexico's northern mountains and points south.[18] Such a large area with so few animals whose genes may already be mixed with those of feral dogs and coyotes generates the following question: What is a "genuine" Mexican wolf? The probability of people generating "wolf revenue" by driving to Chihuahuan or Sonoran deserts for Mexican "wolf howlings" or "watchings" is negligible.

Conclusion

"Wolf watchings" and "wolf howlings" will never generate the revenue it takes to cover the minimum costs required to manage these predators. And unlike elk, moose, deer, and bighorn sheep hunters, who return annually, "wolf watchers" who see a wolf generally are satisfied with a one-time sighting for their "bucket list" and move onto their next quest. "Wolf watchers" returning year after year like traditional hunters exists only in the mind of wolf promoters.

Hunting license sales to take a wolf will make no difference in replacing the revenues from disappearing hunters, who no longer make their annual trips to Western states because the traditional big game populations there have disappeared or declined to the point that it's no longer worth the trip. Attempts by the mountain states to raise the price of hunting licenses only exacerbates the declining hunter shortage because there is no game to hunt, and increasing the cost of hunting licenses cannot make up revenues lost from lack of hunters, most of which historically came from out of state.

When wolf hunting was permitted in Montana in 2009, license sales from wolf licenses brought in $325,916, an amount that did not come close to covering the million dollars a year it costs the state to manage them. Nor did it even move the needle on Montana's $57 million annual wildlife budget, *66% of which comes from the sale of the state's traditional hunting and fishing licenses.* Wolves cannot pay their way like the herds they decimated did.

Today, these imported wolves have spilled out of the Northern Rockies into Washington, Oregon, Utah, and Colorado in the West. In New England, their numbers are growing. Michigan, Minnesota, and Wisconsin have an estimated 6,000 in their area, while the USFWS continues trying to reintroduce the Mexican "wolf" throughout the Southwest.

Some researchers now suggest that the number of Canadian wolves in North America is in the tens of thousands, none of which were wanted or needed for economic or environmental reasons. In their wake lie tens of millions of dollars in unnecessary and unwarranted expenses generated by the USFWS using taxpayer dollars, along with decimated wildlife populations and dead cattle, sheep, horses, guard dogs, and other domestic livestock and pets.

Of what value are these imported Canadian wolves?

🐾 🐾 🐾 🐾 🐾 🐾 🐾 🐾

1 "Hunting and Fishing: A Force As Big As All Outdoors," http://www.nssf. org/PDF/research/bright%20stars%20of%20the%20economy.pdf

2 *Wood Duck Report*. USDA Natural Resources Conservation Service, ftp://ftpfc:sc.egov.usda.gov/WHMI/WEB/pdg/woodduck%281%29.pdf

3 "A Dedication to Wild Turkey Conservation," http://www.nwtf. org/conservation/?utm_source=nwtf.org&utm_medium=conserve-article&utm_campaign=spring-hunting

4 Plugger, Wade, *Fishing the Gulf Coast: Rudy Grigar, W.R. McAffee*, Texas Tech University Press, 1997; http://www.amazon.com/Plugger-Wade-Fishing-Gulf-Coast/dp/0896725103

5 "Texas Ranchers Fight To Breed, Hunt Endangered Antelope," http:// articles.latimes.com/2012/apr/03/nation/la-na-nn-texas-antelope20120403

6 "Abundant Wildlife Society of North America Fact Sheet, T.R. Mader, Research Director," http://www.aws.vcn.com/fact.html

7 "Rocky Barker: After wolf attacks, rancher regrets…-Idaho Statesman," www.Idahostatesman.com/2013/09/02/2738613/after-wolf-att

8 "Benjamin Corbin's advice, The Wolf Hunter's Guide. A first hand picture of tens of thousands of wolves running rampant in the Mountain and Western States in 1900," Internet Archive, archive.org, Ebook and Texts Archive, The Library of Congress, http://archive.org/details/corbinsadviceorw00corb

9 *Canid News*, Vol. 3, 1995, "Hybridization: The Double-Edged Threat," by Ron Nowak, Canid Specialist Group, www.canids.org/publicat/cndnews3/hybridiz.htm

10 *Gibbs v. Babbitt*, 214 F.3d 483 (4th Cir. 2000).

11 Duffield, John, Chris Neher, and David Patterson, "Wolves and People in Yellowstone: Impacts on the Regional Economy," September 2006, http://www.defenders.org/publication/wolves-and-people-yellowstone

12 "Major Yellowstone elk herd continues steep decline," *Missoulian, Associated Press*, March 8, 2013, http://missoulian.com/news/state-andregional/major-yellowstone-elk-herd-continues-steep-decline/article_ea10e4cc-880a-11e2-9ef3-0019bb2963f4.html

13 Retrieved from http://wolfsource.blogspot.com/; http://www. yellowstonepark.com/MoreToKnow/ShowNewsDetails.aspx?newsid=182

14 "Wolf restoration worth millions of dollars to the economies of Idaho," www.forwolves.org/ralph/wolf-economic-impact.htm

15 Yellowstone National Park Visitor Statistics Page; http:www.yellowstone. co/stats.htm

16 "The Human Dimensions of Wolf Ecotourism in North America," by Matthew A. Wilson, Departments of Sociology and Rural Sociology, University of Wisconsin-Madison; http://www.wolf.org/wolves/learn/wow/regions/Canada_subpages/wolf_human_interactions2.asp

17 Schaller, David T., "The Ecocenter as a Tourist Attraction: Ely and the International Wolf Center," Ely, MN: IWC, 1999.

[18] "Feds Make Mexican Wolf Deals," by Center for Biological Diversity, August 29, 2013; http:pinedaleonline.com/news/2013/08/FedsmakeMexicanWolfd.htm

ericlafrancais / Shutterstock.com

Chapter 21

I REST MY CASE

By Ted B. Lyon

Photo West / Shutterstock.com

Closing Statement

May it please the court, ladies and gentlemen of the jury of public opinion, I want to first of all thank you for taking the time to read this book. The case this book presents has been about addressing the crisis of wolf proliferation across North America environmentally and the myths that have been perpetuated by those who believe it is a good thing to reintroduce wolves into the lower 48 states. We, the authors and contributors, do not believe the wolf reintroduction in the western U.S. was done properly or honestly. Some people are opposed to having any wolves in the lower 48 states, and others feel they can be managed as wildlife populations and game animals. Regardless, federal control with misuse of science and abuse of citizens' and states' rights, as described in this book, is not the right way to manage any natural resource, including wolves.

The wolves that were reintroduced into Yellowstone and Idaho are here to stay, according to every wolf expert that has been quoted in popular print from David Mech to Doug Smith to Ed Bangs, all men who have spent their careers studying wolves.

When I started on this journey in 2007, I did not see or understand that wolves are some of the most destructive animals on earth. This book proves—through ongoing scientific research developed over the last 100 years—that wolves devastate elk, caribou, deer, and moose unless they are tightly controlled. From Russia to Canada and from Minnesota to Idaho, the same story repeats itself: where wolves go, big game herds disappear or are severely impacted.

Teddy Roosevelt, our greatest conservation President, once said that wolves were "beasts of waste and desolation." The tragedy that has happened to the elk and moose in Yellowstone National Park and to other herds in Montana, Idaho, and Wyoming is beyond comprehension. It is unbelievable that the elk herd in Yellowstone, which was at 19,000 when wolves were introduced in 1995, now numbers fewer than 4,000 elk as I write this final chapter in 2013. The moose count has gone from an estimated 1,200 in 1995 to almost zero now.

What has been shocking to me is that wolves, while causing all of this destruction, have been elevated to an exalted status by the federal government and by some in the environmental movement. I have been shocked that no one at the top of the Yellowstone National Park hierarchy has come out and told the truth about what was happening to the elk, the mule deer, and the moose. The only statements that have come out

318

of the Park about wolves were made by the chief wolf biologist for the Park, who is an admitted wolf proponent. No one with authority has come forward publicly to defend the elk, the mule deer, the wild sheep, or the moose that are disappearing from the Park. No one!

This book hopefully has proven to its readers that wolves are *not* an economic boon to the economy as has been represented by many environmental groups. When wolves cause huge losses of wild game, there is a resultant loss of license revenues and expenditures by hunters that is dramatic, as has been seen in Idaho and Montana, where millions of dollars in revenue have been lost due to fewer hunters coming in to hunt big game.

Wolves also cause significant losses to ranchers, who have no way of recovering from their losses absent programs that require them to prove beyond a reasonable doubt that wolves caused a kill. Studies cited in this book also show that, for every cow identified as a wolf kill, there are at least seven other cattle that are killed that are not even found because a wolf pack eats almost everything and leaves little or nothing to prove or disprove what killed the animal.

The United States Department of Agriculture reported that, in 2010 alone, wolves killed over 8,000 head of cattle. At an average cost of $1,000 per cow in 2013, that is a direct economic loss of approximately $8,000,000 to the cattle industry in states that harbor wolves. Multiply this number by seven, and you get a real number of at least $56 million since only one in eight kills is actually proven to be a wolf kill. That loss, however, does not even touch the indirect losses suffered by ranchers whose cattle are continually harassed by wolves. Studies done in Idaho, Oregon, and New Mexico demonstrate much greater costs because of the stress that cattle suffer, which results in loss of weight and the failure to get pregnant.

What has also been shown here is that Congress should require each federal agency that pays attorney fees for the enforcement of the Equal Access to Justice Act or the Endangered Species Act to keep an accounting of those payments and what they are for and to send those records to the General Accounting Office, which shall send a report to Congress each year detailing these costs. In addition, each federal district court should be required by statute to send a copy of any judgment awarding attorney fees under the act to the Government Accounting Office as well.

It has been estimated that, since the mid-1970s, the federal government and the states have spent over $200 million to reintroduce Mexican wolves in New Mexico and Arizona and Canadian gray wolves in Idaho, Montana, and Wyoming. Assuming that the government's numbers of wolves are correct, there are roughly 75 wolves in New Mexico and Arizona and a minimum of roughly 1,800 wolves in Montana, Idaho, and Wyoming. Assuming that the numbers in the Northern Rockies are correct, the cost per wolf for reintroduction is roughly $104,000 per wolf.

A reasonable person, when thinking about governmental policy and what is good for all wildlife and the economy, would have to come to the conclusion that further federal expenditures to propagate a species that causes such huge economic losses to people living in the areas affected is not rational.

The chapters by Laura Schneberger and Jess Carey, who have been intimately involved with the Mexican Gray Wolf Recovery Program, call into question whether or not Congress should allow for any further expenditures for this program, which has been going on for over 34 years at a cost of over $20 million. According to the U.S. Fish and Wildlife Service, there are 75 Mexican wolves in the wild. Divide 75 into $26,000,000, the amount that has been spent on the Mexican Gray Wolf Recovery Program, and the astounding number is more than $346,000 per wolf. That is not rational under any sense of the word when one considers that there are 50,000 to 100,000 wolves in Canada; another 1,800 to 4,000 and perhaps more in Montana, Idaho, and Wyoming; and an unknown number that have expanded into Oregon, Washington, and many other states. Additionally, there are several thousand in Minnesota, Michigan, and Wisconsin. Continued expenditure of federal or state tax dollars to promote wolves is *not* sound governmental policy.

I also think people who are part of the environmental movement need to take a close look at the groups that use the wolf as a rallying cry to raise money and that file endless lawsuits over wolf hunts. If you are an environmentalist and believe in the preservation of our wild species, why would you support an organization that continually tries to expand wolves when, in fact, you are destroying other wildlife in order to do so?

Hunting, fishing, and sportsmen groups across the United States should also understand that, when they work together and put aside their differences, they are more powerful than any group in the country and can truly impact public policy. When farm and ranch groups combine with organized sportsmen groups, then they are truly powerful.

By the time people in Oregon and Washington understand what the wolves have done to their big game herds and to their ranching industry, it may be too late. Hopefully someone from those states reads this book and uses this political template for success in changing the laws in those states to allow for the control of wolves before they destroy their big game hunting industry and severely affect ranchers and cattlemen.

I did not pick this fight or this issue to write a book about, but after six years of study, I think that I am on the right side of the preservation of our wild game. Others may disagree, which is their right, but if they look at the research that has been done on wolves for over 100 years, I think they will conclude, as have I, that wolves need to be stringently controlled if we are to have sustainable game herds.

I told my wife that I expected this book to be widely praised by some groups and widely condemned by others who may use sophisticated methodology, even misinformation, to attack every facet of the book. The research and facts contained in this book are irrefutable.

I rest my case.

Ted B. Lyon
September 1, 2013

Holly Kuchera / Shutterstock.com

Epilogue

I had a front-row seat to one of the worst natural resources decisions ever foisted upon the West. I was Idaho's lieutenant governor in the mid-1990s when—like a shotgun wedding—Canadian gray wolves were "reintroduced" to the Idaho backcountry by the U.S. Fish and Wildlife Service against our wishes. And I was a member of Congress in the early years of the 21st century as those transplanted predators grew into packs that began ravaging our elk herds and terrorizing our livestock. So, when I became Idaho's governor in 2007, one of my top priorities was working with our congressional delegation, sportsmen, ranchers, and many others—including Ted Lyon—to overcome the legal hurdles set up by environmental extremists and activist judges to states wresting management of these big marauders from federal bureaucrats.

Ted's book *The Real Wolf* does a compelling job of chronicling that process and why it remains so important to those of us—no matter our political affiliation—who care deeply about states' rights and responsible stewardship of our public lands, wildlife, and other resources. The hard-won experience of too often being at the mercy of that federal fiat we call the Endangered Species Act has taught us the value of on-the-ground collaboration, consensus-building, and counting on those who live with the outcomes of our public policies for the best, real world perspectives.

There have been few issues during my 40 years in public life that have provoked the raw passions of so many people from around the world as the debate over wolves. I was deluged with some of the nastiest, most disparaging, and truly hateful letters, emails, and phone calls from well-meaning but badly misinformed folks, who saw wolves only as big, beautiful dogs harmlessly pursuing their majestic lives in the trackless wild. Wolves are an essential and misunderstood part of the Rocky Mountain ecosystem, many argued, and we owe it to our Western heritage to enable wolves to once again roam freely in the Idaho wilderness.

The problem is that wolves don't stay put. Their enormous range, high reproductive rate, and insatiable hunger for ungulates inevitably draw them out of the woods to interface with man. As their numbers spiraled far beyond expectations, so did the conflicts, and so did my determination to manage wolves as we do any other species—with an eye toward the bigger picture of a balanced ecosystem that includes man.

I'm grateful to Ted and the many good people who feel a strong affinity for Idaho, Montana, Wyoming, and the other states where wolves were another government-imposed challenge to overcome. It was a problem created by "conservationists" who speak floridly about the primal necessity of having wolves in our midst, but for whom the real goal is raising money and disrupting or shutting down such traditional multiple uses of public lands as grazing, logging, mining, and especially hunting. It was a problem created by "conservationists" who consistently moved the recovery targets, forum-shopped for a sympathetic judge, collected millions of taxpayer dollars to pay their lawyers, and looked for any opportunity to abandon their commitment to pay for our ranchers' losses to wolves released in Idaho.

Ted, and many others who recognize that reality, fought tough odds to turn the tide on the wolf issue. Now Idaho, Montana, and Wyoming are managing wolves—wolves that never should have been here in the first place. But since they are, the happy ending to this story is that the people most affected by their presence now are managing them in a way that's far more balanced and reflective of the realities of today's West. They will never be "our wolves," but at least now we have a primary role in controlling their population and impacts.

It's my sincere hope that *The Real Wolf* will help open some eyes to the bigger problems with the Endangered Species Act—a once well-intentioned but incredibly flawed law that undermines the real interests and values of conservation by placing the well-being of humans and their livelihoods far down the food chain.

Idaho Governor C.L. "Butch" Otter
November 1, 2013

Appendices

Holly Kuchera / Shutterstock.com

Will N. Graves
900 Hillen Drive
Millersville, MD
21108
3 Oct 1993

Gray Wolf EIS
P.O. Box 8017
Helena, MT 59601
ATTN: Ed Banks, US Fish & Wildlife Service Project Leader

Dear Mr. Banks,

Thank you for the opportunity to comment on the DEIS regarding
"The Reintroduction of Gray Wolves to Yellowstone National Park &
Central Idaho." I support Alternative 3, the No Wolf Alternative.

1. Diseases, Worms, and Parasites. I was surprised that the
DEIS did not make a detailed study on the impact issue of diseases,
worms, and parasites (page 9). I believe an EIS is not complete
without a detailed study covering the diseases, worms and parasites
that wolves would carry, harbor, and spread around in YNP and in
Idaho. The study should cover the potential negative impact of
these diseases on wild and domestic animals, and on humans. I
believe the potential negative impact of these diseases is a valid
reason not to reintroduce wolves into YNP and to Idaho.

Countless articles about the diseases, worms, and parasites
carried, harbored, and spread around by wide ranging wolves have
been published in a magazine sponsored by the former Soviet
Ministry of Agriculture. For example, a Soviet biologist reported
that gray wolves are carriers of a number of types of worms and
parasites which are dangerous for animals and for humans.
According to this biologist, the main one is cestoda. Over
approximately a ten year period, the Soviets conducted a controlled
study on this subject. They made the following observations. When
and where wolves were almost eliminated in a given research area,
(where almost all wolves were killed by each spring and new wolves
moved into the controlled area only in the fall) infections of
taenia hydatigena in moose and boar did not occur in more than 30
to 35% of the animals. The rate of infections were 3 to 5 examples
in each animal. When and where wolves were not killed in the
controlled areas in the spring, and where there were 1 or 2 litters
of wolf cubs, the infections in moose and boar of taenia hydatigena
reached 100% and up to 30 to 40 examples of infection (infestation)
were in each moose and boar. Each year the Soviets studied 20
moose and 50 boar. The research was documented and proved that
even in the presence of foxes, racoons, and domestic dogs, ONLY THE
WOLF was the basic source of the infections in the moose and boar.
Examinations of 9 wolves showed that each one was infected with
taenia hydatigena with an intensity of 5 to 127 examples. This
confirmed the Soviet conclusions. The damage done by taenia
hydatigena to cloven footed game animals is documented by Soviet

veterinarians. My concern is that if grey wolves in the former USSR carried and spread to game animals dangerous parasites, then there must be danger that gray wolves in YNP and in Idaho would also spread parasites. Why should we subject our game animals, and possibly our domestic animals to such danger?

If wolves are planted in YNP and in Idaho, I believe the wolves will undoubtedly play a role in the epizootiology and epidemiology of rabies. The wolf has played an important role, or perhaps a major role, as a source of rabies for humans in Russia, Asia, and the former USSR. From 1976 to 1980 a wolf bite was the cause of rabies in 3.5% of human cases in the Uzbek, Kazakh, and Georgian SSRs and in several areas of the RSFSR. Thirty cases of wolf rabies and 36 attacks on humans by wolves were registered in 1975-78 only in the European area of the RSFSR. In the Ukraine, wolf rabies constituted .8% of all cases of rabies in wildlife in 1964 to 1978. The incidence of wolf rabies increased six fold between 1977 & 1979. The epizootic significance of the wolf has been shown in the Siberian part of the former USSR. Between 1950 and 1977 a total of 8.7% of rabies cases in the Eastern Baikal region were caused by wolf bites. In the Aktyubinsk Region of Kazakhstan, of 54 wolves examined from 1972 to 1978, 17 or 31.5% tested positive for rabies. During this period, 50 people were attacked by wolves and 33 suffered bites by rabid wolves. This shows that healthy wolves also attack and bite humans. Recent Russian research states that as the numbers of hybrid wolves increases, the likelihood of a healthy hybrid wolf attacking humans also increases. Russian wildlife specialists state that when there is no hunting of wolves, the possibility of wolves attacking humans also increases, as the wolves lose their fear of humans.

Wolves not only have and carry rabies, but also have carried foot and mouth disease and anthrax. Wolves in Russia are reported to carry over 50 types of worms and parasites, including echinococcus, cysticercus and the trichinellidae family.

Prior to planting wolves into YNP and Idaho, I respectfully request a detailed study be made on the potential impact wolves will have in regard to carry, harboring, and spreading diseases.

I believe bringing wolves into YNP and into Idaho will increase the cost of meat and animal products. It will take a lot of work and cost a lot of money for our ranchers to protect their animals from wolves. Why should we subject our ranchers to these increased costs?

2. Prey Animals Normally Killed by Wolves. I do not understand the wide discrepancy in the number of prey animals projected to be killed annually by one grey wolf in YNP/Idaho, and the numbers of prey animals killed annually by one grey wolf in Russia. The Soviets have professionally documented the number of prey animals "normally" killed by one wolf in one year. I realize there are many factors involved. However, Soviet literature states that one grey wolf will kill 50 deer each year (about 1 per week),

328

or up to 90 saiga (Saiga tatarica L.) or 50 to 80 boar, or about 8 to 10 moose. I understand that the US estimate is one wolf will kill one deer every 23 days. Why is there such a difference in these figures?

Russian and Soviet literature is filled with examples reporting that wolves kill extremely large numbers of game and domestic animals. In the Krasnoyarskij Region wolves kill 30 to 40,000 reindeer and 700 to 800 moose each year. In the Putorana Plato, just east of Norilsk, wolves kill about 20,000 reindeer each year.

3. Lustful Killing by Wolves. It is well documented that wolves are lustful killers. Look at some of the figures. Here are some examples from Soviet literature. Eight wolves killed 50 reindeer in about two days. In 1978 one wolf killed 39 reindeer in one attack, and another wold killed 29 in one attack. A pack of wolves killed over 200 sheep in a few short attacks. There are many, many examples of lustful killing of animals by wolves in the former USSR.

In Sweden in 1977 one wolf killed between 80 and 100 reindeer in 19 days. Is there any doubt that a high majority of Swedish reindeer owners (over 70%) reject efforts to protect wolves?

Does the FWS expect that wolves in the US will not carry out lustful killing attacks, especially during periods of crusted or heavy snow? What would these attacks have on the populations of small game herds? Do the US estimates on the number of prey expected to be taken by wolves account for any lustful attacks?

4. Wold Attacks on Humans in the Former USSR. In the former USSR, it was not unusual for wolves to attack or threaten humans. (Details are available.) Some of these attacks were done by non rabid wolves. There are remote villages in Siberia that have been under siege by wolves. The villagers would not dare go out of their houses at night for fear of wolf attacks. I believe bringing wolves into the US will create a threat to humans that is just not necessary. There is extensive information in the world about the behavior of wolves. Gray wolves attack and kill humans in Asia, so I do not support planting them anywhere in the US.

5. Wolves Affect Structure of Prey Populations. Wolves not only reduce the size of prey populations, but also have a marked effect on the structure of the prey population. Detailed research by the Soviets showed that in areas inhabited by wolves, about 50% of the moose and deer calves do not reach the age of six months. In areas where the wolves had been exterminated, the loss of calves of moose and deer was only 7 to 9%. The negative effects of wolves on populations of ungulates is especially apparent in years when there is heavy snow or crusted snow. In just one severe winter, wolves can almost completely destroy the ungulate population in the area. In these conditions, wolves pay little attention to the animals they killed, because it is so easy for the wolves to kill

the almost helpless animals. Documenting these lustful killings is almost impossible due to the conditions.

6. Cost. I do not support spending 6 million dollars to reintroduce an experimental wolf population. In this high deficit period, we need to cut all unnecessary spending, and this is a good area to cut. During the years of the USSR, wolves cost the Soviets about 45 million dollars per year. The US needs to keep our "wolf costs" to a minimum.

7. Reduction in Harvest of Female Elk. Why should we let wolves kill female elk rather than let US hunters bag them? Let us keep the wolves out, and if the ungulate population becomes to large for food and cover, then we should let hunter bag the excess game. Part, or even all of the meat could be given to the poor. I believe wildlife managers should use hunters to control excess populations of game, and not wolves.

8. States Manage Own Land & Resources. I believe each state should have primary authority and rights to manage its own land and resources.

9. The Wolf is the Most Dangerous & Damaging of all Predators of Fauna. Am eminent Soviet professional wrote that if all the predators of fauna were placed in a list according to the degree of damage and danger to humans, the wolf, without a doubt, would be in first place at the very top of the list, and would far outdistance its closest competitors.

I believe it is time to stop idealizing the role of the wolf in nature. The wolf is a powerful, vicious, dangerous, damaging and lustful predator. The wolf is an almost perfectly designed killing machine. Some Soviet researchers state that US wolf specialists generally overestimate the selectivity role of the wolf in nature. Wolves do kill weak and deformed animals, but wolves kill many perfectly fit, healthy animals. Wolves select prey animals based on many circumstances. It is a fact that wolves prefer females that are in late stages of pregnancy and young animals. There are many people who try to emphasize that wolves select only old, weak, sick or deformed animals for prey; the facts do not support this belief.

I recently read a book for children about wolves. The book showed a white tail deer about two years old, and stated that this is a healthy deer and thur does not have to be afraid of wolves since a healthy deer can run faster than a wolf. If you accept this as fact, then how do you explain that thousands upon thousands of perfectly healthy, fit deer are killed each year by wolves. It is time for us to teach the truth about wolves. I still meet people who believe that wolves in the northers regions of Canada do not kill reindeer for food - that the wolves prey only on lemmings. Then they say that a "specialist" on nature proved that by a detailed study about wolves which he published in a book.

Wolves cause some prey to suffer terribly. Some Soviet technical literature about wolves describes in detail how wolves kill their prey. It is not a pleasant subject. It may take several days for a few wolves to kill a large, male moose. During this time the wolves are actually eating the moose alive. Wolves often eat sheep when they are still alive.

After reading Russian and Soviet literature about wolves for so many years, and talking to Russians who have had experiences with wolves, I have come to the conclusion that many American wildlife biologists have become enamoured/infatuated with the wolf. To these American wolf experts, it appears that "the wolf can do no wrong." Although I am not a biologist, I have learned a lot about wolves and their behaviour in Czarist Russia and in the USSR. What I have learned has convinced me that it would be a mistake to plant wolves in the YNP and in central Idaho.

Signed
Will Graves
mailed on 30 Oct 93
to Mr. Bangs

THE STATE OF ARIZONA

GAME AND FISH DEPARTMENT

5000 W. CAREFREE HIGHWAY
PHOENIX, AZ 85086-5000
(602) 942-3000 • WWW.AZGFD.GOV

GOVERNOR
JANICE K. BREWER

COMMISSIONERS
CHAIRMAN, J.W. HARRIS, TUCSON
ROBERT E. MANSELL, WINSLOW
KURT R. DAVIS, PHOENIX
EDWARD "PAT" MADDEN, FLAGSTAFF

DIRECTOR
LARRY D. VOYLES

DEPUTY DIRECTOR
TY E. GRAY

September 9, 2013

The Honorable Sally Jewell
Secretary
United States Department of the Interior
1849 C Street NW
Washington, DC 20240

The Honorable Dan Ashe
Director
United States Fish and Wildlife Service
1849 C Street NW
Washington, DC 20240

Dear Secretary Jewell and Director Ashe:

It is my duty as Chairman of the Arizona Game and Fish Commission to forward the attached Commission Resolution passed by this body at our September 6 Commission meeting in Show Low, Arizona, to express our disappointment at the refusal to hold public scoping hearings in Arizona regarding the proposed expansion of Mexican wolf conservation, an expansion that will almost certainly deliver a preponderance of the program's social, financial, and biological costs to the citizens of Arizona.

As your partners in this effort since the program's birth, the passion of the resolution's language reflects our regret at what seems like an opportunity being missed by the Department of Interior and the U.S. Fish and Wildlife Service to show their respect for those who share the working lands of Arizona with this species and who will live with the consequences of this program, potentially in perpetuity.

The best meetings in Sacramento, Albuquerque, or even the nation's capital, will not suffice as substitutes for meeting on the affected land with the affected parties.

Sincerely,

John W. Harris
Chairman
The Arizona Game and Fish Commission

AN EQUAL OPPORTUNITY REASONABLE ACCOMMODATIONS AGENCY

A RESOLUTION OF THE ARIZONA GAME AND FISH COMMISSION CONCERNING SCOPING HEARINGS ON THE EXPANSION OF MEXICAN WOLF CONSERVATION

WHEREAS, the Arizona Game and Fish Commission and Arizona Game and Fish Department (AGFD) have been partners with the United States Fish and Wildlife Service (USFWS) in Mexican wolf conservation since the effort's inception; and

WHEREAS, the Endangered Species Act (ESA) envisions a significant role for state fish and wildlife agencies in endangered species conservation and recovery efforts; and

WHEREAS, since the Mexican Wolf conservation effort promulgated under section 10(j) of the ESA, Arizona has been the only state in which primary releases of Mexican wolves have been allowed; and

WHEREAS, the primary capacity for the Interagency Field Team has been supplied by personnel from AGFD, and

WHEREAS, the congressional record clearly reflects congressional intent that rules promulgated under section 10(j) of the ESA will reflect an agreement between the USFWS and the affected states; and

WHEREAS, the National Environmental Policy Act requires adequate public notice and opportunity to comment which should involve public hearings near the location of the proposed action; and

WHEREAS, the USFWS has received written requests to hold public scoping hearings in Arizona directly from the AGFD, Graham County, Greenlee County, Navajo County, Apache County, Gila County, and specifically from US Congressman Paul Gosar, US Congresswoman Ann Kirkpatrick, and Senator Jeff Flake; and

WHEREAS, the USFWS has publicly announced its decision to hold scoping hearings in Washington, DC, where there are no plans to conserve Mexican wolves; and to hold public scoping hearings in Sacramento, CA where there are no plans to conserve Mexican wolves; and

WHEREAS, the USFWS has elected to hold a public scoping hearing in Albuquerque, NM as the only state in which Mexican wolf conservation would be involved; and

WHEREAS, the USFWS has specifically not elected to hold any public scoping hearings in Arizona; and

WHEREAS, the citizens of Arizona will unarguably be the most affected by the decisions to be made;

NOW, THEREFORE, BE IT RESOLVED that the Arizona Game and Fish Commission expresses its extreme concern that the USFWS has refused to hold public scoping hearings in Arizona regarding proposals to expand Mexican Wolf conservation efforts;

BE IT FURTHER RESOLVED that the Arizona Game and Fish Commission demands that the Secretary of the Interior and Director of the USFWS reconsider this decision and schedule public scoping hearings in the State of Arizona with at least one meeting to occur in a community geographically located within the proposed 10(j) area.

ADOPTED on the 6th day of September, 2013 by the Arizona Game and Fish Commission.

John W. Harris
Chairman
Arizona Game and Fish Commission

Global Socioeconomic Impact of Cystic Echinococcosis

Christine M. Budke,* Peter Deplazes,* and Paul R. Torgerson*

Cystic echinococcosis (CE) is an emerging zoonotic parasitic disease throughout the world. Human incidence and livestock prevalence data of CE were gathered from published literature and the Office International des Epizooties databases. Disability-adjusted life years (DALYs) and monetary losses, resulting from human and livestock CE, were calculated from recorded human and livestock cases. Alternative values, assuming substantial underreporting, are also reported. When no underreporting is assumed, the estimated human burden of disease is 285,407 (95% confidence interval [CI] 218,515–366,133) DALYs or an annual loss of US $193,529,740 (95% CI $171,567,331–$217,773,513). When underreporting is accounted for, this amount rises to 1,009,662 (95% CI 862,119–1,175,654) DALYs or US $763,980,979 (95% CI $676,048,731–$857,982,275). An annual livestock production loss of at least US $141,605,195 (95% CI $101,011,553–$183,422,465) and possibly up to US $2,190,132,464 (95% CI $1,572,373,055–$2,951,409,989) is also estimated. This initial valuation demonstrates the necessity for increased monitoring and global control of CE.

Cystic echinococcosis (CE) is a condition of livestock and humans that arises from eating infective eggs of the cestode *Echinococcus granulosus*. Dogs are the primary definitive hosts for this parasite, with livestock acting as intermediate hosts and humans as aberrant intermediate hosts. The outcome of infection in livestock and humans is cyst development in the liver, lungs, or other organ system. The distribution of *E. granulosus* is considered worldwide, with only a few areas such as Iceland, Ireland, and Greenland believed to be free of autochthonous human CE. However, CE is not evenly distributed geographically (Figure 1) (*1*). For example, the United States has few cases in livestock and most human cases are imported. The same is true for regions of Western and Central Europe. In many parts of the world, however, CE is considered an emerging disease. For example, in the former Soviet Union and Eastern Europe, the number of observed cases has dramatically increased in recent years (*2–4*). Additionally, in other regions of the world, such as parts of China, the geographic distribution and extent of CE are greater than previously believed (*5*). CE not only causes severe disease and possible death in humans, but also results in economic losses from treatment costs, lost wages, and livestock-associated production losses. To date, no global estimates exist of CE burden (total health, socioeconomic, and financial cost of a given disease to society) in humans or livestock. Such an estimate is imperative since it can be used as a tool to prioritize control measures for CE, which is essentially a preventable disease.

Two methods previously used to assess disease burden are disability adjusted life years (DALYs) and the calculation of monetary losses (*6*). DALYs were first developed in the 1990s and were used in the Global Burden of Disease (GBD) Study to determine the worldwide burden of disease due to both communicable and noncommunicable causes (*7*). Although the application of DALYs is becoming more

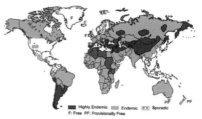

Figure 1. Global distribution of zoonotic strains of *Echinococcus granulosus*. (Adapted from Eckert and Deplazes, 2004 [*1*]. Copyright Institut für Parasitologie, Universität Zürich); used with permission.

*University of Zürich, Zürich, Switzerland

335

commonplace, the use of DALYs and the methods behind the creation of this measure remain debatable (8). The GBD Study was an extensive undertaking; however, echinococcosis was not among the conditions studied. Nevertheless, DALYs have been applied to cystic echinococcosis and alveolar echinococcosis, caused by *E. multilocularis*, on a small scale in western China (9). Likewise, monetary evaluations have been applied to CE infections in humans and livestock only at a local level (10–14). Global burden indicators not only give an idea of the scope of the disease under study, but can also be used to direct limited financial resources to sites where they can be most effective. Because of the magnitude of applying burden of disease measurements on a global scale, this study must be considered a preliminary estimate. Nevertheless, this report should increase awareness of the global impact of CE by both the public health and livestock sectors.

Materials and Methods

CE Incidence in Humans

Data on country-specific annual reported human CE cases were obtained from the Office International des Epizooties (OIE), World Health Organization Handistatus II database for the years 1996–2003 (15). This information was then merged with published case reports from numerous countries and logged into an Excel spreadsheet (Microsoft Corp., Redmond, WA, USA). Type and quality of incidence data varied by country or region; however, most data consisted of annual numbers of detected cases per susceptible population or was converted into this form for analysis purposes. If both an OIE-reported and a literature-based value were available, the larger of the 2 was used. However, if the higher value appeared to be from a survey that evaluated a highly disease-endemic region and was, therefore, not applicable to the entire country, a corresponding adjustment was made. In addition, we assumed that ≈10% of annual cases are not officially diagnosed, and those patients do not receive medical attention because of their socioeconomic status or the subclinical nature of the illness. Based on past studies, this estimate is most likely conservative (12,14). For example, in China, mass ultrasound screening in remote areas has shown high prevalence rates of CE (9). A number of these patients have advanced clinical disease but would not normally have access to treatment because of poverty and distance from medical facilities. Human cases of CE are also systematically underreported by the healthcare establishment, with up to 75% of clinic or hospital-diagnosed cases never recorded in local or national databases or published reports (16,17). Therefore, adjustments were made to account for the substantial underreporting of known treated cases.

CE Prevalence in Livestock

Numbers of annual reported CE cases in slaughtered livestock (sheep, goats, cattle, camels, and swine) for the years 1996–2003 were obtained from the OIE-Handistatus II database (15). This information was merged with abattoir studies performed in numerous countries. If data from both sources were available, the larger of the 2 estimates was used. However, if the higher value appeared to be from a region that was highly disease-endemic and was not appropriate for a countrywide estimate, an adjustment was made. Prevalence per species, for each country, was applied to the estimated number of slaughtered animals per year, with 2004 livestock numbers obtained from the FAO-STAT database (18). The assumption was made that approximately one fourth of sheep and goat populations, one sixth of cattle and camel populations, and the entire swine population would be slaughtered annually, based on estimated average species' lifespan (e.g., approximately one fourth of a country's sheep population would be slaughtered annually, with a typical animal life expectancy of 4 years). Such a general estimate was used because of the large amount of variation in animal production practices between and within countries. As with the human incidence data, the true number that were positive for *E. granulosus* at slaughter is substantially higher than reported. Therefore, a correction factor was used to estimate true prevalence.

Application of DALYs to Human Incidence Data

The DALY formula (shown below) was applied to global human incidence data.

$$-\left[\frac{DCe^{-\beta a}}{(\beta+r)^2}\left[e^{-(\beta+r)(L)}\left(1+(\beta+r)(L+a)\right)-\left(1+(\beta+r)a\right)\right]\right]$$

In this equation, D is a disability weight, β is an age-weighting function parameter, C is an age-weighting correction constant, r is a discount rate, a is age at clinical onset, and L is the duration of disability or time lost because of death (7). Disability weight for CE was assigned a multinomial distribution based on numerous retrospective studies evaluating postoperative outcome (Table 1) (19–24). The percentage of patients projected to improve after surgery was assigned a disability weight of 0.200 (Dutch weight for clinically disease-free cancer) for 1 year, the percentage of patients projected to have substantial postsurgical conditions was assigned a disability of 0.239 (GBD weight for preterminal liver cancer) for 5 years, the percentage of patients projected to have recurrent disease was assigned a disability of 0.809 (GBD weight for terminal liver cancer) for 5 years, and the percentage of patients projected to die postoperatively were assigned a disability of 1 (indicating death) for the remainder of their predicted lifespan (7,25). An assumption was

336

Table 1. Outcome of surgery for cystic echinococcosis in humans

Country (y)	No. patients	Cure (%)	Morbidity (%)	Relapse (%)	Death (%)	Reference
Greece (1984–1990)	56	40 (72)	13 (23)	3 (5)	0	(9)
Italy (1950–1987)	298	244 (82)	27 (9)	15 (5)	12 (4)	(20)
Turkey (1992–1999)	95	32 (34)	38 (40)	24 (25)	1 (1)	(21)
Turkey (1990–1995)	108	88 (81)	19 (18)	0	1 (1)	(22)
Greece (1985–1990)	67	59 (86)	4 (6)	3 (6)	1 (2)	(23)
Italy (1982–1994)	89	70 (79)	17 (19)	1 (1)	1 (1)	(24)
Total	713	533 (75)	118 (17)	46 (6)	16 (2)	

also made that ≈10% of cases are not reported and do not receive medical treatment. These cases were assigned a disability weight of 0.200 (Dutch weight for clinically disease-free cancer) for 10 years (25). For the GBD Study, a standardized life table was used for L (7).

Economic Evaluation of Human-associated Losses

Overall cost per human surgical case was based on findings from previous international studies (Table 2) (11,13,14,26,27). Expenses taken into consideration included diagnostic costs, surgical cost, hospitalization, and postoperative costs. The average cost per surgical patient was shown to be significantly correlated ($R^2 = 0.898$, $p < 0.05$), with the country-specific per capita gross national income (per capita GNI) (Atlas Method) (Table 2). Therefore, the linear regression coefficient was used as a predictor of treatment costs for each disease-endemic country. In addition to medical costs and single-year wage losses, past studies have indicated an average 2.2% postoperative death rate for surgical patients (Table 1). Approximately 6.5% of cases also are assumed to relapse and require a prolonged recovery time (Table 1) (11). Therefore, these outcomes were also taken into account. We assumed that, in addition to surgical cases, ≈10% of cases are not officially diagnosed each year, and those patients never receive treatment. Wage losses for this group were thus taken into consideration. Economic losses in humans were also evaluated, taking into account the nearly 4-fold degree of underreporting of patients who received treatment.

Economic Evaluation of Livestock-associated Losses

Production-based losses attributable to infected sheep, goats, cattle, camels, and pigs were estimated. Losses from liver condemnation, defined as the action of preventing the sale of livers deemed unfit for human consumption (sheep,

goats, cattle, pigs, camels), reduction in carcass weight (sheep, goats, cattle), decrease in hide value (sheep, cattle), decrease in milk production (sheep, goats, cattle), and decreased fecundity (sheep, goats, cattle) were taken into account. Only liver-associated losses in camels and pigs are presented since few studies have evaluated production losses from echinococcosis in these species (28). Losses from liver condemnation are assumed to occur since hepatic pathology is associated with infection in swine and camels (29). Losses from liver condemnation were presumed proportional to those used for the analysis of economic impact of CE in Jordan (12). Decrease in hide value (20%) and decrease in fecundity (11%) were presumed proportional to values suggested by numerous Soviet studies conducted from the 1950s through the 1980s (28). Reductions in carcass weight (2.5%) and milk production (2.5%) were also based on previous reports (30).

Analysis

Spreadsheet models were constructed in Excel to estimate global impact of CE in terms of DALYs and monetary losses. Total disease effects, in DALYs lost or monetary losses, was calculated by summing all of the constituent components. Uncertainty in parameter estimates was modeled by using Monte Carlo techniques (6). Briefly, all parameters were assigned a probability distribution based on the quantity and quality of reported data. Macros were written in Excel to sample across these distributions, with 10,000 iterations of each model calculated. Mean and 95% confidence intervals (CIs) for losses were then determined from these iterations.

Reported global human incidence was assigned a normal distribution, with a standard deviation of 5%. Adjustments were then made to account for the nearly 4-fold degree of underreporting of treated cases believed to occur (16,17). In addition, cases that would not be official-

Table 2. Average cost per surgical case of cystic echinococcosis

Country	Years	Average cost per case (US $)	% of real per capita GNI* per patient	Reference
Jordan	2002	701.50	40	(26)
Spain	1987–2001	10,915.00	76	(27)
Tunisia	2000	1,481.00	71	(11)
Uruguay	2000	6,721.00	110	(14)
Wales, UK	2000	13,600.30	54	(13)

*World Bank Atlas method for converting data in national currency to US dollars; GNI, gross national income.

337

ly acknowledged had to be accounted for, i.e., cases in persons who never receive treatment in a hospital. We therefore assumed that ≈10% (uniform distribution of 8% to 12%) of cases would not be detected. This estimate is conservative compared to other country-specific estimates (12,14).

The DALY formula was applied to worldwide CE cases in a stochastic manner similar to that used to apply DALYs to echinococcosis cases in a region of western China (10). Mean age of clinical onset (a) was allocated a uniform distribution of 30 to 40 years, established on the basis of various studies (Table 3) (4,9,21,31–34). Numerous and varying reports have indicated the sex of CE-positive persons with women tending to be infected at a higher rate than men. Based on these reports, we assigned a uniform distribution of 50% to 60% of infected persons as female (4,35). Number of DALYs lost, using incidence values corrected and uncorrected for underreporting of surgical incidence, was determined.

Human-associated economic losses were applied in a stochastic manner similar to that used for a region of western China (10). Variability in surgical treatment costs, due to CE, was modeled by using a uniform distribution of 50% to 90% of per capita GNI per country and was weighted by each country's contribution to global human CE incidence (36). Lower income, higher unemployment, or both has been associated with a diagnosis of CE (4,10). Consequently, a decrease in wages earned was assumed, at least for the year of initial diagnosis and treatment. Therefore, all patients were assigned a uniform loss of 50% to 90% of country-specific per capita GNI for 1 year (36). Approximately 6.5% of patients were also assigned a 50%–90% wage loss for 4 additional years because of relapse and prolonged recovery time. In addition, 2.2% of patients were assigned a 100% wage loss until the expected retirement age of 65 due to postsurgical death. A standard 3% discounting rate was applied to all income losses (7). In addition to surgical cases, ≈10% of cases (uniform distribution of 8% to 12%) annually were assumed to not be officially diagnosed. A 25% wage loss for 5 years was consequently assigned to this population. This estimate is conservative and does not take into account income losses attributable to undiagnosed cases

with fatal outcomes. Projections were made that assumed the absence and presence of underreporting of surgical incidence (16,17). In addition to using real per capita GNI (Atlas Method), calculations were also performed by using purchasing power parity (ppp) adjusted per capita GNI.

As with human-associated economic losses, livestock-associated losses were applied in a stochastic manner (10). Livestock prices were given uniform distributions of US $30–$60 for sheep, US $15–$30 for goats, US $150–$350 for cattle, US $300–$600 for camels, and US $55–$75 for pigs. Uniform distributions were used because of the large regional variations in prices and assigned in accordance with baseline prices for most affected countries. Production losses were assumed to follow a log-normal distribution; most affected animals were lightly infected, and only a small proportion of animals had severe losses. As with human cases, substantial underreporting of livestock infection was recognized, since official reporting is not mandatory in most countries. Therefore, a uniform correction factor of 1.5 to 2 was used to approximate true economic losses. A large uniform distribution was used because of the lack of information concerning true global prevalence of CE in livestock. This lack will, therefore, be represented in the wide confidence limits obtained.

Results

DALYs

Regional findings for predicted global burden of CE in terms of DALYs lost, with 95% CIs, can be found in Table 4. The most conservative estimate of number of global DALYs lost is 285,407 (95% CI 218,515–366,133), with no consideration for disease underreporting. Estimated number of global DALYs lost, taking into consideration nonreported surgical cases, is 1,009,662 (95% CI 862,119–1,175,654).

Human-associated Economic Losses

Findings for predicted regional burden of human CE in economic terms, with 95% CI, can be found in Table 5. Global losses, assuming no underreporting, are estimated at US $193,529,740 (95% CI $171,567,331–$217,773,513).

Table 3. Average age at ultrasound detection or surgery			
Country	Years	Average age at onset/detection (y)	Reference
China	2001–2003	35*	(9)
Jordan	1994–2000	31–45†	(31)
Kenya (Turkana)	1979–1982	21–30*	(32)
Kyrgyzstan	1991–2000	22†	(4)
Morocco	2000–2001	32*	(33)
Turkey	1992–1999	44†	(21)
Uruguay	1991–1992	45*	(34)

*Age at time of ultrasound detection.
†Age at surgery.

338

Table 4. Estimated global impact of cystic echinococcosis in terms of DALYs lost

Region*	Total unadjusted DALYs lost (95% CI)†	Total adjusted DALYs lost (95% CI)
Western Europe, USA, Canada, Australia, New Zealand	11,842 (8,977–15,722)	41,891 (30,949–55,014)
Middle Eastern Crescent	104,503 (79,291–135,722)	370,056 (275,228–486,353)
Formerly socialist economies of Europe and Russia	17,317 (13,129–22,371)	61,369 (45,800–80,077)
China	112,451 (85,001–145,898)	398,015 (295,922–521,879)
Other Asia and Islands	1,130 (851–1,462)	4,003 (2,971–5,256)
Sub-Saharan Africa	2,639 (1,926–3,518)	9,314 (6,664–12,623)
Latin America and the Caribbean	14,834 (11,252–19,241)	52,693 (38,787–69,380)
India	20,691 (15,666–26,822)	73,364 (54,518–96,263)
World	285,407 (218,515–366, 133)	1,009,662 (862,119–1,175,654)

*Regional breakdown of disability-associated life years (DALYs) lost is based on that used in the Global Burden of Disease study (7).
†CI, confidence interval.

Losses, adjusted for underreporting, are estimated at US $763,980,979 (95% CI $676,048,731–$857,982,275). When ppp adjusted per capita GNI is used instead of real per capita GNI, estimated annual overall losses, without correction for underreporting, are US $484,878,359 (95% CI $432,898,134–US $542,048,125). When corrected for underreporting, annual losses are estimated at US $1,918,318,955 (95% CI $1,700,574,632–$2,142,268,992) (Table 5).

Livestock-associated Economic Losses

Estimated livestock-associated losses, with 95% CI, can be found in Table 6. Minimal annual losses, assuming liver condemnation alone with no correction for underreporting, is estimated at US $141,605,195 (95% CI $101,011,553–$183,422,465). However, when losses from additional production factors (decreased carcass weight, decreased milk production, decreased hide value, decreased fecundity) are taken into account, losses range from US $1,249,866,660 (95% CI $942,356,157–$1,622,045,957), not taking into account underreporting, up to US $2,190,132,464 (95% CI $1,572,373,055–$2,951,409,989), when underreporting is considered.

Discussion

Even without correcting for the underreporting of human and livestock cases, CE has a substantial global disease impact in terms of DALYs and monetary losses. The importance of using both indicators is illustrated by the proportional difference in DALYS lost versus economic losses per region (Tables 4 and 5). If only monetary losses were evaluated, the severity of the situation in poorer regions would be underestimated because of the decreased income and economic value of livestock products relative to more economically prosperous regions. For example, China is responsible for 40% of the world's CE DALYs but only 19% of human-associated economic losses. However, losses based on ppp-adjusted per capita GNI give a better picture of the relative distribution of disease impact (Table 5). When the number of DALYs lost, taking into account the recognized underreporting of human cases, is compared with those of other parasitic conditions evaluated by the World Health Organization (WHO), worldwide losses due to CE are slightly less than those caused by African trypanosomiasis (1,525,000) and more than those caused by onchocerciasis (484,000) or Chagas disease (667,000) (37). Even though estimated number of DALYs lost from

Table 5. Global annual cystic echinococcosis–associated economic losses to humans

Region	Total adjusted economic losses (95% CI)*(US $)	Total adjusted economic losses (95% CI)† (US $)
Western Europe, USA, Canada, Australia, New Zealand	$309,983,585 ($244,256,327–$383,371,741)	$354,460,281 ($277,178,852–$440,438,597)
Middle Eastern Crescent	$197,276,106 ($158,870,204–$240,282,181)	$564,496,304 ($454,402,304–$690,682,060)
Formerly socialist economies of Europe and Russia	$46,896,902 ($37,750,210–$57,355,873)	$143,921,865 ($114,323,294–$176,555,114)
China	$146,129,578 ($114,279,187–$181,937,463)	$663,712,150 ($516,048,103–$826,353,341)
Other Asia and Islands	$1,535,990 ($1,159,946–$1,946,632)	$2,412,386 ($1,826,342–$3,074,240)
Sub-Saharan Africa	$832,295 ($649,915–$1,035,681)	$5,176,229 ($3,710,869–$6,969,680)
Latin America and the Caribbean	$48,396,449 ($38,408,001–$59,672,173)	$120,717,047 ($95,789,339–$148,939,896)
India	$12,930,073 ($9,674,489–$16,499,072)	$63,422,693 ($47,576,673–$80,430,630)
World	$763,980,979 ($676,048,731–$857,982,275)	$1,918,318,955 ($1,700,574,632–$2,142,268,992)

* Income losses based on per capita gross national income (GNI) (Atlas method); CI, confidence interval.
†Income losses based on purchasing power parity–adjusted per capita GNI.

339

Table 6. Global annual cystic echinococcosis–associated livestock production losses

Category	Economic losses (95% CI) (US $)*
Liver condemnation†	$141,605,195 ($101,011,553–$183,422,465)
Decreased carcass weight†	$241,525,979 ($100,335,764–$518,035,773)
Decreased hide value‡	$34,871,148 ($23,965,776 – $46,162,828)
Decreased milk production§	$378,722,717 ($279,048,143–$495,682,356)
Decreased fecundity§	$453,141,617 ($278,287,046–$671,424,319)
Overall cost (no correction factor)	$1,249,866,660 ($942,356,157–$1,622,045,957)
Overall cost (adjusted for underreporting)	$2,190,132,464 ($1,572,373,055–$2,951,409,989)

*CI, confidence interval.
†Sheep, goats, cattle, camels, pigs.
‡Sheep, cattle.
§Sheep, goats, cattle.

CE is greater than estimated losses from multiple members of the tropical disease cluster, CE continues to be excluded from funding associated with conditions related to low socioeconomic status. This exclusion best illustrated by evaluating research and training funding provided by the United Nations Children's Fund (UNICEF)/United Nations Development Programme (UNCP)/World Bank/ WHO-supported Special Programme for Research and Training in Tropical Diseases (TDR). If funding for CE were placed on the same scale as TDR-supported diseases, based on estimated DALYs lost, CE should receive approximately US $1,200,000 annually (Figure 2) (38). For now, however, CE continues to be widely underappreciated by most international agencies. These findings emphasize the need for CE to be taken seriously as a global public health condition, regardless of its economic implications. What makes this disease exceptional, however, is that it is not only a substantial human health problem but also has a considerable economic effect on the livestock industries of some of the most socioeconomically fragile countries.

In addition to reporting the estimated global burden of CE, this study has shown the need for more accurate reporting of infected humans and livestock. Very few country-specific estimations of the true incidence of CE in humans have been made and no studies, to the authors' knowledge, that estimate its true prevalence in livestock (16,17). Presentation of the substantial economic losses for both the public health and agricultural sectors will also, we hope, encourage countries and international organizations to more closely examine potential control programs and cost-sharing methods between the 2 affected sectors (10).

The values presented in this paper are not definitive but instead estimates of the severity of the global situation from human- and livestock-associated CE. Considerable sums of money have been invested in the investigation and control of such parasitic conditions as lymphatic filariasis and onchocerciasis. Although these conditions can result in severe human disease, unlike CE they do not have severe secondary economic implications, such as massive livestock production losses (39,40). In addition, regional control programs that have been implemented and recommended thus far for CE, based on combinations of dog deworming, stray dog culling, sheep and goat vaccination, and education programs, have been shown to be very cost effective (10,27). CE is, therefore, a worthy condition for research and control program implementation, with substantial anticipated return on invested funding.

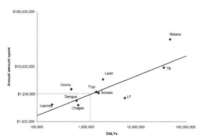

Figure 2. Annual budget (in US $) for diseases included in the United Nations Children's Fund/UNDP/World Bank/World Health Organization-supported Special Programme for Research and Training in Tropical Diseases (TDR) compared to their estimated global disability-associated life years (DALYs). The thinner lines indicate estimated DALYs lost because of cystic echinococcosis (CE) and the recommended funding level based on the TDR 2004-2005 approved program budget (Oncho, onchocerciasis; Tryp, trypanosomiasis; Schisto, schistosomiasis; Leish, leishmaniasis; LF, lymphatic filariasis; TB, tuberculosis). This figure does not take into account the substantial regional variability in both the estimates of DALYs lost and the annual budget for the diseases illustrated.

Acknowledgments

We thank F.-X. Meslin of WHO for encouraging us to undertake this study.

The authors received financial support from the University of Zürich, an Ecology of Infectious Diseases program grant from the US National Institutes of Health, and the National Science Foundation (TWO 1565-02), and the International Association for the Promotion of Co-operation with Scientists from the New Independent States of the Former Soviet Union (INTAS 01-500, INTAS 03-51-5661).

340

Dr Budke was a research scientist at the Institute of Parasitology, University of Zurich, Switzerland, where she studied the transmission and economic effects of echinococcosis. She is currently an assistant professor of epidemiology at the College of Veterinary Medicine and Biomedical Sciences, Texas A&M University. Her research interests include the epidemiology of emerging and zoonotic diseases.

References

1. Eckert J, Deplazes P. Biological, epidemiological, and clinical aspects of echinococcosis, a zoonosis of increasing concern. Clin Microbiol Rev. 2004;17:107–25.
2. Torgerson PR, Shaikenov BS, Baitursinov KK, Abdybekova AM. The emerging epidemic of echinococcosis in Kazakhstan. Trans R Soc Trop Med Hyg. 2002;96:124–8.
3. Todorov T, Boeva V. Human echinococcosis in Bulgaria: a comparative epidemiological analysis. Bull World Health Organ. 1999;77:110–8.
4. Torgerson PR, Karaeva RR, Corkeri N, Abdyjaparov TA, Kuttubaev OT, Shaikenov BS. Human cystic echinococcosis in Kyrgystan: an epidemiological study. Acta Trop. 2003;85:51–61.
5. Chai JJ. Epidemiological studies on cystic echinococcosis in China—a review. Biomed Environ Sci. 1995;8:122–36.
6. Carabin H, Budke CM, Cowan LD, Willingham III AL, Torgerson PR. Methods for assessing the burden of parasitic zoonoses: cysticercosis and echinococcosis. Trends Parasitol. 2005;21:327–33.
7. Murray CJL, Lopez AD. The global burden of disease: a comprehensive assessment of mortality and disability from disease, injuries, and risk factors in 1990 and projected to 2020. Cambridge: Harvard University Press; 1996.
8. Anand S, Hanson K. Disability-adjusted life years: a critical evaluation. J Health Econ. 1997;16:685–702.
9. Budke CM, Qiu J, Wang Q, Zinsstag J, Torgerson PR. Utilization of DALYs in the estimation of disease burden for a high endemic region of the Tibetan plateau. Am J Trop Med Hyg. 2004;71:56–64.
10. Budke CM, Qiu J, Wang Q, Torgerson PR. Economic effects of echinococcosis on a highly endemic region of the Tibetan plateau. Am J Trop Med Hyg. 2005;73:2–10.
11. Majorowski MM, Carabin H, Kilani M, Bensalah A. Echinococcosis in Tunisia: a cost analysis. Trans R Soc Trop Med Hyg. 2005;99:268–78.
12. Torgerson PR, Dowling PM, Abo-Shehada MN. Estimating the economic effects of cystic echinococcosis. Part 3: Jordan, a developing country with lower-middle income. Ann Trop Med Hyg. 2001;95:595–603.
13. Torgerson PR, Dowling PM. Estimating the economic effects of cystic echinococcosis. Part 2: an endemic region in the United Kingdom, a wealthy, industrialized economy. Ann Trop Med Hyg. 2001;95:177–85.
14. Torgerson PR, Carmona C, Bonifacino R. Estimating the economic effects of cystic echinococcosis: Uruguay, a developing country with upper-middle income. Ann Trop Med Hyg. 2000;94:703–13.
15. OIE-Handistatus II, Office International des Epizooties, Paris; 2005. [cited 4 Jan 2006] Available from http://www.oie.int/hs2/report.asp.
16. Serra I, Garcia V, Pizzaro A, Luzoro A, Cavada G, Lopez J. A universal method to correct underreporting of communicable diseases. Real incidence of hydatidosis in Chile, 1985–1994. Rev Med Chil. 1999;127:1075–85.
17. Nazirov FG, Ilkhamov IL, Ambekov NC. Echinococcosis in Uzbekistan: types of problems and methods to improve treatment [article in Russian]. Medical Journal of Uzbekistan. 2002;2/3:2–5.
18. FAOSTAT. FAO statistical databases, February 2004 ed. Food and Agricultural Organization of the United Nations, Rome; 2004. [cited 4 Jan 2006] Available from http://www.apps.fao.org/.
19. Gogas J, Papachristodoulou A, Zografos G, Papastratis G, Gardikis S, Markopoulos C, et al. Experience with surgical therapy of hepatic echinococcosis [article in German]. Zentralbl Chir. 1997;122:339–43.
20. Cirenei A, Bertoldi I. Evolution of surgery for liver hydatidosis from 1950 to today: analysis of a personal experience. World J Surg. 2001;25:87–92.
21. Yorganci K, Sayek I. Surgical treatment of hydatid cysts of the liver in the era of percutaneous treatment. Am J Surg. 2002;185:63–9.
22. Ozacmak ID, Ekiz F, Ozmen V, Isik A. Management of residual cavity after partial cystectomy for hepatic hydatidosis: comparison of omentoplasty with external drainage. Eur J Surg. 2000;166:696–9.
23. Vagianos CE, Karavia DD, Kakkos SK, Vagenas CA, Androulakis JA. Conservative surgery in the treatment of hepatic hydatidosis. Eur J Surg. 1995;161:415–20.
24. Altieri S, Doglietto GB, Pacelli F, Costamagna G, Carriero C, Murigani M, et al. Radical surgery for liver hydatid disease: a study of 89 consecutive patients. Hepatogastroenterology. 1997;44:4 96–500.
25. Stouthard MEA, Essink-Bot ML, Bonsel GJ. Disability weights for diseases: a modified protocol and results for a Western European region. Eur J Public Health. 2000;10:24–30.
26. Nasrieh MA, Abdel-Hafez SK, Kamhawi SA, Craig PS, Schantz PM. Cystic echinococcosis in Jordan: socioeconomic evaluation and risk factors. Parasitol Res. 2003;90:456–66.
27. Jimenez S, Perez A, Gil H, Schantz PM, Ramalle E, Juste RA. Progress in control of cystic echinococcosis in La Rioja, Spain: decline in infection prevalences in human and animal hosts and economic costs and benefits. Acta Trop. 2002;83:213–21.
28. Romazanov VT. Evaluation of economic losses due to echinococcosis. In: Lysendo A, editor. Zoonosis control: collection of teaching aids for international training course vol. II. Moscow: Centre of International Projects GKNT;1983. p. 283–85.
29. Njoroge EM, Mbithi PM, Gathuma JM, Wachira TM, Gathura PB, Magambo JK, et al. A study of cystic echinococcosis in slaughter animals in three selected areas of northern Turkana, Kenya. Vet Parasitol. 2002;104:85–91.
30. Polydorou K. Animal health and economics. Case study: echinococcosis with a reference to Cyprus. Bull Off Int Epizoot. 1981;93:981–92.
31. Al-Qaoud KM, Craig PS, Abdel-Hafez SK. Retrospective surgical incidence and case distribution of cystic echinococcosis in Jordan between 1994 and 2000. Acta Trop. 2003;87:207–14.
32. Macpherson CNL. An active intermediate host role for man in the life cycle of Echinococcus granulosus in Turkana, Kenya. Am J Trop Med Hyg. 1983;32:397–404.
33. Macpherson CNL, Kachani M, Lyagoubi M, Berrada M, Bouslikhane M, Shepherd M, et al. Cystic echinococcosis in the Berber on the Mid Atlas mountains, Morocco: new insights into the natural history of the disease in humans. Ann Trop Med Parasitol. 2004;98:481–90.
34. Cohen H, Paolillo E, Bonifacino R, Botta B, Parada L, Cabrera P, et al. Human cystic echinococcosis in a Uruguayan community: a sonographic, serologic, and epidemiologic study. Am J Trop Med Hyg. 1998;59:620–7.
35. Schantz PM, Wang H, Qiu J, Liu FJ, Saito E, Emshoff A, et al. Echinococcosis on the Tibetan plateau: prevalence and risk factors for cystic and alveolar echinococcosis in Tibetan populations in Qinghai Province, China. Parasitology. 2003;127:S109–20.
36. World Bank data and statistics. World Bank Group; 2005. [cited 4 Jan 2006] Available from http://www.worldbank.org/data/.
37. World Health Organization. The world health report. Geneva: the Organization; 2004. Available from http://www.who.int/whr/2004

341

38. United Nations Children's Fund/United Nations Development Programme/World Bank/World Health Organization Special Programme for Research and Training in Tropical Diseases. Approved Programmed Budget 2004–2005. 2003. TDR/PB/04-05/Rev.1. [cited 4 Jan 2006] Available from http://www.who.int/tdr/ublications/publications/budget_04.htm

39. Ramaiah KD, Pradeep KD. Mass drug administration to eliminated lymphatic filariasis in India. Trends Parasitol. 2004;20:449–502.

40. Remme JH. Research for control: the onchocerciasis experience. Trop Med Int Health. 2004;9:243–54.

Address for correspondence: Paul R. Torgerson, WHO Collaborating Centre for Parasitic Zoonoses, Institute of Parsitology, University of Zurich, Winterthurerstrasse 266A, 8057 Zurich, Switzerland; fax: 41-44-63-58907; email: paul.torgerson@access.unizh.ch

NSW DPI

primefacts

PROFITABLE & SUSTAINABLE PRIMARY INDUSTRIES   www.dpi.nsw.gov.au

FEBRUARY 2007 | PRIMEFACT 475 | (REPLACES AGFACT A0.9.43)

Hydatids – you, too, can be affected

Stuart King

Former Senior Field Veterinary Officer

Dr Gareth Hutchinson

Research Officer, Health Science, Strategic Alliances & Evaluation, Menangle

Introduction

Hydatid disease (also known as hydatidosis or echinococcosis) is caused by a tapeworm which infects dogs, dingoes and foxes. At its intermediate stage, it forms cysts in the internal organs, especially livers and lungs, of a number of animals, including humans. In humans, the disease is so serious that it requires surgery for treatment. Hydatid disease also causes losses in livestock with the downgrading of edible meat by-products because of the presence of the hydatid cysts.

Control of hydatid disease involves the elimination of the hydatid tapeworm from dogs. By carrying out the recommended control measures, the infection of dogs with the tapeworm can be prevented, and the spread of the disease to other animals, including humans, reduced.

With better control of hydatid disease in domestic livestock and dogs, it is the hydatid cycle that occurs in wildlife which is becoming relatively more important as a threat to human health.

The hydatid tapeworm

In Australia, cystic hydatid disease (CHD) is caused by the unilocular hydatid tapeworm, Echinococcus granulosus (see Figure 1), which infects primarily the dog and the dingo. (The term 'unilocular' refers to the fact that the cysts are characterised as having only one bladder.) There have been reports of infection in the fox, but, because of its feeding habits and the fact that it carries fewer worms than dogs, the fox does not pose as serious a threat to humans or livestock.

The multilocular (multiple-cyst-forming) tapeworm Echinococcus multilocularis, which causes very aggressive alveolar hydatid disease (AHD) where the vesicles resemble lung pockets, is found in Japan, Alaska and parts of Europe, but fortunately does not occur in Australia or New Zealand.

The *E. granulosus* tapeworm consists of 3–4 segments (proglottids) and is only about 6 mm long when fully grown. Because it is small, thousands of these tapeworms can inhabit the intestine of a dog without causing any ill effects, and are usually very difficult to see in gut contents because they resemble intestinal villi (the finger-like folding of the gut lining).

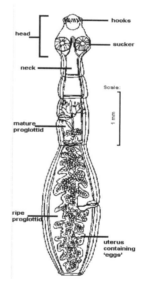

Figure 1. The hydatid tapeworm Echinococcus granulosus

A dog, dingo or fox can become infected with these tapeworms only by eating tapeworm heads, called **protoscolices**, contained in cysts or in food contaminated with cyst fluid that contains protoscolices. (Protoscolices are sometimes called 'hydatid sand' because of their gritty feel when cysts are sliced open.)

When swallowed by a dog, the tapeworm head embeds in the lining of the dog's intestine and begins to grow, taking about 6 weeks to reach maturity. When mature, the last segment of the tapeworm may contain a thousand eggs. A segment containing eggs is shed by the tapeworm every 14 days. A new segment, filled with eggs, then develops to take its place.

Hydatid eggs in the life cycle

After being shed in faeces, the segments rupture, scattering the eggs which can be moved about by wind and water. They are highly resistant to weathering and can remain infective for several months, particularly in cool climates. Contamination of the dog's kennel area, playgrounds, vegetable gardens, pastures and the dog's coat can easily occur (see Figure 2).

The eggs cannot continue their life cycle until swallowed by a susceptible animal – sheep, pigs, goats, camels, deer, cattle, kangaroos and humans. When the egg is swallowed, it hatches to release a small, hooked embryo which burrows its way through the gut wall. It is then picked up by the

the infected dog (tapeworm inside intestine)

the hydatid tapeworm (up to 6 mm)

passed eggs in faeces contaminate pastures

also contaminate kennel area ...

affecting stock by forming cysts in liver, lungs, heart and brain

... and dog's hair

possibly affecting humans by forming cysts

only when a dog eats a cyst can tapeworms develop

Figure 2. The hydatid life cycle. Each time an infected dog passes faeces, thousands of eggs may be released.

bloodstream and transported to the liver, lungs or, less frequently, to another organ such as the brain. Once lodged, the embryo develops into a hydatid cyst, which is a fluid-filled sac.

The hydatid cyst

The outer layers of the cyst (see Figure 3) consist of fibrous tissue laid down by the host and have a characteristic laminated appearance. The cyst is lined by a germinal membrane from which brood capsules grow. Within the brood capsules, the next generation of tapeworm heads forms. Each capsule may contain up to 40 heads or protoscolices (see photo of protoscolices in Figure 4).

Brood capsules which become detached from the germinal membrane and float free within the fluid are referred to as daughter cysts. Brood capsules release the tapeworm heads ('hydatid sand'). If a cyst ruptures within a host, the brood capsules can form new cysts, called secondary cysts.

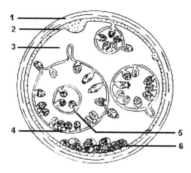

1. Fibrous tissue of host
2. Germinal membrane of cyst
3. Hydatid fluid
4. 'Daughter' cyst or brood capsule
5. 'Grand-daughter' cyst
6. Hydatid 'sand' (protoscolices)

Figure 3. Echinococcus granulosus cyst

Figure 4. Photograph of protoscolices from a hydatid brood capsule. Photo: Gareth Hutchinson

The speed with which the cyst grows varies with the species of the host. In sheep, for example, it takes about 3 months for a cyst to grow to 4–5 mm in diameter, and another 3 months for it to reach 20 mm in diameter.

In other animals, cysts with diameter 5–20 cm, or larger, can develop. A mature, fertile hydatid cyst can contain up to 100 000 brood capsules, and each of these may contain up to 40 tapeworm heads – a total of 4 million tapeworm heads. Not all hydatid cysts are fertile.

As the animal gets older, some of the cysts die and form scars. Often, dead cysts become filled with caseous (cheesy) material, or they become calcified, but the outer laminated layers can still be distinguished microscopically.

Figure 5. Photograph of fluid-filled hydatid cyst in bovine lung. Photo: Gareth Hutchinson

Animals affected by hydatids in Australia

The adult hydatid tapeworm can infect the dog (including wild dogs and feral crosses with dingoes), dingo and fox.

Larval cysts, resulting from ingesting hydatid eggs released from an infected dog, dingo or fox, can develop in humans, sheep, cattle, goats, deer, horses, pigs, camels, kangaroos, wallabies and wombats. In Australia, hydatid cysts have not been reported in alpacas, and have only rarely been reported in horses, although they are common in horses in the UK. No reports of hydatid cysts in water buffalo are known from Australia. The percentage of fertile cysts varies with the species and age of the host animal.

- In Australia, most fertile cysts are found in sheep or certain species of macropod marsupials (kangaroos and wallabies). The main source of domestic transmission of infective cysts in Australia is from sheep to dogs (the domestic cycle – see below).

- Most cysts in cattle, feral pigs or other hosts in Australia are not infective.

Hydatid life cycles in Australia – how the infection spreads

Domestic cycle

Dogs, dingoes or foxes can become infected with tapeworm only if they swallow a tapeworm head. This comes from either fertile cysts contained in the liver, lungs or other organs (offal) they eat or by eating meat or blood that was contaminated by cyst fluid when the animal was killed. (See Figure 6.)

Wildlife or sylvatic cycle

Dingoes and wild dogs usually become infected by eating kangaroos or wallabies carrying cysts. The eggs shed in the faeces of these dogs sometimes contaminate areas grazed by sheep or cattle, causing them to be infected with cysts. In comparison with the dingoes and wild dogs, very few foxes are likely to be infected and their tapeworm burdens are usually low (<50 worms), so their contamination of pastures will be insignificant in the overall disease pattern. (See Figure 7.)

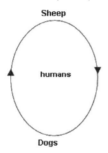

Sheep

humans

Dogs

Figure 6. Hydatid tapeworm domestic cycle

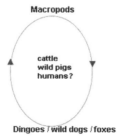

Macropods

cattle
wild pigs
humans?

Dingoes / wild dogs / foxes

Figure 7. Hydatid tapeworm sylvatic (wildlife) cycle

At one time it was thought that the strains in the domestic life cycle and the sylvatic life cycle were distinct, but it now appears that they are genetically the same.

Cysts in accidental intermediate hosts, such as feral pigs, cattle and humans, are usually not fertile and/or not involved in the maintenance of the life cycle (i.e. a 'dead end' host).

Hydatid strains

Field and laboratory studies over almost 50 years have revealed a wide variety of morphological, biological and biochemical differences in *Echinococcus granulosus*. Variations have been reported in:

- fertility of cysts;
- rates of development in the dog (this is important as it can affect timing of treatment for adult tapeworms before they have the opportunity to pass eggs in faeces);
- rostellar hook size;

Such worms have been recognised as 'strains', and even as separate species. See Thompson & McManus 2002 for details and discussion on the case for separate *Echinococcus* species in the horse (*E. equinus*) and cattle (*E. ortleppi*).

Biochemical and recently genetic (molecular biology) characterisation of different isolates/strains has revealed at least nine genetically distinct strains (see Table 1).

- In Australia, only two genotypes, G1 (common sheep strain) and G2 (Tasmanian sheep strain), have been recorded, but on the Australian mainland, only G1 occurs. The current status of G2 in Australia is uncertain since it has never been reported on the mainland (DJ Jenkins, pers. comm., 2005), and the success of the Tasmanian eradication program is likely to have made this strain extinct in its place of origin. Cattle in Australia are frequently infected with hydatid cysts, but in all cases these are the common sheep strain (G1) and the cysts are usually not fertile.

- In overseas countries, where there are other genetic strains of hydatids present, 94% of hydatid cysts in sheep, 88% in pigs, 70% in goats, and 10%–20% in cattle, may be fertile and able to infect dogs.

Cattle case study

Cattle may be infected with a number of strains (although, in Australia, only with the G1 common sheep strain).

346

Table 1. Strains of hydatids

	Genotype (strain)	Other hosts	Location in Australia	Other countries
Found in Australia	G1 (common sheep strain)	sheep	mainland	UK, Europe, Africa, Middle East
		cattle	mainland	UK, Europe, Africa, Middle East, China
		human	mainland and (formerly) Tas.	Europe, Middle East, Africa, China, South America
		buffalo	NR	India
		pig (domestic)	NR	China
		pig (feral)	mainland	NP
		camel (incl. alpaca)	?	China
		kangaroo	mainland	NP
		wallaby	mainland	NP
		dog	mainland & (formerly) Tas.	Africa
		dingo	mainland	NP
	G2 (Tasmanian sheep strain)	sheep	?Tas.	Argentina
		human	NR	Argentina
Not found in Australia	G3 (buffalo strain?)	buffalo		India
	G4 (horse strain)	horse, donkey		UK, Europe
	G5 (cattle strain)	cattle, sheep, buffalo, human		Europe, India, Nepal, Brazil
	G6 (camel strain)	camel, cattle, human, sheep, goat		East Africa, Middle East, China
	G7 (pig strain)	pig, wild boar, beaver, cattle, human		Eastern Europe, Argentina
	G8 (cervid strain)	moose, human		USA
	G9?	pig, human		Poland

NR = not recorded, NP = not present, ? = current status not certain. Adapted from McManus & Thompson 2003

In Europe there is a well characterised 'Swiss' cattle strain (G5), also found in India, Nepal and South America:

- This strain has precocious development in the dog, taking only 5 weeks to produce eggs (i.e. 5 weeks to patency).
- It has a predilection for the lungs for cyst development.
- The cysts are frequently fertile.
- This strain can also infect humans.

Camel strain (G6) is reported in cattle in Iran (Fasihi Harandi et al. 2002).

Cysts caused by other larval tapeworms

Hydatid cysts need to be distinguished from **cysts caused by other larval tapeworms** found in cattle and sheep: see Table 2, below.

Hydatid disease in humans

Hydatid disease in humans can result only from swallowing eggs released from an infected dog, dingo or fox. Early in the 20th century in some areas of New South Wales, particularly the Southern and Northern Tablelands, infection in humans was a common problem and death as a result of the disease was all too frequent (Dew

Table 2. Cysts of larval cestodes of sheep and cattle

Cysts/larval stage	Intermediate hosts	Location	Size	Appearance	Tapeworm, length, location	Definitive host
Hydatid cyst (*Echinococcus granulosus*)	Sheep, cattle, goat, pig, wallaby, kangaroo, human, deer, camel, wombat.	Liver, lung, kidneys, spleen, heart, brain, bone.	4–5 mm at 3 months; 20 mm at 6 months.	Viable cysts enclosed within laminated fibrous capsule and embedded in substance of affected organ. If fertile, contain many protoscolices ('hydatid sand'). Degenerated cysts contain caseous material that 'shells out'.	*Echinococcus granulosus*, 4–6 mm (4–6 segments), small intestine.	dog, dingo, fox
Sheep measles (*Cysticercus ovis*)	Sheep, goat.	Heart, diaphragm, masseter muscles, oesophagus, all striated muscle.	3–6 mm at 7 weeks. Oval shape, up to 10 mm long.	Viable cysts contain fluid and a single protoscolex. Dead cysts become calcified.	*Taenia ovis*, 2 m, small intestine.	dog
Beef measles (*Cysticercus bovis*)	Cattle, buffalo, deer, giraffe.	Heart, tongue, masseter muscles, diaphragm, all striated muscle.	Variable in size: 2–20 mm; average 5 mm. Fully developed in 16 weeks.	Viable cysts contain fluid and a single protoscolex. Degenerated cysts become caseous and calcified.	*Taenia saginata*, 4–10 m, small intestine	human
Bladder worms (*Cysticercus tenuicollis*)	Sheep, cattle, goat, pig.	Liver, abdominal cavity.	Average 50 mm; range 1–60 mm.	Cysts loosely attached to surface of viscera. Contain clear, jelly-like fluid and a single large protoscolex.	*Taenia hydatigena*, 3 m, small intestine.	dog, dingo

Note: *Taenia pisiformis*, *T. serialis* and *Dipylidium caninum* are common tapeworms of dogs, foxes and dingoes that need to be differentiated from *T. ovis* and *T. hydatigena*. The intermediate hosts of *T. pisiformis* and *T. serialis* are the rabbit and hare. The flea and possibly the biting louse are the intermediate hosts for *D. caninum*. Modified from Love & Hutchinson 2003, adapted from Cole 1986.

1928). Although human hydatidosis was a notifiable disease in most states until 2001, it was frequently under-reported (Schreuder 1990).

A study of hospital records in New South Wales and the Australian Capital Territory (Jenkins & Power 1996) covering the period 1987–92 found:

- in 15 shires in the north-east, a range of 0.3 to 17.7 cases per 100 000 population;
- in 24 shires in the south-east, a range of 0.5 to 3.5 cases per 100 000 population.

In the 5 year period covered by the study, 321 patients were treated in NSW and ACT hospitals, but only 17 of these cases had been notified. Of these, 195 were new cases and 117 recurrent cases.

The mean annual human prevalence in rural NSW was calculated to be 2.6 cases per 100 000. Four cases were reported in Aboriginal persons (mean annual prevalence of 1.1 cases per 100 000 Aboriginal population of NSW).

The age at which hydatids was found varied depending upon the place of birth. Most Australian-born patients were 31–40 years of age, whereas those born outside Australia were mostly 41–50 years of age. Because the disease appears as a result of the slow growth of the cyst, infection picked up in childhood often does not appear until later. However, there is evidence that a proportion of new infections would have occurred as adults, so no person should consider themself immune from this disease. An earlier study of the disease in

348

humans found that the average stay in hospital for hydatid disease patients was 23 days.

Children are the group at greatest risk of hydatid infection because of their close association with dogs and their sometimes casual approach to personal hygiene. Any dog that has had access to fresh offal, including livers from butchers, could be infected with this tapeworm.

Hydatids in humans is a serious disease. With improved diagnostic techniques, in particular computerised axial tomography (CT) and magnetic resonance imaging (MRI), whereby cyst lesions can be more easily identified, the danger of the disease has been reduced. However, the formation of cysts in the body is always dangerous and their surgical removal is never straightforward. A major concern during surgery to remove cysts is that brood capsules can float free within the cyst. If a cyst ruptures, the brood capsules can spread through the body and secondary cysts can grow wherever they come to rest. This contributes to the high level of recurrence (37.5% of patients). Steps taken to sterilise the contents of the cyst during surgery, and post-operative treatment with albendazole, will greatly reduce this risk. Deaths from hydatid disease still occur both before and after surgery.

The real tragedy of hydatid disease is that it happens at all, as the infection can result only from swallowing eggs released from an infected dog, dingo or fox. This is an event that can be prevented by following the simple measures outlined below.

The control of hydatids

The control of hydatids involves ensuring that dogs are not infected with the tapeworm, either by preventing the dog from eating tapeworm heads or by cleaning out any tapeworms infecting it, preferably before they have had a chance to mature. Effective control methods include the following:

1. Feed only manufactured dog foods.

These foods contain materials cooked in such a way that there is no risk of contamination by tapeworm heads in or on the materials used such as liver, lung or meat by-products. Also, the majority of commercial rations have been prepared under the direction of nutritionists specialising in such foods, and so contain all the nutrients essential for a balanced diet.

Feeding unprocessed meats or meat by-products to dogs carries with it the risk that contamination from ruptured cysts may have occurred during the slaughtering process, and, as a result, the dog could become infected. It is very difficult to cook meat at home adequately because of the time required to thoroughly heat the centre of a piece of meat. Commercial companies use pressure cookers which ensure that adequate temperatures are reached.

2. Prevent access to dead stock or fresh offal.

Access by dogs to dead stock or fresh offal can be from scavenging or by the deliberate feeding of meat, liver, lungs or even blood from the slaughtering process. Hydatid cysts, which can be in the carcase or have contaminated the meat or other material, may be eaten by the dog, causing it to become infected with the tapeworm.

- **Rural dogs.** Dogs should be kept in kennels, in runs or on leads when not working or under supervision. Unsupervised dogs can find carcasses or, worse still, attack and kill lambs, sheep or other small stock. Sheep provide the greatest risk because they have the greatest proportion of viable cysts when infected.

- **Suburban dogs.** Legislation requires that all dogs in public places be under the direct control of a responsible person.

If you know that what your dog eats is safe and you know where your dog goes, you can make sure that no risk of eating tapeworm heads can occur and your dog cannot become infected with the hydatid tapeworm.

3. Wash your hands after handling dogs.

Even though you take every precaution to keep your dog free from hydatids, you cannot be sure of the attention other people give to their dogs.

Before handling food, smoking a cigarette or, in fact, doing anything where your hands could transfer eggs from the dog's coat to your mouth, you should wash your hands thoroughly. Teach your children to do the same, particularly after playing with dogs and before eating.

4. Treat dogs which may have infection.

If dogs have not been fed safe foods or if they have wandered away and there is a risk that they might have eaten hydatid cysts by scavenging, then treatment with a drug that is 100% effective against the hydatid tapeworm should be carried out. The only drug which is this effective at the moment is praziquantel. (There are 86 different registered products contain praziquantel, many of which are 'all-wormers' with activities against other worm types.) This drug is very safe to give to dogs and can be served in their food.

When treating potentially infected dogs, take great care in disposing of the droppings for 2–3 days after treatment. Numerous eggs from dead and dying tapeworms will be present in the droppings.

Dispose of them by deep burial or burning. These eggs pose a health hazard because they are not affected by the treatment.

Where a constant risk of infection occurs, treat the dog every 6 weeks to eliminate tapeworms before they grow to maturity and begin to release eggs.

Eradication programs

As far as human health is concerned, the infectivity of various strains to humans is of primary importance for the development of control programs. Several countries have undertaken programs to either eradicate or control hydatid disease in their human populations, including Iceland, New Zealand and Cyprus, as well as in Tasmania.

Tasmanian program

The Tasmanian program started in 1962 with public meetings and the formation of action committees to create public awareness of the serious health risk being posed by hydatids. In 1960 a survey had shown a human incidence of 92.5 cases per 100 000. A further study in 1963 identified 537 new cases in the decade 1953–62. Also, 60% of sheep carried cysts and about 12% of rural dogs carried the tapeworm.

The program was planned along the lines of the successful New Zealand program but with one important difference. In essence, the program was based on the regular testing of dogs for infection with the tapeworm, combined with an educational program emphasising that preventing dogs from obtaining offal would prevent hydatid infection.

However, the whole test was carried out on the spot rather than having the samples sent to a central laboratory for examination. Later, abattoir monitoring of sheep was introduced to trace properties with infected dogs.

In 1966, the voluntary program became compulsory because a large part of the population favoured the initiative. The success of the program can be measured by the fact that, from its commencement in 1966, the number of new cases in humans fell from 18 per year to 4 per year in 1983. The prevalence of the tapeworm in dogs fell from 12% to 0.04% (1 dog), and the prevalence of hydatid cysts in sheep older than 3 years of age fell from 60% to 0.8%.

Tasmania was favoured in this program because the domestic dog is the only host of the adult tapeworm on the island. There are no dingoes, and so the wildlife (sylvatic) cycle does not occur (recent deliberate introductions of foxes into Tasmania notwithstanding).

By 1996 the last known infected sheep groups had been slaughtered, and Tasmania declared itself provisionally free from hydatids disease in dogs and sheep (Middleton 2002). If this status is maintained, it is likely that the G2 Tasmanian sheep strain will have been eradicated from its source, although it is still found in Argentina (Kamenetzky et al. 2002).

The difficulty on mainland Australia

On mainland Australia, the dingo, wild dog and, to a very minor extent, the fox provide alternative hosts, and so the dingo–wallaby cycle can continue the infection. This makes the eradication of the tapeworm impossible and the wildlife cycle of increasing importance in human infections, especially in areas bordering National Parks, such as Kosciuszko, where control of wild dogs/dingoes is difficult (Jenkins & Morris 2003, Jenkins & McPherson 2003), and in suburban areas fringing regional towns (Brown & Copeman 2003).

New Zealand program

In New Zealand, *Echinococcus* has been 'reduced to the point of extinction', with the last reported fertile cysts occurring in three sheep on an island in the Marlborough Sound in 1995. The country was declared provisionally free from hydatids disease in 2002 (Pharo 2002).

The cost of the disease in animals

Estimating the cost of the disease in animals is very difficult because of the wide range of animals affected. However, large quantities of edible meat by-products, such as liver and lungs, are downgraded to 'meatmeal' because of the presence of hydatid cysts at slaughter. One estimate of the loss in 1980 from this source put the figure at about $650 000 annually.

In some lines of sheep, particularly old ewes from the Tablelands, up to 80% may be infected. There has been no estimation of the potential loss from ill-thrift or early culling of sheep and goats. From abattoir reports on the extent of infection with cysts in individual animals, economic loss of this kind could be possible.

Evidence exists that there has been an improvement in the situation in areas where efforts in controlling hydatids are being put into effect. In the majority of areas, hydatid disease can be controlled and the prevalence of the infection reduced by following the recommended methods.

Only in those areas where the infection results solely from contamination of pastures by dingoes will there be little impact on the disease in livestock. However, protection of the family by following the recommendations with regard to dogs is still applicable.

Further information

For further information, contact your local veterinarian or NSW Department of Primary Industries.

Acknowledgments

The author of the first edition (Stuart King, former Senior Field Veterinary Officer) acknowledged the work of the Goulburn and District Hydatid Eradication Committee in the preparation of the hydatid life cycle illustration, and the New Zealand National Hydatids Council and the Tasmanian Hydatid Eradication Council for the material they produced.

Further reading

Brown, B, & Copeman, DB 2003, 'Zoonotic importance of parasites in wild dogs caught in the vicinity of Townsville', *Australian Veterinary Journal*, 81: 700–2.

Cole, VG 1986, *Animal Health in Australia*, Volume 8 'Helminth parasites of sheep and cattle', Department of Primary Industries, Australian Government Publishing Service, Canberra, p. 255.

Dew, HR 1928, *Hydatid disease. Its pathology, diagnosis and treatment*, Australasian Medical Publishing Company Ltd., Sydney, p. 427.

Eckert, J, Gemmell, MA, Meslin, F-X & Pawlowski, ZS (eds) 2001, *WHO/OIE Manual on Echinococcosis in humans and animals: A public health problem of global concern*, World Organisation for Animal Health, Paris, France, p. 265.

Fasihi Harandi, M, Hobbs, RP, Adams, PJ, Mobedi, I, Morgan-Ryan, UM & Thompson, RCA 2002, 'Molecular and morphological characterisation of *Echinococcus granulosus* of human and animal origin in Iran', *Parasitology*, 125: 367–373.

Jenkins, DJ & Macpherson, CNL 2003, 'Transmission ecology of *Echinococcus* in wild-life in Australia and Africa', *Parasitology*, 127: secs 63–72.

Jenkins, DJ & Morris, B 2003, '*Echinococcus granulosus* in wildlife in and around the Kosciuszko National Park, south-eastern Australia', *Australian Veterinary Journal*, 81: 81–85.

Jenkins, DJ & Power, K 1996, 'Human hydatidosis in New South Wales and the Australian Capital Territory, 1987–1992', *Medical Journal of Australia*, 164: 18–21.

Kamenetzky, L, Gutierrez, AM, Canova, SG, Haag, KL, Guarnera, EA, Parra, A, Garcia, GE & Rosenzvit, MC 2002, 'Several strains of *Echinococcus granulosus* infect livestock and humans in Argentina', *Infection, Genetics and Evolution*, 2: 129–36.

Love, SCJ & Hutchinson, GW 2003, 'Pathology and diagnosis of internal parasites of ruminants' in *Gross Pathology of Ruminants*, Proceedings 350, Post Graduate Foundation in Veterinary Science, University of Sydney, Ch. 16, pp. 309–38.

McCullagh, PJ 1996, 'Hydatid disease: medical problems, veterinary solutions, political obstacles' (editorial), *Medical Journal of Australia*, 164: 7–8.

McManus, DP & Thompson, RCA 2003, 'Molecular epidemiology of cystic echinococcosis', *Parasitology*, 127: secs 37–51.

Middleton, M 2002, 'Provisional eradication of hydatid disease in Tasmania – the final campaign stages', Proceedings, Australian Society for Parasitology, Annual Scientific Meeting, Hobart, p. 19.

Pharo, HJ 2002, 'New Zealand declares provisional freedom from hydatids', Proceedings, Australian Society for Parasitology, Annual Scientific Meeting, Hobart, p. 19.

Schreuder, S. 1990, 'Survey of hospital admissions for hydatidosis in New South Wales and Australian Capital Territory, 1982–1987', *Australian Veterinary Journal*, 67: 149–51

Thompson, RCA & McManus, DP 2002, 'Towards a taxonomic revision of the genus *Echinococcus*', *TRENDS in Parasitology*, 18: 452–7.